An Approach to Physical Science

An Approach to
PHYSICAL SCIENCE

Physical Science for Nonscience Students

PSNS Project Staff

JOHN WILEY & SONS, INC.
New York   London   Sydney   Toronto

PREFACE

This text has been written and revised cooperatively by a group of college teachers—about half by physics teachers and half by chemistry teachers. The entire effort was stimulated by the Commission on College Physics and the Advisory Council on College Chemistry, which had identified a need for better materials for college physical science courses taken by nonscience majors. During 1963 and 1964 these two groups held conferences that led to the organization of the PSNS Project (Physical Science for Nonscience Students) in April, 1965. With the support of the National Science Foundation, the PSNS staff produced this new course, *An Approach to Physical Science*. It was first used at eight colleges in the academic year 1965–1966. Interest in it has grown rapidly since that time.

Preliminary studies indicated that many of you who take a single physical science course in college have little previous training in science and mathematics, and much of what you did have, you did not enjoy and have forgotten. We hope that this approach to physical science will be a pleasant and profitable experience for you.

This text and the course it presents has several novel features. First, the text serves as a laboratory manual as well as providing discussion of the science that the experiments reveal. Experimentation is central to physical science, and it seemed to us that it should be central to a physical science course. Only by personal involvement in the phenomena of science can you develop a significant feeling for the nature of science.

Another unusual feature of this course is that it deals with only a small number of the topics usually treated in physical science courses. This is not a survey course. In a survey course one learns a large number of facts in a number of related fields of science, but there is always a very much larger number of facts that have to be left out. A premise on which this course is based is that, in the limited time available, more real science can be taught if less broad coverage of subject matter is attempted. Attention is focused on the nature of solid matter and how we find out about it. You will be encouraged to observe carefully the nature of the physical world about you; you will be invited to wonder about it and suggest hypotheses consistent with your observations, just as those who engage in scientific research for a living do. Such activity is closer to the nature of scientific investigation than is the memorizing of many facts.

Some of you will later become teachers of children. A teacher who thinks that his or her responsibility is to know all the facts of science so as to be able to produce answers to all the questions that children can ask is not only wrong, but is also doomed to many hours of anxiety and insecurity. A teacher who looks upon the science class as a place to encourage observation, wonder, and questions which suggest more observations and experiments will not only be more truly teaching science, but will find that most children, naturally curious and anxious to experiment, are scientists at heart.

What of those parts of physical science that are often touched upon in a survey course and are not found in this text? As with any one-year course that is not followed by more advanced courses in the same field, this course should be a key to open the door to a variety of related fields. No course is complete. If, in *An Approach to Physical Science,* you discover the scientist's approach to the world around him, you will be better equipped to learn from the many excellent elementary books on science available today than if you had not known the experience of personally investigating one area of science in some depth.

To all you students who are about to start *An Approach to Physical Science,* we who prepared this book hope that this course will not be the end of your experience with physical science, but the beginning of a lifetime of enjoyment in observing and wondering about the world.

July, 1968   *PSNS Project Staff*
Rensselaer Polytechnic Institute
Troy, New York

PSNS PROJECT STAFF

S. ARONSON
Nassau Community College
Garden City, New York

J. J. BANEWICZ
Southern Methodist University
Dallas, Texas

L. G. BASSETT
Rensselaer Polytechnic Institute
Troy, N.Y.

S. C. BUNCE
Rensselaer Polytechnic Institute
Troy, N.Y.

W. E. CAMPBELL
Rensselaer Polytechnic Institute
Troy, N.Y.

E. L. CARLYON
State University of New
York College at Geneseo

M. T. CLARK
Agnes Scott College
Decatur, Georgia

SISTER BERNICE PETRONITIS, O.S.F.
Alverno College
Milwaukee, Wisconsin

T. H. DIEHL
Miami University
Oxford, Ohio

W. E. EPPENSTEIN
Rensselaer Polytechnic Institute
Troy, N.Y.

D. F. HOLCOMB
Cornell University
Ithaca, New York

H. B. HOLLINGER
Rensselaer Polytechnic Institute
Troy, N.Y.

S. J. INGLIS
Chabot College
Hayward, California

J. L. KATZ
Rensselaer Polytechnic Institute
Troy, N.Y.

H. M. LANDIS
Wheaton College
Norton, Massachusetts

S. H. LEE
Texas Technological University
Lubbock, Texas

A. LEITNER
Rensselaer Polytechnic Institute
Troy, N.Y.

W. J. McCONNELL
Webster College
Webster Groves, Missouri

H. F. MEINERS
Rensselaer Polytechnic Institute
Troy, N.Y.

E. J. MONTAGUE
Ball State University
Muncie, Indiana

L. V. RACSTER
Rensselaer Polytechnic Institute
Troy, N.Y.

A. J. READ
State University of New
York College at Oneonta

R. RESNICK
Rensselaer Polytechnic Institute
Troy, N.Y.

F. J. REYNOLDS
West Chester State College
West Chester, Pennsylvania

R. K. RICKERT
West Chester State College
West Chester, Pennsylvania

R. S. SAKURAI
Webster College
Webster Groves, Missouri

J. SCHNEIDER
St. Francis College
Brooklyn, New York

R. L. SELLS
State University of New
York College at Geneseo

L. SMITH
Russell Sage College
Troy, New York

A. A. STRASSENBURG
State University of New York
at Stony Brook and
American Institute of Physics

P. WESTMEYER
Florida State University
Tallahassee, Florida

S. WHITCOMB
Earlham College
Richmond, Indiana

E. A. WOOD
Bell Telephone Laboratories
Murray Hill, New Jersey

E. WRIGHT
Montana State University
Bozeman, Montana

CONTENTS

Chapter 1 You and Physical Science

Chapter 2 When, Where, and How Much?

Chapter 3 A Look at Light

Chapter 4   Interference of Light

Chapter 5   Crystals In and Out of the Laboratory

Chapter 6   What Happened in 1912

Chapter 7   Matter: A Closer Look at Differences

Chapter 8 Matter in Motion

Chapter 9 Energy

Chapter 10 The Kinetic Theory of Gases

Chapter 11 Bonding Forces within a Crystal

Chapter 12   Electric Charges in Motion

Chapter 13   Models of Atoms

Chapter 14   Ions

Chapter 15   *The Nature of an Ionic Crystal*

Chapter 16   *Molecules*

Chapter 17 Nonionic Materials

Chapter 18 What It Is All About

An Approach to Physical Science

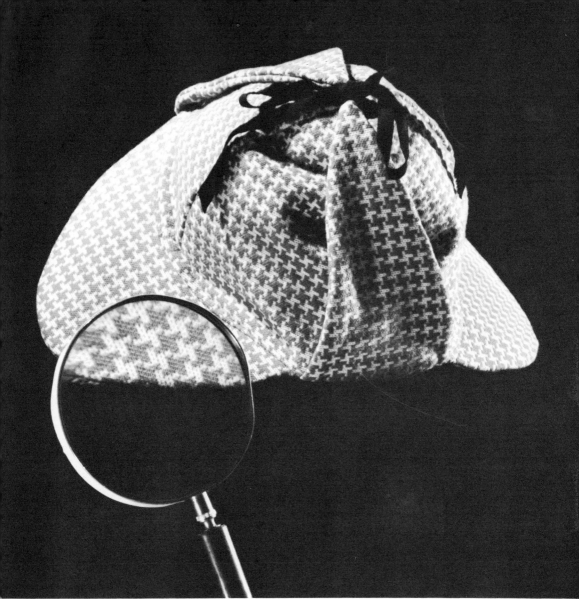

CHAPTER 1

YOU AND PHYSICAL SCIENCE

1-1 *About this course*

Why does a person become a scientist? Ask one and he will probably tell you that he gets more enjoyment, excitement, and intellectual stimulation from working in science than from anything else he can think of doing. Nature is a great mystery that interests him. Like a detective working to solve a crime, a scientist tries to understand nature by piecing together his observations and the observations of others into a coherent whole. Yet many students not majoring in science find these courses boring and not at all enjoyable. Must this be true? We don't think so. We think that it is possible for the scientist to impart some of the fun he finds in his work to the nonscience student.

In this course we hope that you will experience some of the excitement and enjoyment we find in science by joining us as we undertake a scientific adventure. We hope that you will encounter the thrill of exploration and discovery, the disappointment when things don't seem to fit together properly, and the exhilaration of success when they do. We also expect you to share in the hard work of interpreting your own observations about nature. You will be able to make sense out of your observations only after you have some background; building this depends, in large part, on you. It will require that you study and memorize important material. As the course proceeds, you will gain more familiarity with this material and will be able to search more deeply into the subject.

In science, as in the rest of life, interpretations are subjective. This means that many people may observe the same thing and yet describe or explain it in different ways. For example, several people may see the same automobile accident and yet not agree on exactly what happened. Hopefully, when differences occur in this course, we will be able to choose the best interpretation after further discussion and observation. We can then use this information to plan the next steps in attacking the problem.

In making interpretations scientists have found that certain techniques and tools are more helpful than others. Probably the most useful tool of all is mathematics. Mathematical symbols are a form of shorthand often used to represent, in a simple way, relations between the physical properties of substances. Every mathematical equation can be written out in words, but in its verbal form it would be clumsy and consequently more difficult to understand. In this course we will use mathematics to help clarify our thinking and express our ideas. However, we will be much more concerned with insight into the nature of the physical world around us than with quantitative mathematical analysis.

Each of the various fields of science has a specialized vocabulary, as does every trade and profession. Every area of human activity has words that come to have special meaning, understood fully only by the people in that field. Your family and group of friends probably have abbreviated ways of talking to one another, use certain private expressions, and have little jokes that are understood only by members of the group. In the same way, people who are in a specialized field use a specialized vocabulary among themselves, and it is sometimes hard for them to remember that people who are new to the field may give their words a different meaning. In this text we will try to define any new words we use and to explain their special meaning when they commonly have a more general or different one. You will also find that many terms you use every day, such as "work" and "speed," will have a much more restricted use. In part, your understanding of the concepts in this course will depend upon how familiar you become with the terms used.

What part of science shall we study in this course? Since many scientists have spent their entire lives working to comprehend even a small part of the universe, we know that we must limit our field of investigation. The limitations are really quite arbitrary. One of the purposes of this course is to give an understanding of how a scientist works; therefore, the subject matter we choose must illustrate this. Since another purpose of this course is to impart some knowledge of the physical world about us, we have chosen to ask the question, "What is the nature of solid matter?" This is only one of many possible choices.

What do you already know about solid matter? If you were asked to consider your desk, you would probably say that it has volume, weight, and color. But to ascertain volume, we need to learn about measurements. Why does it have weight? For this, we need to investigate gravity. Why does it have color? To answer this, we need to talk about light. And just why is it solid? Why doesn't the table flow like water? Why does it change if you put it into fire? All of these are questions we will investigate during the course.

If we had chosen to study a different part of science, such as outer space or the oceans, much, but not all, of the subject matter of the course would have been different. However, it would still have been science, and many of the ways of thinking about it and of proceeding to find out about it would have been very much the same.

Professor Albert Einstein and his colleague, Professor Leopold Infeld, made a similar choice when they wrote *The Evolution of Physics.*° The following paragraphs from their introduction to that book fit remarkably well as an introduction to this text.

Here is no systematic course in elementary physical facts and theories. Our intention was rather to sketch in broad outline the attempts of the human mind to find a connection between the world of ideas and the world of phenomena. We have tried to show the active forces which compel science to invent ideas corresponding to the reality of our world. But our representation had to be simple. Through the maze of facts and concepts we had to choose some highway which seemed to us most characteristic and significant. Facts and theories not reached by this road had to be omitted. We were forced, by our general aim, to make a definite choice of facts and ideas. The importance of a problem should not be judged by the number of pages devoted to it. Some essential lines of thought have been left out, not because they seemed to us unimportant but because they do not lie along the road we have chosen.

Whilst writing the book we had long discussions as to the

° A. Einstein and L. Infeld, *The Evolution of Physics*, Simon and Schuster, New York, 1942 (now available in paperback).

characteristics of our idealized reader and worried a good deal about him. We had him making up for a complete lack of any concrete knowledge of physics and mathematics by quite a great number of virtues. We found him interested in physical and philosophical ideas and we were forced to admire the patience with which he struggled through the less interesting and more difficult passages. He realized that in order to understand any page he must have read the preceding ones carefully. He knew that a scientific book even though popular, must not be read in the same way as a novel.

The book is a simple chat between you and us. You may find it boring or interesting, dull or exciting, but our aim will be accomplished if these pages give you some idea of the eternal struggle of the inventive human mind for a fuller understanding of the laws governing physical phenomena.

1-2 The scientist as a detective

In a real sense a scientist is a detective, for he arrives at answers to his questions by employing an approach much like that used by a detective. One of the keys to success, both in detective work and in scientific investigation, is observation—scrupulously honest observation. To observe what actually happens—not what we think ought to happen—is very difficult. Even the professional scientist doesn't always succeed.

Through observation, we discover the dependability of the physical world. If you hold a rock in your hand and release it, it will fall; if you hold a flame under a dry newspaper, it will catch on fire; days and nights follow each other in dependable succession, as do the seasons. But it is the keen observer who can do more than just see the objects and events about him. The scientist must observe the inherent order—the regularity in the observed events.

1-3 Observation and Sherlock Holmes

Sir Arthur Conan Doyle created one of the greatest of all detectives, Sherlock Holmes, and made it clear that Holmes owed a good deal of his success to his ability to make critical observations. For example, in "A Scandal in Bohemia," his faithful assistant, Dr. Watson, protested that he never noticed the things that Holmes did ". . . and yet I believe that my eyes are as good as yours." Holmes answered, "Quite so. You see but you do not observe. The distinction is clear. For example, you have frequently seen the steps which lead up from the hall to this room."

"Frequently."

"How often?"

"Well, some hundreds of times."

"Then how many are there?"

"How many! I don't know."

"Quite so! You have not observed. And yet you have seen. That is just my point. Now I know that there are seventeen steps, because I have both seen and observed."

Dr. Watson was too polite to ask Mr. Holmes why it made any difference whether there were seventeen steps or twenty steps. If he had, the detective might have told him that the reason he observed the number of steps was to keep his powers of observation active; to train himself to notice things, all sorts of things, so that he could discover subtle clues to help him solve mysteries.

Most of us are in Dr. Watson's position. We see, but we do not observe. You might test yourself here on a few things which scientists observe and find interesting. Undoubtedly you have seen a rainbow, but can you describe one? Which color is on the outside, and in what order do the colors appear? Probably you have heard the change in pitch in the whistle of a passing train. Does the pitch become higher or lower?

When scientists observe an effect, they wonder about its cause. They try to think of a reasonable explanation, which they call a hypothesis. They don't know right away whether it is the true explanation, and it sometimes takes years of work to find out.

We can illustrate how observation leads to the formation of a hypothesis by returning to Sherlock Holmes, this time using an episode from "The Adventure of the Speckled Band." A young lady came to visit Holmes. "You have come by train this morning, I see," said Mr. Holmes. The lady replied, "You know me, then?" "No," said Holmes, "but I observed the second half of a return ticket in the palm of your glove. You must have started early and yet you had a good drive in a dogcart along heavy roads before you reached the station."

The lady was startled and bewildered. "There is no mystery, my dear madam," Sherlock Holmes reassured her, smiling, "The left arm of your jacket is spattered with mud in no less than seven places. The marks are perfectly fresh. There is no vehicle save a dogcart which throws up mud in that way, and then only when you sit on the left-hand side of the driver." (The driver, of course, sits on the right in England.)

The astonished lady assured him that he was absolutely right. She had, in fact, started from home early that morning. It had been a long drive by dogcart over muddy roads to reach the station where she bought a round-trip train ticket for her visit to Sherlock Holmes.

Holmes observed the mud spots on the lady's left sleeve and developed the hypothesis that she had been splattered while riding in a cart over muddy roads. He tested his hypothesis by telling it to her as if he knew it to be true. When she admitted that it was so, he had the satisfaction of knowing that he had produced an accurate hypothesis.

It is interesting to speculate what Holmes would conclude if he met a mud-bespattered woman on the twenty-fifth floor of a modern office building. It is doubtful that he would hypothesize that she had taken a trip by dogcart. Like Holmes, scientists also make interpretations in terms of what is familiar to them. We now have a large number of observations and interpretations that serve as a background against which modern scientific observations are made and interpreted. Observations made today are interpreted in terms of past experience. However, if these same observations were made twenty-five years from now, they might be interpreted somewhat differently.

In this course we will try to see exactly how a scientist proceeds to satisfy his curiosity. In fact, we will try to proceed in the same way—that is, by observing, wondering, asking questions, and trying to think of a way to get at the answers. By doing our experiments the way a scientist would, we will understand better how scientists have proceeded in those cases that would take us too long to work out for ourselves.

We will begin just as Sherlock Holmes began on his case in "The Speckled Band," by observation. Objects will be handed out in class, and we will make observations on them during this course. We will need to record these observations in a notebook so that we can refer to them later. The records should be accurate, complete, and carefully kept. Every entry in the notebook should be dated. In scientific research laboratories, dated records of observations in notebooks often become very important. For example, suppose the observer needs to know how long it took for some change in his experiment to occur. If his notebook was accurately kept, he can check back and find out. He may not expect any change when he first records his observations. However, if he makes the record anyway, rather than trusting his memory, he may save himself much time and energy later.

A record book should not contain a lot of trivial observations. However, it is not always easy to determine what is pertinent and what is trivial. Experience and intuition seem to be the best guides. For the beginner, it may be well to remember that it is better to record too many observations than to omit one that may be needed later.

EXPERIMENT 1-1   The salol experiment

To illustrate the detective-like nature of the scientist's work and to give you some indication of the subject matter covered in this course, you will perform the first of several "chair-arm" experiments, that is, an experiment done on the arm of your chair in the lecture room. You will be given a hand magnifying glass, some matches, and a microscope slide with two blotches of white salol crystals on it; one blotch is bigger than the other. Examine the two blotches of crystals with your magnifying glass and describe, as best you can, what you see. Then carefully, and without losing any of the pieces, loosen a bit of crystal from the small blotch and leave it there.

Now hold a lighted match under the big blotch so as to heat the glass without cracking it or depositing soot on it. What happens to the crystals when they are heated?

Let the slide cool to room temperature. Did you notice any change in the material of the big blotch during cooling?

Finally, take the little bit of crystal you broke off from the small blotch and put it in the material that remains after heating and cooling the big blotch. Use your magnifying glass to watch *carefully* what happens when you add the bit of crystal.

Describe what happens.

Explaining what happens during this experiment is not as easy as describing what you observe. In fact, a good part of this course is directed toward that explanation.

EXPERIMENT 1-2   Solution of powders

> *Caution.* When handling unfamiliar substances, it is prudent and customary to assume they are unsafe, that is, that they are poisonous or will stain fabrics. During this course you may be asked to perform some "take-home" experiments, and since children may not be as aware of dangers as you, this caution is especially pertinent. Therefore, place the uncovered jars where they will not be accidentally spilled and will be out of the reach of children.

You will be given a jar containing a powder. Examine the powder carefully by noting its properties: color, size, shape of grains, etc. Record every detail of your observations. As we have said, observations are subjective. Therefore, you should compare yours with those of other students to see whether theirs agree with yours or whether they noticed something you missed. If someone has made an observation that you can verify on your own sample, add it to your list. If you do not agree, it may be because the other student's specimen is really different from yours or because one of you is mistaken in your observations. To eliminate the latter possibility, you should observe your samples again. If problems arise, you may find it helpful to discuss them with others.

In this particular experiment you are simply following instructions. In addition to the observations you've just made, we are concerned about the very broad question, "What will happen?" Some of the other experiments you will perform also answer broad questions; others are designed to answer specific questions.

This experiment is quite easy. Let hot water run from the tap until it is so hot that you cannot hold your hand under it. Then carefully add hot water to the jar containing powder until the water level is about $\frac{1}{4}$ inch from the top. Screw the cap tightly onto the jar and shake it until all of the powder has dissolved.

Now choose a place where the jar can sit undisturbed for several weeks and not be subjected to large changes in temperature. A corner of a bookcase away from a radiator would be a good place. Remove the cap from the jar so that the water may evaporate slowly.

Observe the contents of the jar periodically. While making your observations, don't even pick up the jar. If you should, and you subsequently get unexpected changes, you won't be able to repeat the experiment in exactly the same way. To reproduce it exactly, you would need to pick the jar up at exactly the same point in the experiment, hold it for the same amount of time, etc.

As time passes, observe any changes you see with the close attention of a Sherlock Holmes. Be sure to record what you observe, as well as the date of your observations. Later in the course we will need the results of this experiment; but in the meantime, think about what is happening in that jar.

As the water in the jar evaporates, the water level will drop. Do not let the water level fall below the solid material that appears in the jar. Before that happens, stop the experiment: pour the water off and carefully dry the solid material with a cleansing tissue. Dispose of the water carefully, and wash your hands; some of the material is poisonous.

In contrast to Experiment 1–2, the following experiment will require your close, immediate attention and can be performed within one laboratory period. Be sure to record your observations in a notebook.

EXPERIMENT 1–3 Formation and dissolving of crystals

Part A   Put about 1 inch of cold water into a beaker and an equal amount of very hot water into another beaker. After the water stops swirling, drop a small piece (about the size of a pinhead) of potassium permanganate into each beaker. Observe the manner in which the coloration spreads. Be sure to observe the spread of color from above to detect the horizontal distribution and from the side to see the vertical distribution.

Record your observations and compare them with those of other observers. Comparison of results obtained by different observers performing the same experiment is one way of checking the results for reliability. Another way would be for the same experimenter to repeat his experiment, but this requires much more time.

Propose an explanation for both the spread of color and the differences between the spread of color in the hot and cold water. Is the coloration a result of visible particles? Approximately how long is solid potassium permanganate visible? Would that crystal remain as long or longer if the water were stirred? Why? Test your answer by putting a small piece in water and swirling it about.

Part B   This experiment involves a piece of metal and a vial containing powder. Fill the vial with water (distilled water if available) until it is almost full, screw the cap on, and shake for 10 to 15 seconds. What happens to the powder? Now put the piece of metal in the vial, set the vial down so that it won't be disturbed, and observe the contents with your magnifier. What is happening to the metal? Continue your observations for several minutes and record them. Compare your results with those of others. Try to suggest an explanation for the results of this part of the experiment.

1-4 *Asking answerable questions*

The history of science makes it clear that some of the most important contributions have consisted of asking the right questions—questions which went to the heart of the problem. Questions that can be answered by a specific experiment or sequence of experiments are better than questions that do not lend themselves to an experimental or logical search for the answer. Children often ask very complicated questions that can be answered fully only by a highly specialized expert. "How can we go to the moon?" "Why is grass green?" "How high is the sky?" "What makes water so wet?"

Some questions are a matter of semantics; that is, they involve the particular meaning you ascribe to a word. If, to the question, "What do you mean by wet?" your answer is "Behaving like water," then the question becomes, "Why does water behave like water?" The obvious answer to this is, "Because that's what it is." If, however, you mean by "wet" the ability to spread out over a surface and make close contact with every hump and hollow, then the question is a deeper one: "What are the chemical and physical properties of water that give it this surface-covering ability, unlike mercury, for instance?"

Most scientists perform experiments to find out about the physical world. Those who do not include the astronomer, who cannot experiment with the distant galaxies but only observe them; and the geologist, who can only observe the mountains as they are now and cannot make new ones to verify his mountain-building theories. But the chemist, the physicist, and the biologist can experiment with the objects of their study.

Let us ask a simple question that can be answered by an experiment: "At what temperature does water boil?" That sounds easy. Can't you find out simply by putting a thermometer in a container full of water, heating the water until a lot of bubbles break on the surface, and then reading the temperature? The answer is no. If you live in New York City, you will find the boiling temperature of water is approximately 212°F; in Denver, you will find that it is close to 200°F. When you place the thermometer in the container of water, you could let it rest on the bottom or suspend it in the water near the top. Would you expect the temperature to be the same in both cases? You could use tap water, which has substances dissolved in it, or you could use chemically pure water. Would you expect the temperature to be the same in both cases?

We have implied that the boiling temperature of water depends on a number of factors; for example, it depends on the air pressure and on the purity of the water. In addition, during the performance of an experiment to determine temperature, the experimenter must be careful to place the thermometer where it will indicate the temperature of the water, not of the container, for the two may not be the same.

Were we to make measurements of the boiling temperature of water, how would we indicate our results so that someone else could repeat the experiment and achieve the same or a comparable result? We would have to state exactly how we made our measurements, how we determined the purity of the water, and what the air pressure was when we made our measurements. Only then could someone else interpret our results.

1-5   *Classification*

Science is a combination of observing, formulating hypotheses or models to account for the observations, predicting, and testing. Observations are the experimental data; they often become more meaningful when classified. (Classification is the process of identifying groups such that all members of a group have at least one characteristic in common.)

What is it that we classify? The properties of things. For example, some of the criteria by which we could classify automobiles are color, make, number of doors, body style, and year of manufacture.

In the following questions and experiments you will be asked to establish your own criteria for classification. Frequently, scientists will agree to use certain selection criteria. Often these are chosen on the basis of usefulness, but sometimes the choice is quite arbitrary. In these questions several equally correct answers are possible, depending upon your classification scheme.

Question 1-1   In the photograph which opens Chapter 2 there are many different objects. Classify these objects by grouping them into sets. A set may consist of only one item, although it may have 5, 10, 1000, or more. After carefully observing the objects in the photograph, establish a criterion that differentiates some of them as members of one set, distinct from the others. Indicate what your criterion is and then list the items that fall within that set. Establish several other criteria, and classify these same objects into still other sets. How many sets can you establish?

Question 1-2   Following this question are five objects (Fig. 1-1). Classify these figures into several different sets by establishing criteria that distinguish some from the others.

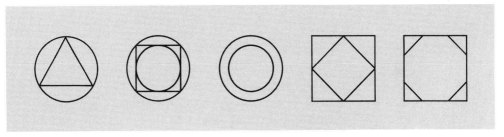

FIGURE 1-1

Question 1-3   Choose the pattern from the three patterns within the gray area (Fig. 1-2) that best fits into the empty square. Explain why it is best.

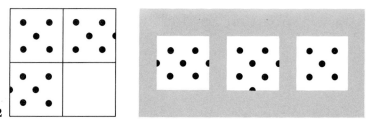

FIGURE 1-2

Question 1-4   Choose the pattern from the two within the gray area (Fig. 1-3) that fits in the empty square.

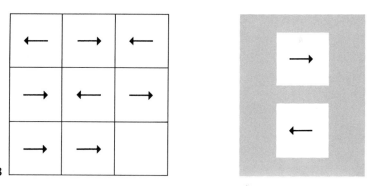

FIGURE 1-3

Question 1-5   Organize or classify the words in List I into sets, then those in List II.

List I	*List II*	
steam	record	water
gasoline	experiment	kilometer
salt	blue	rainbow
oxygen	Fahrenheit	hearing
water	model	light
mercury	violet	yard
carbon dioxide	crystals	waves
wood	vehicle	question
waves	puzzle	measure
doughnuts	solids	sky
alcohol		
ice		

EXPERIMENT 1-4 Classification of objects

You will be given small and approximately equal quantities of nine different substances. Observe these and then establish criteria that permit you to classify them into two sets. Now establish other criteria that let you group them into different sets. Record in your notebook the various criteria and sets you create.

1-6 *Summary*

In the first chapter of this course, there are not many facts to be learned. However, as we look back on what has been done, we find several ideas that will be useful as the course continues. Some of these are listed here to help you review the chapter.

1. Careful observation is essential to an understanding of the physical world.

2. A hypothesis is based upon observations and can be tested by making additional observations.

3. Hypotheses grow out of that which is familiar to the observer and reflect his past experience.

4. Experimental results carry more weight when repetition of the experiment gives the same result.

5. The conditions under which an experiment is performed frequently affect the results obtained.

6. Classification is made on the basis of criteria that may be arbitrarily selected.

QUESTIONS 1-6 Using a dictionary or textbook, define each of the following terms: classify, criterion, hypothesis, intuition, phenomenon, property, qualitative, rational, solid.

1-7 Suppose you were given a piece of solid, metallic material (you have just learned two of its identifying properties). Could you determine easily whether it is iron or aluminum? List as many ways as possible in which these two metals differ.

1-8 Classify the following substances as solid, liquid, or gas, and specify the criteria you used to make your decisions: air, chocolate pudding, glass, granite, ice, jello, mercury, milk, oxygen, sand, steam, tar, wood.

1-9 Devise a means of classifying buildings.

1-10 In Experiment 1–2, you used very hot water to dissolve the powder in the jar. Why? Can you think of any analogous everyday situation that justifies your answer?

1-11 Suppose you were asked to explain the physical principles involved in the following phenomena. Try to do this as well as you can, using only terms you can readily define. Later, after further study on your part, you will be asked this same question. It will be interesting to compare your answers then with your answers now.

(a) It is usually found that after a car has been driven for a considerable distance at high speeds, the pressure in the tires changes.

(b) Water condenses on the outside of a glass. Before explaining this, it may help to consider the conditions under which this occurs as well as the conditions under which it does not occur.

References 1. Chem Study, *Chemistry, an Experimental Science*, W. H. Freeman & Co., 1963, Chapter 1, Section 1–1, pages 1 to 8.
   A discussion of some of the activities of science, with some simple, everyday examples.
2. Christiansen, G. S., and P. H. Garrett, *Structure and Change*, W. H. Freeman & Co., 1960, Chapter 1, pages 1 to 13.
   A consideration of the language and structure of science.
3. Feynman, R. P., R. B. Leighton, and M. Sands, *The Feynman Lectures on Physics*, Vol. I, Addison-Wesley Publ. Co., 1963, Chapter 2, Section 2–1, pages 2–1 to 2–2.
   A great physicist introduces physics to his students. Also, Chapter 22, pages 22–1 to 22–10, gives an appropriate treatment of the way elementary mathematics is put together.

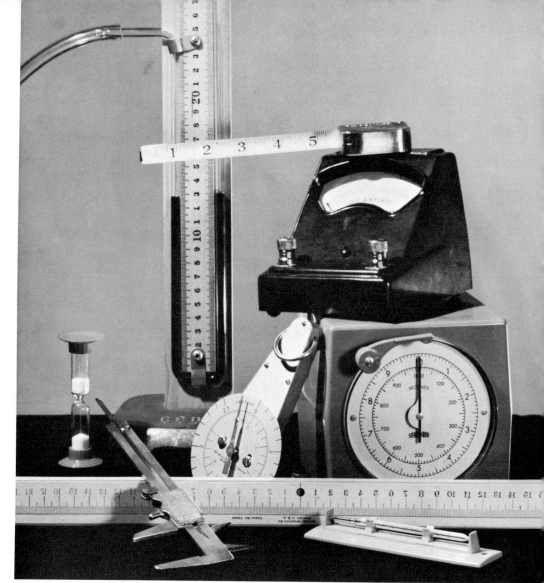

CHAPTER **2**

WHEN, WHERE, AND HOW MUCH?

2-1 *A controlled environment, the laboratory*

Man's explanations of the phenomena of nature were considerably different in previous ages from what they are today. The early explanations were qualitative; that is, measurements were not made to help explain the observed phenomena. More often than not, the early observers failed to recognize the causes of the phenomena they tried to explain. As a consequence, they ascribed "natures" and "humours" to materials in order to explain their actions. For example, it was thought that an apple fell from the tree to the ground because of its "earthy nature." The Moon moved in a circular orbit about the Earth supposedly because of its "heavenly nature." The apple and the Moon presumably had nothing in common; their motions were unrelated.

But a revolutionary approach to the study of the world about us developed in the 17th century. The principal stimulus for this new approach was the idea of using experiments and of making measurements, that is, explanations became quantitative. Explanations which had previously been acceptable became inadequate if they could not be supported by experimentation. Experiments became significant, not only because they helped develop new and better hypotheses, but because other experiments could in turn be used to test those hypotheses. Not only did the resulting explanations become more meaningful, they became more inclusive. Phenomena which seemed as unrelated as the fall of an apple and the motion of the Moon could be accounted for by the same general principles.

The belief that many phenomena, on the Earth and in the heavens, can be explained by the same principle is called *the universality of physical laws.* The inclusion of many phenomena within the realm of one principle increased the ability of the scientist to understand the world about him. A few general principles which explain a vast variety of phenomena are much simpler to understand and to handle than a vast variety of unrelated explanations, one for each phenomenon. In addition, since universal laws cover such a wide variety of phenomena, they lend themselves to experimental testing. For example, the law of gravitation was the first of the great universal laws to be developed. It was proposed by Sir Isaac Newton in 1686 to account for both the fall of the apple and the motion of the Moon. Many experiments have been successfully performed to verify the law of gravitation and many observations vouch for its validity.

Performing an experiment involves focusing your attention on a small and isolated portion of the world. Usually, this is most conveniently done in a laboratory where conditions such as temperature, pressure, and movement of air can be *controlled;* that is, either kept constant or varied at will. Any condition or property of the environment in which an experiment is being performed is called a *variable.* However, the term usually refers to only those conditions which actually affect the experiment.

The experimenter must know in advance—or guess— which variable or variables will influence the object of his studies. He then designs an experiment so that all of the variables except one are kept constant. For example, suppose you wish to study whether the boiling temperature of water depends on the air pressure, the purity of water, or both. If you make repeated measurements of the boiling temperature while you vary both the air pressure and the purity of the water, you will not be able to ascribe any change in the boiling temperature to either variable. On the other hand, if you vary only the air pressure and keep the purity of water (and all other variables) constant, then any change in the boiling temperature can be ascribed to the change in air pressure. A cause and effect relationship will have been observed.

When each of the possible variables is explored in turn, enough bits of information (or data) can be obtained to establish a pattern. That pattern should lead to a relationship between the variables. When found, that relationship constitutes a hypothesis. The hypothesis is then systematically tested by additional experiments, ones that go beyond those used to establish the hypothesis in the first place.

In announcing the hypothesis to others, it is always necessary to state precisely the conditions under which the experiments were carried out, for only then can others check your results. If the experimental results of others are consistent with your hypothesis, the relationship is incorporated into the formal body of science; if those results do not continue to be consistent with your hypothesis, it may be necessary to modify the hypothesis. It may even be necessary to reject it altogether.

One of the most important aspects of the 17th century scientific revolution was the realization that careful observations were not enough—precise measurements should be made. The data become a collection of numbers. The relationship between those various numbers is most easily and precisely stated in quantitative mathematical forms. For example, Newton's law of gravity can be stated as an algebraic equation (see Chapter 11).

2-2 *How much?*

We indicated in Chapter 1 that the focus of this course is the study of solid matter and that classification of information is fundamental to all scientific work. In investigating solids, much of our information comes from making measurements of properties that can be used to distinguish one substance from another, such as color, volume, mass, melting point, solubility, ability to conduct heat or electricity, and density. You are undoubtedly familiar with all of these words in a general way; precise definitions will be developed as we go along.

Suppose you were to pick up a piece of iron. What are the first three things you notice? Probably they would be: what color it is, how big it is (its volume), and how heavy it is. For the moment we are concerned with these last two properties.

You all know something about computing volume—to do so you need the dimensions of the object. Let us, therefore, take a look at how we measure length.

There is no evidence to indicate that there is such a thing as a natural or fundamental unit of length. We could set up our own system quite arbitrarily if we wanted to, by using the length of a paper clip, an umbrella, and a city block as standards. There need be no special relationship between these three standard units of measurement. In fact, if you think for a moment, you will note that the units in our English system are not related to each other in a very convenient way. The foot is 12 inches long, the yard is 3 feet long, and the mile is 5280 feet long; in order to convert from one unit to another, we have to memorize the multiples that are used. Since we are accustomed to these units through constant use, we seldom think about how inconvenient they are.

How did our commonly used units of length originate? They were initially defined in terms of local rather than universal standards. The foot of the English king was used to define a unit of length; but not all kings had the same size feet. Nevertheless, each king's foot, in turn, became a primary standard, and all meaningful measurements somehow had to be compared with that standard of length. However, since the king's foot was not readily accessible to most people, it was necessary to make another and more readily available standard of the same length. Such a standard is called a secondary standard; in this case, it took the form of a foot ruler. Eventually, one of the secondary standards became established as the primary standard, and the king's foot was relieved of this duty. The present primary standard does not change with succeeding kings.

However, not all secondary standards are equally good. Some are not the same length as the primary, a fact you can verify by comparing a number of different kinds of rulers with one another. Of course, using a secondary standard that is not exactly the same as the primary standard systematically introduces an error into all measurements made with that secondary standard; that error will permeate all the calculations made from those measurements. Such an error is called a *systematic error.*

Your measurements in this course will not be so precise as to cause us to worry about systematic errors resulting from rulers of the wrong length. Systematic errors are, however, very important in many fields. For example, many parts of an automobile engine are made five to ten times more accurately than rulers. In situations such as this, an exceptionally accurate secondary standard is needed. So that accurate measurements may be made when necessary, good secondary standards may be obtained from the U.S. Bureau of Standards.

In all scientific measurements we use the metric system. In this system the basic unit of length is the meter. It is much simpler than the English system: here all the units are related by multiples of ten.

$$1 \text{ millimeter} = 0.001 \text{ meter} = 10^{-3} \text{meter}$$
$$1 \text{ centimeter} = 0.01 \text{ meter} = 10^{-2} \text{meter}$$
$$1 \text{ meter} = 1 \text{ meter} = 10^{0} \text{ meter}$$
$$1 \text{ kilometer} = 1000 \text{ meter} = 10^{3} \text{meter}$$

The international primary standard meter, kept at Sèvres, France, is a platinum-iridium bar with two marks etched on it. The distance between these marks is called 1 meter. This is what we mean by defining the meter. Congress has recently established 1 inch as exactly 2.54 centimeters. Note that the inch becomes a secondary unit defined in terms of the meter bar in Sèvres.

All primary standards of measurement are chosen arbitrarily. For example, the meter was first defined in 1799 as one ten-millionth of the distance between the Earth's equator and the north pole. The marks on the platinum-iridium bar in France were carefully made so that they correspond to this definition. However, later measurements of the size of the Earth revealed that the distance between the north pole and the equator is not 10.000 million meters, but 10.002 million meters. Because of the arbitrary nature of primary standards, however, the marks were not changed to correspond to the new measurements of the Earth.

The problem of providing an independently reproducible and accessible standard of high precision might still be with us if it weren't for the adoption of a new International Standard of length in 1960. This standard is the wavelength of one of the narrow bands of light which krypton gas can be made to emit. (We will discuss how this can be done later in the course.) Since this wavelength has so far been found to be constant at all times under essentially all conditions, and since it can be determined to a precision of better than 1 part in 10,000,000 in any well-equipped scientific laboratory in the world, it is precise and accessible to everyone who might need it.

Turning now to the calculation of volume , we always multiply three numbers that represent lengths. For a rectangular box these are depth, width, and height. If each of these measurements is expressed in centimeters, then the volume is expressed as cm $\times$ cm $\times$ cm, or cm^3, or cc (cubic centimeters). Other appropriate units of volume are $meter^3$ (cubic meters) and ft^3 (cubic feet). Another unit of volume, the liter, is arbitrarily defined so that 1 liter is very nearly equal to 1000 cm^3 (1 quart is about 0.9 liter). One thousandth of a liter, called a milliliter (ml), is very nearly the same as 1 cm^3. We will use milliliters (ml) and cubic centimeters (cm^3) interchangeably in volume measurements, even though there is a small difference between them.

The concept of weight is not as simple as the concept of length. We are all used to weighing things by measuring how much they push down on a scale. However, it is found that the same object (e.g., a gold brick) weighs less on top of a high mountain than it does at sea level. On the surface of the Moon it would weigh only one-sixth of what it does on the Earth. Yet the brick is the same brick; no bits of it were lost when we took it from sea level to the top of the mountain or to the Moon. The amount of gold in it is the same in all three locations. Consequently, its weight apparently depends not only on how much gold there is in the brick but where that brick is located; that is, its weight is not an unchanging property. The property which is expressed by the unchanging quantity of matter in the brick is called *mass*. The weight of the brick changes because on the mountain top, it is farther from the great bulk of the Earth than it is at sea level and so the attraction between the brick and the Earth is less. On the surface of the Moon it weighs still less because there it is attracted by the mass of the Moon, which is much less than the mass of the Earth.

How can we determine the mass of an object? Simply by comparing, directly or indirectly, the mass of the unknown object with some standard mass. The standard unit of mass, chosen arbitrarily, is a kilogram. A 1-quart carton of milk has a mass of about 1 kilogram (abbreviated kg). A gram (abbreviated g), which is one-thousandth of a kilogram, is often a convenient unit to use in the laboratory. A paper clip has a mass of 1 or 2 grams.

The primary standard for 1 kilogram is, by international agreement, the mass of a carefully preserved piece of platinum-iridium metal. Other objects are said to have a mass of 1 kilogram if they exactly balance the primary standard when placed on the opposite pan of an equal-arm balance. The condition of balance would be maintained at any location; this is consistent with the statement that mass is an unvarying property of an object. If two 1-kilogram masses placed in one pan of an equal-arm balance are balanced by a single object in the opposite pan, the mass of that single object would be called 2 kilograms. Such comparisons enable us to determine the mass of any object.

Are mass and weight related? We find that at a given location, an object with a mass of 2 kilograms presses down on any surface on which it rests with twice the force that an object of 1 kilogram would exert. Thus we discover experimentally that weight is proportional to mass.

We know that all material objects have mass and volume. Recall the piece of iron we mentioned at the beginning of this section. It too has a certain mass and volume. Half the original piece will have half the original mass and occupy half the original volume; one-tenth the mass will occupy one-tenth the volume; ten times the mass will occupy ten times the volume, etc. No matter how much the original volume is divided, the mass is divided in the same proportion. We can see, therefore, that there is a constant relationship between mass and volume. When this is expressed mathematically, the ratio mass/volume is constant. This constant is called the *density*. The density is independent of the amount of a substance. It is a characteristic property of a substance and can be used as an aid in identification.

Question 2–1 A particular iron horseshoe has a volume of 50 cm³ and a mass of 390 g. An aluminum scoop has a volume of 55 cm³ and a mass of 150 g. A large nail has a volume of 0.4 cm³ and a mass of 3.1 g. Some nails are made of iron; others of aluminum. Which metal is this nail made of?

Question 2–2 On the basis of your experience try to arrange the following in order of increasing density: an iron nail, a piece of glass, a feather, mercury, a copper wire, a styrofoam cup, water, an acorn, the human body.

Question 2–3 The density of dry air at normal atmospheric pressure and at 20°C is given as 1.20×10^{-3} g/cm³. What is the mass of air in 1 meter³?

EXPERIMENT 2–1 Making measurements

Part A   You will be given a piece of wood sawed from a log 100 cm in diameter and 10 meters long. From an examination of your sample, determine the mass of the original log. Remember that you should not mix units, such as centimeters and meters. You can convert the 10 meters to centimeters.

Part B   You will be given a piece of a tombstone which measured 50 cm by 100 cm by 150 cm before it was broken into fragments. From an examination of the fragment you have, determine the mass of the original tombstone.

2-3 When?

What is time? How can we measure it? Although we all think we know what the word time means, it is difficult to define. Try it. Note that your definition of time should not depend on measurements, although a definition of its units must. After you have written down your own definition, check it with a dictionary.

Since ancient eras, man has been concerned with quantitative measurement of time. He has observed many processes which are repetitive: for example, the motions of the Earth and other planets around the Sun, the motion of the Moon around the Earth, the rotation of the Earth about its own axis, and the swing of a pendulum. When a repetitive process occurs and we can assume that the interval between repeats is constant, this interval can be used as an arbitrary standard of time.

The basic unit of time, the *second*, was originally defined as 1/86,400 of a day.° This seemingly strange number derives from a conversion of units from 24 hours per day to 60 minutes per hour and, finally, to 60 seconds per minute

$$\frac{24 \text{ hr}}{\text{day}} \times \frac{60 \text{ min}}{\text{hr}} \times \frac{60 \text{ sec}}{\text{min}} = 86,400 \text{ sec/day}$$

Question 2–4 Think about the possible use of the following phenomena as units of time.

(1) The motion of the Moon around the Earth.
(2) The rotation of the Earth about its axis.
(3) The beating of your heart.
(4) The dripping of water.
(5) The time between sunrise and sunset.
(6) A freely swinging pendulum.

2-4 How hot?

There are various ways of determining how hot things are. We will explore some of them by experiment, and

° In October 1967, the 13th General Conference on Weights and Measures defined the second in terms of the frequencies of two energy levels of cesium-133.

describe others that we will not take time to explore.

In the first part of Experiment 2-2 the substances you will investigate change temperature in a way that may surprise you. You will use thermometers and your hands to detect these changes.

EXPERIMENT 2–2   Observations of dissolving solids

Place 5 ml of water in a test tube. This will be your reference tube for future comparison. Use a spatula to place anhydrous sodium carbonate into a second test tube, filling it to a depth of $\frac{1}{2}$ in.

> *Caution.* Since most laboratory chemicals are poisonous when taken internally, remember always to use a spatula (not your fingers) when you transfer solid substances from one container to another. Also, remember that taste is *never* used in observations unless specifically requested.

Place a thermometer in each tube. One partner holds the two tubes while the other adds about 5 ml of tap water to the tube containing the solid and immediately stirs it with the thermometer until all the solid dissolves. After feeling the tubes and reading the thermometers, answer the following questions: What observations can you make about the solutions in the two test tubes? Did you need the thermometers to tell which was hotter? What did you learn from the thermometers that you could not learn by feeling with your hands? Wash and dry the test tube. Change places with your partner and do this part of the experiment again.

Now repeat the experiment with ammonium chloride instead of anhydrous sodium carbonate; however, use the same amount, $\frac{1}{2}$ in. in the bottom of the test tube. Answer the questions as before.

Question 2–5   What is the significant difference in your observations with the anhydrous sodium carbonate and the ammonium chloride? In what ways were you able to determine this?

The following experiment will give you further insight into the reliability of your hands for measuring temperature.

EXPERIMENT 2–3   A thermal illusion

Put cold tap water in one container, lukewarm water in a second, and comfortably hot water in a third. Soak one hand in the cold water and the other hand in the hot water for a few minutes; then transfer both hands to lukewarm water. How do your hands feel? What does this tell you about using your hands to measure temperature?

Question 2–6   In view of the experiment with the three bowls of water, are you justified in relying on the observations made with your hands when you wish to determine temperature? Why?

Measure the temperature of the water in all three containers with a thermometer, placing it first in the hottest, then in the lukewarm water, then in the coldest, then in the lukewarm again.

EXPERIMENT 2–4   The temperature sensitivity of the hand

How small a change in temperature can your hand detect? To determine this, place a thermometer in the same container with your hand and gradually change the temperature of the water by adding either cooler or warmer water.

If we are interested only in a qualitative judgment, the senses can be satisfactory detectors under some circumstances. However, if we want to perform experiments with a controlled temperature so that other experimenters can verify our results, it is necessary to find a way to designate temperature quantitatively. For example, if we wished to find out which of two test tubes was warmer, we could do so by touching them (a qualitative judgment). However, we would not be able to tell whether the difference in temperature between the two was 2 degrees or 8 degrees (a quantitative judgment).

When we speak of measuring temperature, the liquid

thermometer is probably the first thing that comes to mind. Its operation is simplicity itself since it makes use of the property of expansion with increase of temperature. This is common to most materials. It turns out, however, that the percent of expansion for moderate temperature changes is so small for most liquids that it must be magnified to be visible. This is effected by having a relatively large volume of liquid in the bulb of the thermometer which, when it expands, is forced into a tube with a very, very small diameter. A slight expansion of the liquid in the bulb results in an appreciable motion of liquid into the narrow tube. It is along this tube that the temperature scale is placed.

Another common type of temperature-sensing device that also makes use of thermal expansion is the room thermostat. The active part of this device is not the small liquid-in-glass thermometer which is often found on the front. Rather, it is a bimetallic strip made up of thin layers of two different metals bonded tightly and permanently together, as illustrated in Figure 2–1. The two metals are chosen so that when there is a change in temperature, one expands

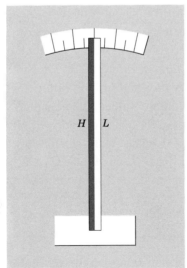

FIGURE 2–1

A bimetallic strip thermometer. The H indicates the high-expansion side; the L, the low-expansion side.

more than the other. If we let *H* represent the high-expansion metal and *L* the low-expansion, which way will the strip bend when it is heated?

If you affix a scale behind the bimetallic strip, as indicated in Figure 2–1, it can readily be used as a direct-reading thermometer. Such thermometers are more useful than liquid-in-glass thermometers for measuring high temperatures; for example, most oven thermometers are bimetallic. However, even the bimetallic strip will melt, so for very high temperatures, it is replaced by other instruments.

One of the most interesting of the high temperature thermometers—the *optical pyrometer*—makes use of the color of light radiated by a very hot object. This color is largely determined by how hot the object is and does not depend greatly on the material or surface of the body itself. The optical pyrometer compares the color of an incandescent object with a color standard and thus determines its temperature. You can make use of a simple rule-of-thumb for estimating high temperatures: if an object just barely glows a dull red, its temperature is about 500°C; if it glows bright or cherry red, its temperature is about 850°C; if it glows yellow, its temperature is roughly 1000°C. When the temperature of an object is over 1500°C, it looks white.

The thermometer scales and units are no less arbitrary than those for length or mass measurements. The Fahrenheit scale is based on the freezing point of saturated salt water (0°F) and what was thought to be man's normal body temperature (100°F). Not only does this choice make the freezing and boiling points of pure water come out to inconvenient values (32°F and 212°F), but the body temperature of a healthy man isn't 100°F. Scientists generally use the Celsuis scale (formerly called centigrade in the United States), on which 0° and 100° correspond to the freezing and boiling points of pure water at standard atmospheric pressure. Other scales are possible; all that is needed are two different reference temperatures which are reproducible.

Question 2–7   Make a diagram of a thermometer showing the following points by marks and labels (in degrees) for the Fahrenheit scale on the left side and the Celsius scale on the right.

(a) The boiling point of pure water.

(b) The freezing point of pure water.

(c) Some particular temperature that would be comfortable as a room temperature.

(d) The freezing point of mercury $(-40°C)$.

2-5   *How much heat?*

Imagine that we have two containers of water which are alike and contain the same amount of water at $20°C$ or approximately at room temperature. We also have two rocks that have been heated to $95°C$, almost the boiling point of water. One rock is the size of a tennis ball; the other is the size of a walnut. If we put one rock in each of the containers of water and test the temperature of the water with a thermometer, what would you expect to find? Within a period of a few minutes, the temperature of the water in the container with the large rock will be higher than that of the water in the container with the little rock, although both started at the same temperature and both received objects at the same temperature.

Clearly, in this case the temperature of the two rocks did not tell us which would raise the temperature of the water more. There was more of something transferred from the larger rock than from the smaller rock. This something we call *heat*.

Now imagine two rocks that are of equal mass and volume, and are made of the same stuff. Let us heat one of these to a higher temperature than the other. Place the rocks in the two containers as before. What do you think will happen? Won't the water in the container receiving the hotter rock rise to a higher temperature than the water in the container receiving the cooler rock?

But what if the second pair of rocks had been made of different materials? Let us perform a laboratory experiment to find out if the heating effect of an object depends upon the material of which it is made.

EXPERIMENT 2-5   Heat transfer from different substances

In order to determine whether the nature of a substance influences its ability to transfer heat to water, we will have to replace the rocks of our previous experiments with objects made of different materials. It will be important to control the variables in this experiment so that any difference in the outcome can be ascribed to the difference in the substances from which the objects are made. Therefore, we will want to have two objects of the same mass, volume, color, shape, etc. In addition the objects should not be soluble in water.

However, it is difficult to satisfy all these requirements, so let us compromise and select objects of equal mass. We see no reason why the volume and color should make any difference, anyway, so we will use equal masses of glass marbles and metal balls. To determine that they are equal in mass you will balance them on an equal-arm balance.

Put the marbles in one heat-resistant plastic bag and the balls in another, and suspend the two bags in the same beaker of boiling water. Leave them there long enough to make sure that each marble and each ball is the same temperature throughout as the boiling water (10–15 minutes should be ample).

Meanwhile, prepare two containers of cool water, each with the same temperature. The amount of water in the two containers should be as nearly the same as you can make it and just enough to cover the marbles. Place a thermometer in each container and observe the temperature. When the marbles and metal balls are ready, quickly transfer the marbles to one container and the metal balls to the other by pouring them out of their plastic bags. Is there a difference in the maximum temperature attained by the water in the two containers? Be sure to catch it at the maximum before it starts to cool again to room temperature.

Question 2-8   What do you deduce from your experiment concerning the effect of the nature of material on heat transfer?

To measure the amount of heat in the last experiment, you used equal amounts of water and noted the change in temperature when heat was added. Since the amount of water was the same in both containers, we assumed that more heat was delivered to the water which suffered the greater change in temperature. In effect, you used water as an instrument for measuring an amount of heat. This is what is done in actual practice; the amount of water and the resulting change in temperature define the unit of heat. The *calorie* is the amount of heat necessary to raise the temperature of one gram of water one degree Celsius. A thousand calories, a kilocalorie, is the "calorie" of those watching their weight.

We have now explored three properties of objects that affect the transfer of heat to water in which those objects are immersed: mass, temperature, and kind of material. What are some properties of heat transfer that we have not yet explored?

EXPERIMENT 2-6   A study of temperature change during cooling

Everyone knows that as water is heated, its temperature increases, and eventually it boils. Conversely, as water is cooled, its temperature falls, and it turns to ice. In this experiment we will observe the temperature of a substance as it changes from a liquid to a solid. Water doesn't freeze until it is far below room temperature; thus, for convenience, we have selected a substance that undergoes this transition at a higher temperature.

You will need to use your alcohol burner, thermometer, test tube, and pegboard, arranged as shown in Figure 2–2. A clock with a sweep-second hand, and a plastic bag containing about 15 grams of the selected substance will be provided. For this experiment, you will need to work in pairs— one partner observing and the other recording. Before beginning, the recorder should prepare a table with the column headings TIME and TEMPERATURE in his notebook. A carbon copy should be made for the observer.

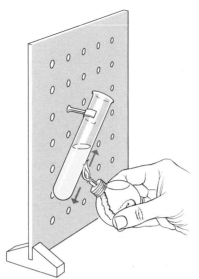

FIGURE 2-2
*Method of heating the sub-
stance in Experiment 2–6.*

Pour the solid substance into a Pyrex test tube and clamp the tube in a sloping position to the pegboard. Slowly heat the test tube until all the solid has just barely melted. Use a low flame and continually play the flame over the side and bottom of the test tube to insure even heating of the entire sample. If you heat the substance unevenly by holding the flame in one spot, part of it may be violently ejected out of the tube.

As soon as the substance has melted, remove the flame, extinguish the burner, place a thermometer in the liquid, and observe the temperature. The observer should stir the liquid continuously and read its temperature every 30 seconds; the recorder should record times and temperatures. Continue taking readings at 30-second intervals until 15 minutes after the substance has solidified. Any change in appearance should be reported and recorded as part of the data. It is important to record the time and temperature at which any interesting phenomenon occurs.

Observations are an important part of any experiment, but it is equally important to analyze and interpret these observations. In this experiment, as in many others, it will be helpful to graph your data. A graph conveys a great deal of information to the reader in such a form that he can detect and analyze a relationship between variables. In this experiment your variables were temperature and time. It is convenient and customary to let the horizontal axis (*abscissa*) represent time and the vertical axis (*ordinate*) represent temperature.

The first step is to scale your graph correctly. This means you will need to choose a suitable spacing along the ordinate to represent temperature intervals and a suitable spacing along the abscissa to represent time intervals. A tiny graph is difficult to construct and interpret, yet the graph should not be so large that it cannot be contained on one page. Plot the points and draw the best smooth line representing your data. (See Appendix B for further instructions on graphing.)

Question 2–9 You used about 15 g of solid substance. How would the curve have changed if you had used 30 g?

Question 2–10 What would the curve look like if the test tube containing the melt had been placed in cold water? Would you expect the solidification temperature to change or the plateau to be shorter? Would the temperature of the water change? Give reasons for your answers. Then check them by the following experiment.

EXPERIMENT 2–7 **A closer look at the plateau**

In Experiment 2–6 you collected data and plotted a cooling curve. Did you notice throughout the experiment that the temperature of the substance was greater than that of the room? If you did, you were not surprised to see the temperature of the substance drop as it lost heat to its surroundings. But during one part of the experiment the temperature of the substance stayed constant; on the graph this appeared as a *plateau*. Why did this plateau appear? Did the substance lose heat during this time?

The last question can be answered by performing another experiment that involves placing the test tube in a cup of water. If the substance loses heat, the temperature of the surrounding water will increase. It is important to control conditions as much as possible by reducing the exchange of heat that might occur between the air and your experimental setup. Therefore, it is helpful to use a heat-insulating vessel, called a calorimeter, to hold the water. A styrofoam drinking cup will do. The cup itself does not absorb much heat, so that almost the only heat lost from the setup will be through contact with the air at the surface. During the experiment you should be prepared to observe and record the time and temperature for both the water in the calorimeter and the substance in the test tube.

Now repeat the procedure of Experiment 2–6 except that the test tube will be cooled in water rather than in air. You will need a second thermometer to both stir and read the temperature of the water. Melt 15 grams of the substance in the test tube, and place a thermometer in the melt. When the substance is melted, put the test tube into the calorimeter, making sure that the level of the melted substance in the test tube is just slightly below the level of the water in the cup.

Stir both the water and the melted substance, and determine the temperature of each every 30 seconds. Continue taking readings for 3 to 5 minutes after the substance has completely solidified. Plot two curves: the temperature-time curve of the substance and the temperature-time curve of the water in the calorimeter.

Question 2–11   Is heat continuously evolved by the substance in the test tube during the time of the experiment?

Question 2–12   How does the temperature-time curve of the water relate to the cooling curve of the substance in the test tube?

Question 2–13   Many people say that temperature is a measure of heat. Do the temperature curves of this experiment support that suggestion?

Question 2–14   If you put ice in a glass of tap water, it will melt. Where does the heat come from to melt that ice? How could you verify your answer?

2-15 If a rectangular object has the dimensions 4.0 cm × 2.0 cm × 6.0 cm and has a mass of 144 g, what is its density?

2-16 If the density of an object is 0.50 g/cm³ and its mass is 144 g, what is its volume?

2-17 Do you think it would make any difference in the results if the density of a cube of copper metal were determined out-of-doors on a hot day in August (95°F) or on a cool day in November (45°F)? Why?

2-18 Scientists are faced with the problem of measuring very small, as well as very large, distances. How would you measure the thickness of a page in this book? What difficulties might be encountered? What experimental procedures take care of these difficulties?

2-19 Calculate the approximate weight per cubic foot of your body. In order to do this, you will need to make some approximations. Try to estimate your body volume by substituting for it a rectangular box having a length equal to your height. Make a reasonable guess as to your average width and thickness. Use these three dimensions to calculate your volume in cubic feet and divide that volume into your weight. The result should be close to the corresponding value for water, 62.4 lb/ft³. What independent evidence can you give in support of your result?

2-20 The approximation suggested in Problem 2–19 is a very crude one. It can be improved by substituting appropriately sized geometrical solids for various parts of your body, in succession, as follows.

(a) Consider your trunk from hips to neck to be a cylinder. Measure its approximate length and radius and compute its volume. (The volume of a cylinder is $\pi r^2 l$.)

(b) Consider each leg to be a cylinder. Approximate its dimensions and find the volume.

(c) Each arm is cylindrical. Approximate the dimensions and find the volume of both arms.

(d) What is the approximate shape of the neck? Approximate the dimensions and find the volume of your neck.

(e) Your head is approximately spherical. Determine its average diameter. (The volume of a sphere is $(4/3)\pi r^3$.) Find the volume of your head.

(f) Now find the total volume of your body and again divide it into your weight. Compare this result with that for water, 62.4 lb/ft^3.

2-21 At one time someone proposed using the honeycomb made by honeybees as a standard of length. What factors must be considered in determining whether or not this is a good standard?

2-22 Compare the interval between successive noons with the interval between successive sunrises as a basis for determining a unit of time.

2-23 Imagine that you have a small heavy object tied to one end of a string. The other end is held stationary. The object swings with a period (time for one complete back-and-forth motion) of one second. What would you predict would have to be changed to change the period to two seconds? Try it. Work with your pendulum until you have discovered how to change the period from one second to two seconds. Compare as quantitatively as possible the 2-second pendulum with the 1-second pendulum.

2-24 Why does heating the bottom of a test tube, while the top is still cool, sometimes cause violent ejection of the contents of the tube?

2-25 Draw a heating curve for the substance used in Experiment 2–6. Will there be a plateau? If so, is heat entering the substance during that time? If so, where did this heat come from?

2-26 In the experiment on cooling curves, Experiment 2–6, is the slope (steepness) of your graph before the plateau greater than the slope after the plateau? Is the slope related in some way to the rate of heat transfer from the substance to the surrounding air? Compare the difference in temperature between the substance and the surrounding air at the beginning and at the end of the experiment. Is there some relationship between rate of heat transfer and temperature difference?

2-27 Try to think of a situation in which you could control an experiment in some way. Examples might be (a) a variation of a cooking recipe, as in changing the amount of baking powder; (b) checking gasoline mileage by changing only the octane rating of the gasoline, or the average driving speed, etc. Show how the problem is isolated and how the variable which you chose can be controlled.

2-28 Just as density is one of the characterizing properties of a particular substance, so is solubility. For example, carbon dioxide, one of the constituents in soda water, has a solubility in water of 3.48 grams per liter at a temperature of 0°C and at atmospheric pressure. That is, at this temperature and pressure no more than 3.48 grams of carbon dioxide can be dissolved in one liter of water. As with boiling temperatures, solubilities depend on the physical condition of the substance, so these conditions must be stipulated whenever solubilities are given. In this instance, the temperature and pressure of the water are given.

As a test of your ability to observe and to plan an experiment, outline a simple procedure to find out whether the solubility of carbon dioxide in water increases or decreases when the temperature is raised.

References 1. Baez, A. V., *The New College Physics*, W. H. Freeman & Co., 1967, Chapter 7, pages 81 to 93.
A discussion of time.
2. Feynman, R. P., R. B. Leighton, and M. Sands, *The Feynman Lectures on Physics*, Vol. 1, Addison-Wesley Publ. Co., 1963, Chapter 5, pages 5-1 to 5-10.
A discussion of length and time.

ALSO OF INTEREST:

3. PSSC, *Physics*, D.C. Heath & Co., 2nd Edition, 1965, Chapter 2, pages 7 to 18.
A discussion of time.

4. Booth, V. H., *The Structure of Atoms*, Macmillan, 1964 (paperback), Chapter 9, pages 44 to 55.

A discussion of the way scientists organize their activities. The roles of facts, concepts, laws, theories, and hypotheses are introduced.

5. Holton, G., *Introduction to Concepts and Theories in Physical Science*, Addison-Wesley Publ. Co., 1952, Chapter 13, pages 234 to 256.

A consideration of the growth of scientific understanding.

Francis Laping, D.P.I.

CHAPTER **3**

A LOOK AT LIGHT

3-1 *Light and solid matter*

To learn something of the nature of solids, we must make observations. Some of the observations we made in Chapters 1 and 2 were qualitative, such as observing the color of a crystal; others were quantitative, such as measuring the mass and volume of a particular crystal and then calculating its density. However, both of these particular observations, i.e., determining the color and density, were *passive* —nothing was done to the sample to alter it in any way. What we can learn from such passive observations is limited.

To learn still more about a particular substance we must do something to it. For example, you heated the crystals in Experiments 2–6 and 2–7 until they melted; then you observed the temperature as each cooled and eventually solidified. You know more about those crystals for having done things to them; and later in the course you will see that still more can be learned from these same experiments.

By adding heat to a crystal you caused it to change; would adding light also cause a change? Let us answer this question by means of an experiment. Initially, the crystals in this experiment will be dissolved in water.

EXPERIMENT 3-1   A chemical reaction produced by light

You will need a solution of silver nitrate and a solution of sodium chloride (table salt) for this experiment.

Place about 10 to 20 ml of the sodium chloride solution in a shallow container, such as a watch glass or small dish. Pour a few milliliters of the silver nitrate solution into a small container so that it can be easily withdrawn with a medicine dropper. Cut out a small piece, approximately 3 cm × 8 cm, of absorbent white paper, preferably filter paper, and place it on a paper towel. Thoroughly moisten about half of this paper with the silver nitrate solution, using the medicine dropper. (If, by accident, you spill any silver nitrate on your hands, wash it off immediately with water, or a dark stain will result that will take a few days to disappear.) Pick up the piece of paper by the dry end and submerge the wet portion in the sodium chloride solution for about 20 sec. Do you notice anything happening on the surface of the paper? Is there any change in the appearance of the solution?

Remove the paper and place it between two paper towels for about 10 minutes. Then take off the top towel and quickly place some object such as a coin, key, or paper clip on the piece of treated paper. Place the paper and object in direct sunlight or close to a bright light bulb. What happens to the exposed portion of the paper? The exposure time under the Sun should be about 5 minutes, but under a light bulb 10 or 15 minutes will be needed. Remove the object and record your observations. Keep this paper for a day or two and record your further observations. Have you produced a permanent record of the shape of the object? If not, can you think of ways to make it permanent?

The experiment you have just performed is a photographic experiment. The prefix "photo" refers to light; the suffix "graphic" refers to the record on the paper. The action of the light on the substance with which you impregnated the paper caused it to become dark, and thus the record was made.

It is clear that light does indeed have an effect on the substance you exposed to it. This tells us something about the substance, but it also tells us something about light. Light produced a chemical change; therefore it has energy and can transmit that energy to objects in its path. Let us see what knowledge we can add to what we have just learned.

We see objects because they either reflect light or, like the Sun, emit their own light. From each visible bit of an object, rays of light reach our eyes, bringing us information about each particular spot on the object. The light has intensity—one spot may be brighter or fainter than others; light has color—that spot may be red; and light has direction —we can see the object so we know where it is. The mind pieces together all these bits of information to form a concept of the object from which the light comes. Light is essential to our investigation of solids and of the world in general. Since our investigations sometimes depend upon its less familiar properties, we need to examine light and the nature of its interaction with solid matter in more detail.

3-2   *Color*

Suppose we are watching an American flag at the top of a long white flag pole near the edge of a well-kept lawn. It is a bright sunny day, near noon, and there is a slight breeze stirring the flag. The scene is a colorful one, yet all of the light reaching our eyes (producing the sensation of color) originates in the white light of the Sun. The grass, for example, generates no visible light of its own; if it did, it would glow at night. How, then, does the white light from the Sun become green light from the grass, blue light from the blue field of the flag, and red light from some of the stripes, while it is still white light from the other stripes and from the flag pole? The colored objects have taken the Sun's white light and done something to it, something which is different for each differently colored object. What is it that they have done?

At one time, some people thought that a colored object added colored light to white light. Others thought that the object removed something from white light, leaving it colored. Let's examine these two hypotheses and see whether either is in agreement with our observations. First, as we have already pointed out, green grass produces no light of its own. Therefore, it cannot add green light to the Sun's white light. Apparently the first hypothesis is incorrect.

Let us consider the second hypothesis. Could colored objects absorb part of the white light? If they all absorbed the same part, then they would all be the same color. Therefore, if the hypothesis is correct, they must absorb different parts. Could it be that the grass is green because the blades reflect the green light and absorb the remainder of the Sun's white light? Let us test this hypothesis by performing an experiment.

EXPERIMENT 3-2   Colored objects

Make red, blue, and black marks on a piece of white paper. Use the red and blue crayons which are supplied for you and a regular black pencil. Now look through the red filter (the piece of red transparent material) at the marks on the sheet of white paper.

Question 3-1   Describe the appearance of each mark as viewed through the red filter.

Question 3-2   Is the first hypothesis in agreement with your experimental results? The second? Discuss the evidence.

Question 3-3   Can you explain why there is so little contrast between the red mark and the white paper when both are viewed through the red filter?

This experiment seems to show that white light is made up of many colors, some of which are absorbed by colored objects while others are reflected or transmitted. If this is really true, why doesn't white light look many colored? The answer to this involves the combined operations of the eye and the mind which enable us to see a many-colored beam of light as white. If white light is composed of many colors, can it be separated into those colors? Have you ever seen white light broken up into its many colors? Have you ever seen a little rainbow formed by light passing through a piece of glass? You can demonstrate this for yourself.

EXPERIMENT 3-3   Breaking up white light

Part A   When the Sun is low in the sky, a glass of water standing in the sunlight on the edge of a table will cast a fine rainbow onto a piece of white paper placed in the right position on the floor below it (Fig. 3–1). Try this at home. You will see that the light rays are bent as they enter the

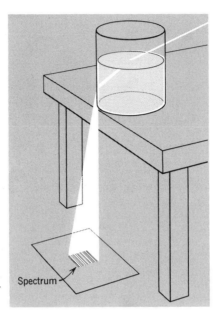

FIGURE 3-1
The glass-of-water experiment.

Spectrum

water; yellow rays are bent more than red, green more than yellow, and blue more than green, so that the various colors are spread out. (If the sun is too high the light will be totally reflected back into the water.)

Part B   (A demonstration)

A glass prism, or a prism-shaped container filled with liquid, shows the effect even better. You can use a plastic pie-wedge box filled with water for an experiment. How can we tell that the prism is not adding color to white light? If this were so, a second prism should add more color. However, with two prisms oriented correctly the spread-out spectrum of white light can be brought back together again to give white light, as in Figure 3–2. The band of parallel rays will appear as white light if the eye encompasses all the rays as shown in the figure.

When you look at this spectrum of white light, note that

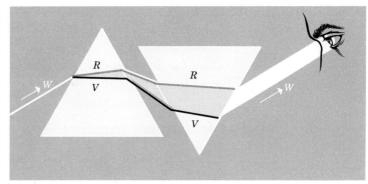

it is continuous. You cannot say exactly where blue stops and green begins, or where the boundary lies between green and yellow, or yellow and red. There is quite a range of colors which can be called blue, another which can be called green, and so on. You can see by this experiment that white light from the sun is not made up of "red, yellow, and blue" or any particular combination of discrete (separate) colors. It is comprised of the whole continuous range of hues that you observed in Experiment 3–3.

All the evidence we have uncovered is consistent with the description of white light as comprising a range of colors. A red substance is, according to this model, one which reflects or transmits red rays and absorbs the others. The fact that some solids absorb some colors and other solids absorb others may provide a clue to the differences in their internal structure. Since we are attempting to understand these structures, we must further pursue the interaction of light with solid matter. But in order to do so, we need to form a better idea of what light is, including the way in which one color of light differs from another.

To help understand the puzzling nature of light, scientists have invented mental models, designed to be simplified ideas of real things and situations. In making a mental model of light, it is sometimes useful to think of it as a stream of particles; at other times, it is useful to think of light as being made up of waves. These two very different models are not as incompatible as you might think. However, for the present, we are interested only in those properties of light which can be understood on the basis of the wave model. In order to talk about this model, we need a vocabulary for describing waves, which we will develop by first considering familiar waves.

3-3   *Mechanical waves*

A good vantage point from which to study waves would be a small boat offshore on a day following a storm at sea. Imagine the situation. The wind has died down, but great

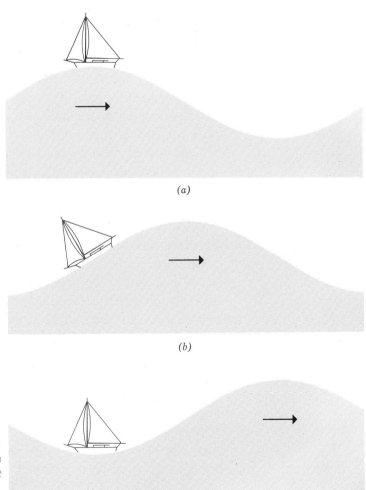

(a)

(b)

FIGURE 3-3
(a) The boat rides high on the crest. (c) The boat sinks low into the trough.

(c)

waves are moving majestically over the open sea. Our little boat rises high on each crest (Fig. 3–3*a*) and then sinks low into the following trough as the crest passes on toward the right-hand side of the picture (Fig. 3–3*b*).

Our little boat is unpowered, and there is no wind to blow it. It simply bobs up and down as the waves pass. (Imagine how your stomach feels.) Although it moves up and down, the boat stays in the same place; the waves do not push the boat along the surface. This can result only from the fact that the water itself does not travel along with the waves; like the boat, it travels only up and down. It is the disturbance that travels across the ocean's surface.

Each wave that passes lifts the boat through a distance of several feet. In order to lift the boat, the waves must have energy. Correspondingly, since the light in Experiment 3–1 produced a chemical change it must also have energy. Both the water waves and light have the ability to transmit their energy to objects.

Not all water waves are the same; some are higher than others, some are spaced more closely together, some travel faster than others. A number of terms are used to describe these differences. For example, the *frequency* of our bobbing boat is the number of times it bobs up and down in a specified interval of time. The boat will bob up and down once as one adjacent crest and trough pass. So the frequency is the number of crests (or troughs) passing any given point in a specified interval of time. The frequency is often designated by the letter *f* and given as so many oscillations per second, or cycles per second (often written *cps*).

The vertical movement of the boat from the level of the calm sea is called the boat's *displacement*. The maximum displacement, that is, the distance from the level of the calm sea to the very top of a crest or to the very bottom of a trough, is called the amplitude *A* (see Fig. 3–4).

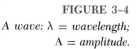

FIGURE 3-4
A wave: λ = *wavelength;*
A = *amplitude.*

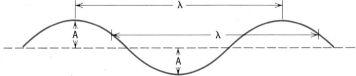

The distance from one crest to the next crest (or one trough to the next trough, or any point to any exactly similar point on the next wave) is the *wavelength* (see Fig. 3–4). In Figure 3–3 the wavelength is nearly eight times the width of the sailboat. The Greek letter for *l*, lambda (λ), is commonly used as a symbol for wavelength. (As you may have noticed, the symbols which scientists use are very often the first letters of the words being symbolized. Our alphabet is very much overworked for that purpose because there are many more quantities to be represented than there are letters. For this reason scientists often resort to using their Greek equivalents.)

If a passenger in the boat counts the number of crests passing him each second (this is the frequency, *f*), and if he also observes the distance between crests (the wavelength, λ), he can then calculate the horizontal distance that a particular crest travels each second. This is the *velocity* of the wave, the distance traveled in a given amount of time.

For example, if two crests passed him each second, *f* was 2, and if the wavelength, λ, was 10 ft, then 2×10 ft of waves went past in one second. Therefore, the waves must be traveling 20 ft/sec. In other words, if we call the speed *v*,

$$f\lambda = v \qquad (3\text{–}1)$$

This relationship holds true for all kinds of waves, including light waves.

EXPERIMENT 3–4   Water waves

Put some water in a container so that the bottom is well covered. A large flat container like a pie plate will do, but a sink is better, a bath tub still better, and a quiet swimming pool or lake best of all.

Now drop a small object such as a pebble or drop of water into the still water. A wave will spread from the place where the object fell. Is there one crest or more than one? How was this wave caused? (Imagine a slow-motion picture of the surface of the water being deformed into a wave shape.) What happens to the wave when it gets to the edge of the water? How do you know?

Try dropping the object in different places. How does this affect what you see? If your container is circular or elliptical, you will find it easier to study these effects.

Can you change the velocity of these waves? Can you change the distances between their crests? Their amplitude?

Water waves are a good example of wave motion, but they are not the only example. Waves can be made to travel along a rope, or a hose, or down a long spring as you will see in the following experiment.

EXPERIMENT 3-5 **Waves in a spring**

Fasten one end of the spring securely to a firm support, or have one of two lab partners hold it firmly. Let the second lab partner pull the spring fairly taut and hold it so that it is free to vibrate up and down without striking anything. Steady the spring, and let the second lab partner pluck the spring near the end he holds. This single disturbance, called a *wave pulse,* travels down the spring. What happens to it when it strikes the end of the spring? Pull the spring tighter and again pluck the spring.

Question 3-4 If the wave pulse travels down the spring as a crest, does it reflect and travel back as a crest or as a trough?

Question 3-5 Does the speed of the wave pulse increase or decrease as the spring is pulled tighter?

What will happen if two wave pulses are sent down the spring in such a manner that the second wave pulse meets the reflection of the first one before that second pulse has a chance to reflect? Try it. Do the two wave pulses stop each other? or do they pass through each other? That is, do they pass through the same spot in the spring going in opposite directions? How do the water waves in Figure 3-5 interfere with one another?

FIGURE 3-5
*Water waves interfering
one with another. (Jacques
Wallach, Photo
Researchers)*

Suppose, for a moment, that a crest is transmitted down the spring and is permitted to reflect back as a trough. Furthermore, suppose that you had sent a trough down the spring at such a time that it intercepted the crest just as it was reflected as a trough. Let us imagine that we photograph those two troughs just as they meet at the far end of the spring. What would they look like? It seems reasonable at this time to suppose that those two troughs would meet and together form a deeper trough, but we will investigate the interference of wave pulses in more detail in Chapter 4.

In the meantime we should extend our discussion of simple wave pulses a little further. If we had a very long spring, so long that we would not need to bother with the reflected wave pulses, we could send repeated wave pulses down the spring by moving the free end up and down. Those wave pulses would follow one after another so as to form a series of wave pulses as seen in Figure 3–6. This is called a *wave train*. The wavelength of the waves in a wave train depends upon two factors: (1) the frequency with

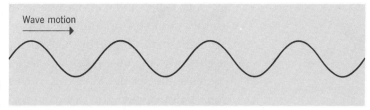

Wave motion

FIGURE 3-6
A wave train.

which we initiate the waves by moving the end of the spring up and down and (2) the speed of those waves.

Question 3-6 If you were actually to set these waves traveling down a very long spring, and if you were to hold the tension in the spring constant, the wave speed would be constant. What would happen to the wavelength of those waves if you were to increase their frequency?

Question 3-7 If you were to hold the frequency constant but increase the wave speed, would the wavelength increase or decrease?

With the spring you have, however, you do need to be concerned with the waves reflecting back. Therefore, any continuing on-going waves you set up in the spring must contend and interfere with the reflected waves coming back. You have already found that the on-going waves and the reflected waves pass right through each other, and we have supposed that when two troughs meet they will make a deeper trough. Correspondingly, two crests should interfere to make a still higher crest.

It might be possible to send waves down the spring with a speed and a frequency such that the crests reflecting as troughs always meet on-going troughs, and troughs reflecting as crests always meet on-going crests. When these conditions are met, the region within one-half wavelength of the end of the spring from which reflections take place should simply oscillate up and down between a crest and a trough. Try it; send waves down the spring so that these conditions are met. It may take some practice before you establish the correct frequency, but you'll know it when you have.

The action you set up in the spring when these conditions

are met is like a wave that doesn't go anywhere; the spring simply oscillates up and down. Consequently, these waves are called *standing waves.* You can set up different standing waves by changing the frequency with which you oscillate the spring. Make the spring oscillate with one loop (one-half wavelength), two loops (one wavelength), and then three loops (one and one-half wavelengths). You can also change the tension in the spring and, consequently, the wave speed.

Standing waves are set up in a guitar string when it is plucked, in a violin string when it is bowed, and in a piano string when it is hit. They are set up in the air of a soda pop bottle when you blow over the top of it and in a clarinet when it is blown correctly. Standing waves can be set up in a tub of water by sloshing it back and forth with the right frequency.

3-4   *The wave model of light*

As we will see in Chapter 4, the behavior of light indicates that it is wavelike, and the vocabulary we have developed for water waves is equally useful for light waves. The amplitude of light waves is related to the intensity of the light; the color depends on the wavelength. For example, the wavelength of red light ranges from about 6.4×10^{-7} meter to 7.0×10^{-7} meter; at the other end of the spectrum of white light, the range of violet light is from about 4.0×10^{-7} meter to 4.3×10^{-7} meter. Light of a single wavelength is called *monochromatic* (from *monos*, single, and *chroma*, color). Monochromatic red light might have a wavelength of 6.6×10^{-7} meter; monochromatic violet light might have a wavelength of 4.1×10^{-7} meter.

Question 3–8   How many wavelengths of monochromatic orange light of wavelength $\lambda = 6.0 \times 10^{-7}$ meter are there in one meter? in one centimeter? How many wavelengths of monochromatic green light of wavelength $\lambda = 5.0 \times 10^{-7}$ meter are there in one centimeter?

Just as it is more convenient to measure little things in millimeters, middle-sized things in centimeters, and big things in meters, those who study light have found it convenient to use an even smaller unit of measurement for measuring wavelengths. The unit is the *Ångström* unit, named for the Swedish physicist Anders Jonas Ångström. It equals 10^{-8} cm. ($1\text{Å} = 10^{-10}$ meter.) Table 3-1 gives the

Table 3-1   Spectrum of White Light

Color	Approximate Walvelength Range in Ångstrom Units ($1\text{Å} = 10^{-10}$ Meter)
Violet	4000–4300
Blue	4300–4900
Green	4900–5700
Yellow	5700–5900
Orange	5900–6400
Red	6400–7000

wavelength ranges of various colors of the spectrum of white light.

Actually, light is only part of a very broad spectrum of waves called the *electromagnetic spectrum,* a term derived from the fact that these waves are composed of electric and magnetic waves coupled together. Other examples of the electromagnetic spectrum are radio waves and x rays. The electromagnetic spectrum includes a vast range of wavelengths. Radio waves may have a wavelength of 1000 meters; gamma rays may have a wavelength shorter than 10^{-13} meter, that is one-thousandth of an Ångström. Figure 3-7 indicates the range of the various regions of the

FIGURE 3-7
The electromagnetic spectrum.

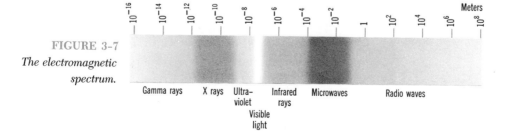

electromagnetic spectrum, and Table 3–1 gives the wavelengths of the various colors in the visible part of the spectrum. But it should be noted that the boundary between any two regions of the electromagnetic spectrum is no more definite than the boundary between any two colors in the visible spectrum. The various regions of the electromagnetic spectrum actually overlap and the naming of the various parts depends upon the origin of the waves as well as on their wavelength. Regardless of their wavelength, however, all parts of the electromagnetic spectrum travel in a vacuum with the speed of light, 3.0×10^8 meters/second or 186,000 miles/hour.

The question now arises: since light is an electromagnetic wave, are we justified in using the analogies of waves in a coiled spring and waves on the surface of water? The answer to this question is yes; for although there are some fundamental differences between electromagnetic and mechanical waves, the characteristics that we wish to study are analogous. This discussion and that in succeeding chapters may convince you of this.

QUESTIONS

3-9   Suppose you are wearing dark-green sunglasses as you drive along a street in town. You come to a traffic light in which the lights are side by side. On what basis would you make the decision to drive through this intersection?

3-10   Another hypothesis concerning colored objects might be that a colored object changes all white light into its particular color. How could you show that this hypothesis is untenable?

3-11   Draw a diagram of the light rays in Experiment 3–3, using a side view. Show how the water in the glass acts like a prism; that is, diagram the light-ray path through the prism and also through the water, and show how these diagrams are similar. Can you explain the fact that the spectrum from the glass of water is curved?

3-12   We have stated that all wavelengths of light have the same speed in a vacuum. The colors have different wavelengths, however, as shown in Table 3–1. What can you say about the frequencies of the various colors? That is, which color has the greatest frequency and which has the lowest? Is the frequency of an x ray greater or less than that of visible light?

3–13 In any equation, the units of the left-hand side of the equation must equal the units of the right-hand side. Show that this is true for the wave equation $v = f\lambda$. (Note: In formal terms, this is a *necessary but not sufficient* condition for the validity of any equation.)

3–14 The wavelength of orange light is about 6000 Å, or 6.0×10^{-5} cm; the speed of light is 3.0×10^{-10} cm/sec. What is the frequency of light of this color?

3–15 Determine the wavelengths of the radio waves which are transmitted by radio stations having the following frequencies.

(a) A standard (AM) broadcasting station at 810 kilocycles per second (kilo means thousand).

(b) An FM broadcasting station at 90 megacycles per second (mega means million).

3–16 Analogous to the period of a pendulum, the period of a wave is defined to be the time for one complete cycle or oscillation. Thus, for the sailboat example in Figure 3–3, the period is the time required for the boat to bob from a crest to a trough and back to a crest. For that example, and using the data given in the text, answer the following questions.

(a) What is the period of oscillation of the boat? (Hint: How are period and frequency related? Reason it out.)

(b) How far did the boat travel during one period?

(c) What is the average speed of the boat going up and down?

(d) Compare this with the speed of the wave going horizontally.

3-17   (a) It is possible to perform Experiment 3–5 quantitatively by actually measuring the wave speed, the frequency, and the wavelength of the standing wave. Suppose that you had set up a standing wave with a frequency of 4 cps and found that two loops had formed in your spring, which was stretched to 3 meters long. What would the speed of the waves be in that spring?

(b) If the tension of the spring in part (a) were held the same, with what frequency would you need to oscillate it to set up a standing wave with three loops?

3-18   The speed of sound in air is about 1100 ft/sec. If the frequency of middle C on the piano is 256 vibrations per second, what is the wavelength of the sound wave?

References   1. Baez, A. V., *The New College Physics,* W. H. Freeman & Co., 1967, Chapter 13, Sections 13.1 through 13.5, pages 159 to 168.

Introduction to some of the language of vibrations and waves.

2. Feynman, R. P., R. B. Leighton, and M. Sands, *The Feynman Lectures on Physics,* Vol. 1, Addison-Wesley Publ. Co., 1963, Chapters 35 and 36, pages 35-1 to 36-12.

Discussions of color vision, physiology of the eye, the mechanism of seeing, etc. Not mathematical. These sections go much beyond our treatment, but they should be of interest to many students nonetheless.

3. PSSC, *College Physics,* Raytheon Educ. Co. (D. C. Heath), 1968, Chapter 3, Sections 3–1 and 3–2, pages 43 to 45.

A discussion of light and color.

ALSO OF INTEREST:

4. Bragg, Sir Wm., *The Universe of Light,* Macmillan, 1933 (also Dover Publications paperback, 1959), Chapter 3, pages 85 to 110.

A popular lecture on color.

5. Garrett, Alfred B., *The Flash of Genius,* D. Van Nostrand, Princeton, New Jersey, 1963, Chapter 20, "The Discovery of Photosensitive Plates."

This is a very brief account of the accidental discovery by Louis Daguerre.

John Shelton

CHAPTER 4

INTERFERENCE OF LIGHT

4-1 *Interference*

Although there had been investigations and hypotheses concerning the nature of light from the time of the ancient Greeks, there was a resurgence of interest in the study of light in the late 17th century. This new interest resulted from the studies of Newton. His experimental evidence convinced him that light is composed of particles which leave the source to travel outward at a very high speed. During the century following the publication of his *Opticks* in 1704, no one seriously questioned either Newton's experiments or his logic that established his particle model of light.

During Newton's lifetime, however, an eminent Dutch scientist, Christian Huygens, put forth the idea that light might be a form of wave motion. Huygens' idea was that light is a periodic wavelike phenomenon consisting of a series of crests and troughs radiating outward in all directions from a light source (see Section 3–4). Huygens was as successful in describing the known properties of light on the basis of a wave theory as Newton was on the basis of a particle theory. In the opening years of the 19th century, some scientists tried to determine, by experiment and observation, which of the two theories better described the behavior of light.

Probably the most important experiments in this effort were performed by an Englishman, Thomas Young,* in 1801. Young designed an experiment that would enable him to decide between the wave and the particle models of light. You will perform a similar experiment which, in general, will proceed something like this.

Narrow slits S_0, S_1, and S_2 are cut in opaque slides A

---

* Thomas Young (1773–1829) was one of the true "natural philosophers" of his age. Trained as a physician, he preferred experimenting in physics to doctoring the sick. An enthusiast and expert in many exotic languages, he helped decipher the text of the Rosetta Stone.

and *B*, and light is allowed to fall on *A* from the left (see Fig. 4–1). We may ask,“ How does the light passing through the slits of the otherwise opaque slides *A* and *B* distribute itself on screen *C*?” You will see this distribution by replac-

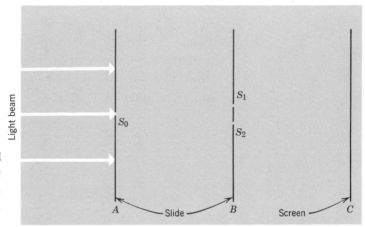

FIGURE 4-1

Young's double-slit experiment. Light from slits S_1 and S_2 overlap on screen C.

ing screen *C* with your eye. Having observed the distribution of light, you can then ask, “Which model of light is consistent with the observed distribution of light on screen *C*?”

The answer to the first question can come only from the observations you make during the performance of Experiment 4–1.

EXPERIMENT 4-1   Young's double-slit experiment

First, let us simplify the experiment by replacing the incident light beam and screen *A* by an incandescent lamp with a long straight filament. This filament will provide us with a simple line source of light at S_0.

Second, take a microscope slide that has been painted with a suspension of graphite in alcohol. You can easily make a single narrow slit in this opaque slide by drawing one razor blade lightly over the graphite surface. For this experiment, however, you'll need two slits very close together, yet each as narrow as the single slit; the two slits must be parallel to each other. The simplest way to make these two slits is to hold two razor blades tightly together and (using a second microscope slide as a guide) draw the razor blades lightly across the graphite-coated surface, so that the slits are parallel to the short edge of the slide. Scratch several of these double slits to make sure you get one good pair. Use your magnifier to look at the slits and estimate the ratio d/w of the distance d between centers of the slits to the width w of each slit opening.°

Now place your eye close to the slits and look at the lamp, as shown in Figure 4–2; make sure the slits are

FIGURE 4-2
Light from a line source passes through two parallel slits and into the observer's eye.

Not to scale

S_1
S_2

Observer

parallel to the filament of the lamp. Choose a pair of slits that gives a distinct, clear pattern when you look at the lamp. Carefully observe the pattern.

If the light passes directly through each of the slits, why do you see several bands of color? The bands of light spreading out to the right and to the left can result only if the light does not pass through a narrow slit as a single narrow beam, but fans out in either direction. This will result in the light from one slit overlapping the light from the other slit. It is possible that these two beams of light will interfere with each other in some way.

The pattern you see in looking at the light through the two slits is called an *interference pattern*. Pay particular attention to the central region of this pattern, and sketch what you see. Indicate the order of colors.

° Note how the expression d/w helps clarify the meaning of this sentence. This is an example of the usefulness of mathematical notation.

Observe the lamp bulb through the red cellophane so that only red light passes through the slits. Again sketch the interference pattern. Now cover the upper half of the light source with a red filter and the lower half with a blue filter. Describe the difference between patterns formed by the blue and the red light, and compare these with that formed by the white light. Record your results.

The interference pattern is formed by the two slits separated by an arbitrary distance. Would it be reasonable to assume that the interference pattern depends upon the distance separating the two slits? To test this assumption you should draw another pair of slits which are separated by a greater distance. This can be done by using a third razor blade as a spacer between the other two. While cutting the double slits, be careful that the middle blade does not touch the graphite. Look at this new pair of slits through the magnifying glass. How does the separation distance between these slits compare with that used earlier? Using this new double slit, observe the red light and compare the interference pattern with the pattern observed with the closer pair of slits. Record your results. We will refer to these results after we have investigated the wave model further. Save these double slits; they will be of use in Experiment 4–2 to measure the wavelength of light.

The second question concerning Young's experiment was, "What model of light is consistent with the observed distribution of light on screen *C*?" That is, does the pattern suggest that light is composed of particles or that it is wavelike?

Question 4–1 What type of pattern would you expect to see if the light were considered to be a large number of small particles coming through the two slits? Does this agree with your observations?

Young considered both the particle and the wave theories, and decided that only the wave model of light was consistent with the interference pattern observed. To see what led to this conclusion, we must take a closer look at the nature of waves. Thus we shall return to our earlier discussion of waves on the surface of water and waves passing along a taut spring, and study both of these examples in greater detail.

4-2   *Superposition of waves*

In Experiment 4-1, it seems clear that the light must have spread out after passing through the two slits. But something else must also have happened because we saw colored bands of light. It is conceivable that these colored bands of light result from the light passing through one slit, spreading out, and overlapping with the light passing through the other slit. But how could overlapping of light waves produce colored bands? We can't answer this question until we know more about waves and how they interact with one another.

First, consider a single upward wave pulse moving along a horizontal spring, like the one you produced in Experiment 3-5. In that experiment, you could not follow its motion carefully; but the motion of such a pulse can be studied in more detail by taking a motion picture of the coil spring and looking at the individual frames.

The pictures in Figure 4-3 show the shape of the coil spring at time intervals of 1/24 sec. It is obvious that the wave is moving from right to left. For ease in observing the motion of a given point on the coil spring, a ribbon has been tied on at the position indicated by the arrow.

As you can see, the ribbon bobs up and down as the pulse goes by but does not move in the direction of the pulse. Because the points on the spring transmitting the wave move in a direction perpendicular to or transverse to the direction of motion of the wave, the wave is called a *transverse* wave. Its motion along the coil spring is similar to the wave motion you observed on the surface of water in Chapter 3. Notice that the shape of the wave does not change appreciably as it moves along.

RIBBON

TIME
(sec)

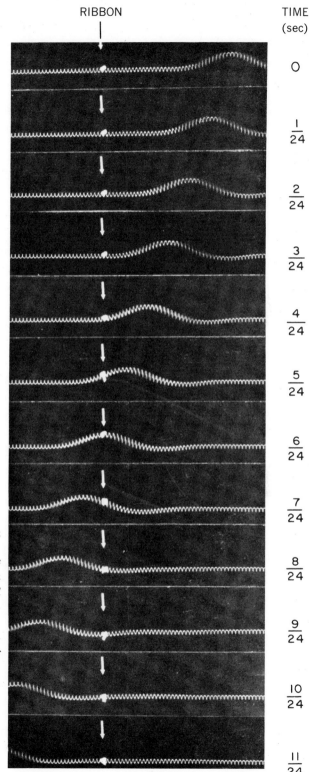

0

$\frac{1}{24}$

$\frac{2}{24}$

$\frac{3}{24}$

$\frac{4}{24}$

$\frac{5}{24}$

$\frac{6}{24}$

$\frac{7}{24}$

$\frac{8}{24}$

$\frac{9}{24}$

$\frac{10}{24}$

$\frac{11}{24}$

FIGURE 4-3

The motion of a pulse from right to left along a spring with a ribbon tied to it. The ribbon moves up and down as the pulse goes by, but does not move in the direction of motion of the pulse. The time at which each frame was taken is indicated on the right. (From Physics, Physical Science Study Committee, D. C. Heath and Company, 1965.)

Question 4–2   Assume that the pictures in Figure 4–3 are one-tenth actual size. Determine the speed of the wave pulse for various time intervals. Record your results. Is the speed constant?

Question 4–3   Measure the amplitude of the wave at time 0 and at time 8/24 sec. Does the amplitude change with time? If so, how?

Now consider what happens when wave pulses meet and interfere with each other. What happens, for example, when two different wave pulses, one from each end of the spring, travel through the same point at the same time?

Once again we make use of the motion picture camera (Fig. 4–4). Here there are two pulses: one, with a larger amplitude, starting from the right and traveling to the left; the other, with a smaller amplitude, starting from the left and traveling to the right. The frame taken at 6/24 sec shows the maximum overlap of the two pulses, with the result that the amplitude is greater than the amplitude of either of the original pulses. Finally, in the last three frames, we see that the wave pulses pass through one another unchanged in shape; the one which began on the right continues on to the left and vice versa. This is a very important observation, and the phenomenon demonstrates one of the fundamental properties of wave pulses: two wave pulses can pass through one another without being altered.

Question 4–4   Assuming that Figure 4–4 is again one-tenth actual size, determine the speed of each pulse and compare the two.

Compare the amplitudes of the individual pulses with the amplitude of the combined pulses when they overlap, as in the 6/24 sec frame. The amplitude can be measured from the edge of a piece of paper that is lined up with the undisturbed spring. The edge serves as a base line. Measure the amplitude of each pulse in the 3/24 sec frame and the amplitude of the combination of these two pulses in the 6/24 sec frame. Your measurements will show that the sum of the amplitudes of the individual pulses equals the amplitude of the combination of those two pulses.

What is the result when more than two pulses meet at the same place at the same time? Many observations indi-

TIME
(sec)

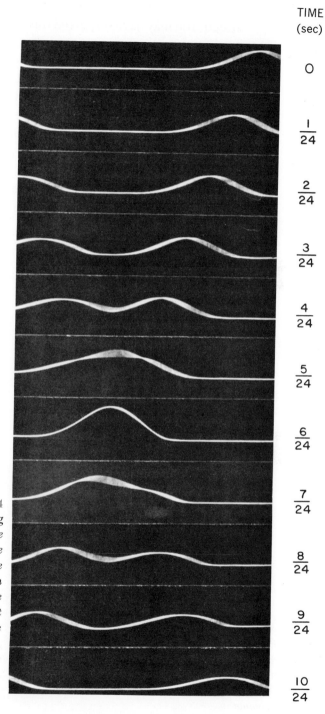

0

$\frac{1}{24}$

$\frac{2}{24}$

$\frac{3}{24}$

$\frac{4}{24}$

$\frac{5}{24}$

$\frac{6}{24}$

$\frac{7}{24}$

$\frac{8}{24}$

$\frac{9}{24}$

$\frac{10}{24}$

FIGURE 4-4
*Two pulses passing
through each other. Notice
that the two pulses have
different shapes. Thus we
can see that the one which
was on the left at the
beginning is on the right
after crossing and vice
versa. (From* Physics,
Physical Science Study
Committee, *D. C. Heath
and Company, 1965.)*

cate that the displacements of any number of pulses can be added. You may have suspected this in Experiment 3-5 when you set up a standing wave in a spring.

In comparing the amplitudes of two separate wave pulses with the amplitude of the combined pulses, we found that these amplitudes are additive. But the amplitude in each pulse is only the maximum displacement, and we considered only the case when the two crests coincided. Now we want to see whether this additivity can be applied to displacements of each point along the overlapping waves. If it can be, then we can find the shape of the total wave displacement in a medium at any time by merely adding to-

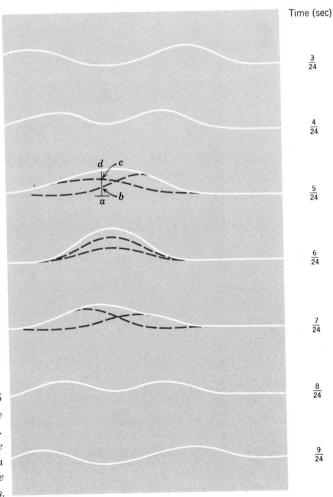

Time (sec)

$\frac{3}{24}$

$\frac{4}{24}$

$\frac{5}{24}$

$\frac{6}{24}$

$\frac{7}{24}$

$\frac{8}{24}$

$\frac{9}{24}$

FIGURE 4-5

The superposition of the two pulses in Figure 4-4. The displacement of the combined pulse is the sum of the separate displacements.

gether, at each point in the medium, the displacement belonging to each pulse that passes through that point.

The solid lines in Figure 4–5 are traced from frames 3/24 sec through 9/24 sec of Figure 4–4. Each solid line represents the displacements of all the points on the spring from their undisturbed positions at the particular time indicated. The dashed lines are our best guess of the shape of each individual pulse.

In one of these frames, a vertical line is drawn and marked with *a*, *b*, *c*, and *d*. Point *d* is a point on the spring which has been displaced from its undisturbed position, marked *a*. The distance *ad* is the amount of this displacement. Distance *ab* would be the displacement if only one pulse were present. Distance *ac* would be the displacement if only the other pulse were present. Measurements of these three displacements will show you that *ad* = *ab* + *ac*; the resultant displacement of a given point on the spring equals the sum of the displacements of the individual waves superposed at that point. This is called the *superposition principle*.

Question 4–5   Examine the same point on the spring in all other frames of Figure 4–5. Does the superposition principle hold in each frame?

Question 4–6   Make measurements of any other point where the waves cross to check whether the superposition principle applies in general.

What if one of the pulses is a trough rather than a crest? The displacements of the two pulses are then in opposite directions, as in Figure 4–6.

In frame 6/24 sec the spring is almost in the undisturbed position as the two pulses pass through each other.* How can we apply the superposition principle to the behavior of the spring in this frame? If this near-zero displacement is to be the sum of the two individual displacements, then one of them must be considered a negative displacement. By

---

* The cancellation in frame 6/24 sec is incomplete because the pulses are not exactly equal and because the picture was taken a moment too soon.

TIME
(sec)

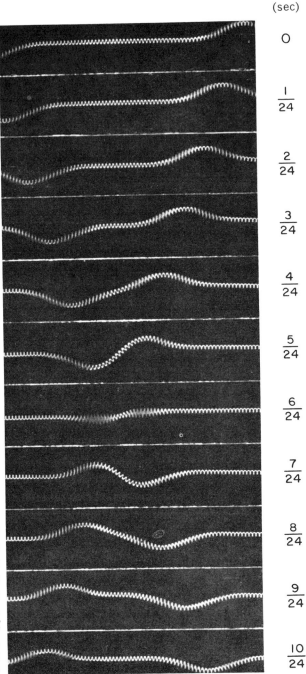

0

$\frac{1}{24}$

$\frac{2}{24}$

$\frac{3}{24}$

$\frac{4}{24}$

$\frac{5}{24}$

$\frac{6}{24}$

$\frac{7}{24}$

$\frac{8}{24}$

$\frac{9}{24}$

$\frac{10}{24}$

FIGURE 4-6
*The superposition of two
equal and opposite pulses
of a wire coil spring. In
the 6/24 sec frame they
almost cancel each other.
(From* Physics, Physical
Science Study Committee,
*D. C. Heath and
Company, 1965.)*

convention, the downward displacement is called negative. The superposition principle can now be verified, as in Figure 4–5, for both positive and negative pulses. This principle, developed to describe the behavior of wave pulses traveling along a coil spring, is applicable to all wave phenomena.

4-3   *Interference revisited*

Let us continue to examine the characteristics of wave motion in general, so that we can apply them to light in particular. We will illustrate the action of light waves in Young's double-slit experiment by studying water waves as seen in a ripple tank—a tank made expressly for the study of water waves. In a ripple tank the waves are generated by disturbing the surface of the water with an object, as you did in Experiment 3–4, and can be observed in the patterns which result from light passing through the water. Light patterns are similarly caused on the bottom of a swimming pool by light passing through the rippled surface of the water. If a series of waves is wanted, the disturbing object is vibrated up and down repeatedly. If the object is very small, it can be considered as a *point source,* and waves will radiate out in circles. From a vibrating straight-edged object like a ruler, the waves proceed with straight-line crests as shown at the left of Figure 4–7.

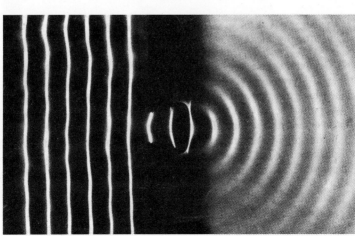

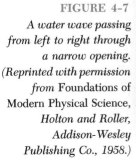

FIGURE 4-7
A water wave passing from left to right through a narrow opening. (Reprinted with permission from Foundations of Modern Physical Science, *Holton and Roller, Addison-Wesley Publishing Co., 1958.)*

In this photograph the bright bands represent the crests of the waves. The distance from the center of one bright band to the center of an adjacent one represents the wavelength. The broad, dark vertical band in the center represents a barrier with a central opening. The straight-line waves passing through this opening from left to right spread out as circular waves beyond the opening. The photograph at the beginning of this chapter is further evidence that waves spread out after passing through a narrow opening.

You may recall that one of the conclusions drawn from your double-slit experiment was that the light spread out after passing through each slit. If light were composed of particles, would you expect it to spread out in the manner you observed?

The spreading out of waves after passing through a narrow opening is an example of *diffraction*, the term used to

FIGURE 4-8   *Ocean waves diffracting around a rock. (John Shelton)*

denote the effects that result from obstructing a portion of a wave. Another example of diffraction is a large rock in the path of the ocean waves. The rock permits most waves to pass, but the waves beyond the rock are altered by its presence (see Fig. 4–8). That alteration is diffraction.

Light waves, water waves, and sound waves all diffract in much the same manner. If the width of the opening through which any of these pass is about the same as the wavelength of those waves, they will spread out as if that opening were a point source of waves.

Let us now consider a water-wave experiment that is analogous to Young's double-slit experiment. Figure 4–1 has been redrawn as Figure 4–9, but with the addition of semi-circles to represent the diffracting water waves. The slit openings are assumed to be very small so that each slit can be thought of as a point source of waves. The lines representing the crests (positive displacements) of the waves are

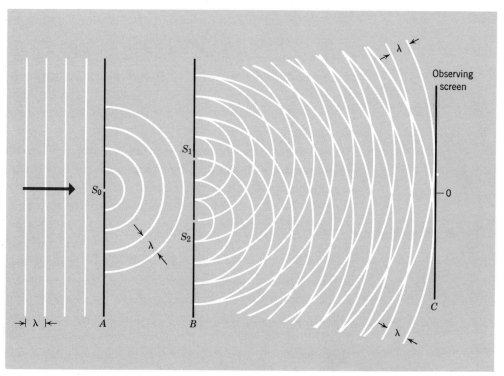

FIGURE 4-9   *Propagation of waves through the slits in Young's double-slit experiment.*

shown in the figure coming in from the left, as in Figure 4–7. The troughs (negative displacements) are not marked on the diagram but lie midway between the crests.

When the waves, spreading out from slit S_0, pass through slits S_1 and S_2 they emerge from each slit as circular waves. Using the principle of superposition for the waves traveling outward from slits S_1 and S_2, we know that when two crests are superimposed (or two troughs are superimposed), a reinforcement occurs that is due to the addition of waves (in a manner similar to the addition in the 6/24 sec frame of Figure 4–4). This reinforcement is known as *constructive interference*. When crest and trough superimpose, cancellation is known as *destructive interference*. These reinforcements and cancellations can be observed in Figure 4–9. Hold the page with your eye close to the middle of the right-hand edge of the figure and look along the figure at a grazing angle. You will see dark streaks (reinforcement) with lighter, less distinct streaks (cancellation) alternating between them.

In the actual physical experiment performed with light, our interpretation is that the bright bands of light are rays of reinforcement while the regions devoid of light are rays of cancellation. Thus, a screen placed anywhere across the superimposed waves will reveal alternate bright and dark bands. In Experiment 4–1 the retina of your eye served as the screen.

Figure 4–10 shows the interference pattern of water waves produced by simulating Young's classic experiment in a ripple tank. However, instead of using slits as the sources, two vibrators, tapping the water surface in unison, are used. Compare Figure 4–9 with Figure 4–10c. The rays of constructive interference in Figure 4–9 lead outward from the region between S_1 and S_2 as dark streaks caused by the crossing of the white lines, which represent crests. Where the white lines cross, larger crests appear; halfway between these crests, but still along a ray of constructive interference, troughs meet to form deeper troughs. These crests and troughs are visible in Figure 4–10c as successive dark and bright bands more or less concentric with

2

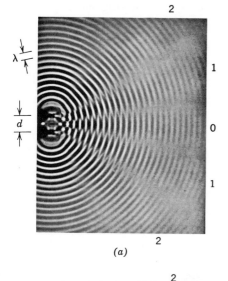

1

0

1

2

λ d

(a)

3

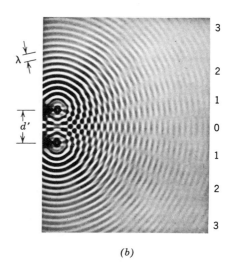

2

1

0

1

2

3

λ d'

(b)

2

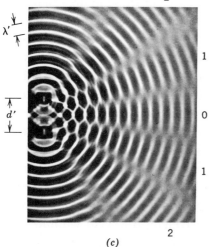

1

0

1

2

λ' d'

(c)

FIGURE 4-10 *Ripple tank photographs of the analog of a double-slit experiment. (a) Vibrator separation* d, *wavelength* λ. *(b) Vibrator separation* d' = 2d, *wavelength* λ. *(c) Vibrator separation* d' = 2d, *wavelength* λ' = 2λ. *(From* Physics, Physical Science Study Committee, *D. C. Heath and Company, 1965.)*

the region between the two vibrators.

The regions of destructive interference are seen in Figure 4–10c as narrow, blurred rays extending outward from the region between the two vibrators. Here the crests meet troughs so that the light and dark bands disappear.

Question 4-7   Describe the motion of a bit of cork placed on the water's surface in the ripple tank shown in Figure 4–10c at either of the positions marked 1. Describe the motion of the cork if it were placed midway between 0 and 1.

The ripple tank is useful for studying wave phenomena because the experimental conditions can be varied, and the resulting wave behavior can easily be observed. For example, the distance between the vibrators can be varied, the frequency of vibration (the rate of tapping) can be changed, and various combinations of openings and obstacles can be placed in the path of the waves.

In Figure 4–10a the two vibrators are separated by a distance d, the wavelength of the water waves is λ. This wavelength is determined by the frequency f and the wave speed v according to Equation 3–1, $v = f\lambda$.

In Figure 4–10b the wavelength is the same as that in Figure 4–10a, but the distance between the vibrators is twice that of Figure 4–10a. In Figure 4–10c *both* the separation distance and the wavelength are twice those in Figure 4–10a. It can be seen from these figures that the pattern of the reinforcement rays depends, in some way, upon the wavelength and the separation distance between the vibrators. Note also that in all three figures there is always a central ray of reinforcement radiating from a point midway between the two vibrators to the point 0 (along the perpendicular bisector of the line joining the two vibrators). The other rays of greatest reinforcement are located symmetrically above and below this one and numbered 1, 2, 3. Note that the angles made by the similarly numbered rays with the central ray are not the same in all photographs of Figure 4–10. They depend on the spacing d between the vibrators and on the wavelength λ. Can we find out how the location of rays of reinforcement in Figure 4–10 is related to d and λ?

Reinforcement occurs when wave crest coincides with wave crest (giving a doubly high wave crest) and when wave trough coincides with wave trough (giving a doubly deep wave trough). Let us examine Figure 4–9 again from the edge of the page and look at the darker paths, those along which constructive interference takes place. In this diagram we have momentarily stopped the waves and can count the number of wavelengths between slit S_1 and the point 0 in Figure 4–9. Here the farthest crest of the waves from slit S_1 coincides with the farthest crest from slit S_2. You will find that this point zero is just 12 wavelengths from slit S_2 and also 12 wavelengths from slit S_1.

Proceeding back from the zero point toward a point halfway between the two slits, we pass through points that are successively 11 wavelengths from each slit, 10 wavelengths from each slit, 9, 8, 7, and so on.

Question 4–8   Now pick a point in Figure 4–9 where two crests coincide on one of the rays immediately adjacent to the central ray and along which constructive interference takes place. Mark it lightly by circling it with your pencil. Count the number of wavelengths from this point back to slit S_1 and also the number back to slit S_2. What do you find? Now using a ruler, measure (in either inches or centimeters) the distance back to slit S_1 and to slit S_2. The difference in the distances is called the path difference.

Repeat these same counts and measurements for other points along this same ray and along the other one adjacent to the central ray. What statement can you make about the points along these rays? Can you establish a relationship between the wavelength and the path difference?

Question 4–9   Now do the same for each of the second rays from the central ray. What statement can you make about the points on these rays along which constructive reinforcement takes place? Make a statement about the third rays.

Question 4–10   You are now in a position to sum up all your results in a general statement that says what must be true about a point where constructive interference takes place between the waves from S_1 and the waves from S_2. Write this statement referring to Figure 4–9.

Each of the radiating rays of constructive interference makes an angle with the central ray. Let us call that angle ϕ (phi). The angle the first ray makes with the central ray can be called ϕ_1, the angle the second ray makes ϕ_2, etc. Compare the angles in Figure 4–10a with the corresponding angles in Figure 4–10b. Is ϕ_1 in *a* larger or smaller than ϕ_1 in *b*? Since the wavelength of the waves in *a* is the same as the wavelength in *b*, any differences in these angles must result from the fact that the spacing *d'* in *b* is twice the spacing *d* in *a*.

Question 4–11   Refer to your record of Experiment 4–1. How did the spacing between the bright bands change when you doubled the distance between the slits in that experiment? Compare this result with the effect of doubling the spacing between the vibrators in Figure 4–10*a*. In Figure 4–10*c* the wavelength is twice as large as that in Figure 4–10*b*, but the spacing is the same in each. Has doubling the wavelength increased or decreased the size of the angles the radiating rays make with the central ray?

Question 4–12   Changing the angle between each of the rays of constructive interference and the central ray of the interference pattern in the ripple tank (Fig. 4–10) is equivalent to changing the spacing between each band and the central band you saw during the performance of Experiment 4–1. Refer to your notes of that experiment. Did increasing the wavelength increase or decrease the spacing of the bands?

In Chapter 3 we stated that, in the wave model of light, the color is related to the wavelength. Now we have seen that the results of our experiment are consistent with this interpretation of color. The relation between the interference pattern of red light and that of blue light appears to be quite comparable with the relation between the interference pattern of long-wavelength water waves and that of short-wavelength water waves. Later, in Experiment 4–2, we will use the interference pattern to determine the wavelength of a particular color of light.

Perhaps the most interesting comparison in Figure 4–10 is the one that remains to be made: the comparison between Figures 4–10*a* and 4–10*c*. In 4–10*c* the spacing between the vibrators is twice that in 4–10*a* ($d' = 2d$), and the wavelength is also twice that in 4–10*a* ($\lambda' = 2\lambda$). What do you find when you compare the angles between each ray and the central ray?

Write down the results of all your comparisons in some organized way, and see whether you can form a hypothesis that describes qualitatively the way the angle ϕ depends on d and λ.

As a result of your analysis of Figure 4–9, you discovered the general rule that constructive interference occurs when the path difference is an integral (whole) number of wavelengths. We will now examine the geometry of the interference pattern and attempt to express this general rule in a quantitative way.

In Figure 4–9 the crests of the water waves are represented by semicircles. As the waves move outward, each semicircular crest grows bigger. It is inconvenient, however, to show the actual motion of each semicircle on a drawing. Therefore we resort to the representation of waves by straight lines proceeding in the direction of wave travel. Such a straight line is called a *ray*. In Figure 4–11

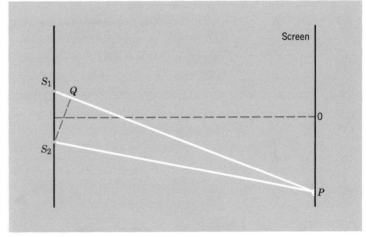

FIGURE 4-11
Diagram illustrating conditions for constructive interference of two rays. In Experiments 4–1 and 4–2 the screen is replaced by the retina of your eye.

two of the paths are represented coming from the two slits and meeting at a point P where they interfere constructively. You determined in Question 4–8 that for constructive interference to occur, the path difference from the slits to the point must equal some integral number of wavelengths $n\lambda$ (where $n = 1$, or 2, or 3, or . . .). For the first ray on either side of the central ray the difference is 1 wavelength, λ (i.e., $n = 1$); for the second rays the difference is 2λ, (i.e., $n = 2$). Now, in Figure 4–11 the distance from slit S_2 to the point P is S_2P, and the distance from slit S_1 to the point P is S_1P. For constructive interference to occur at P, S_1P must equal $S_2P + n\lambda$. The point Q is located so that the distance QP equals the distance S_2P; consequently the distance S_1Q is the path difference, $n\lambda$.

In Experiment 4–1, the distance from the slits to your eye was more than one hundred times the distance between the slits and more than a million times the wavelength of light. We will redraw Figure 4–11 to represent this situation more realistically.

It is not possible to include point P on the paper if we represent the distance between the slits by a convenient length. The lines converge to a point at least a meter away, and consequently, we can consider the lines S_1P and S_2P to be practically parallel. The dashed line S_2Q is practically perpendicular to the two nearly parallel lines in Figure 4–12.

Now let us see if we can find a quantitative statement relating λ, d, and the angle ϕ that the reinforced rays (represented by S_1P and S_2P in Figures 4–11 and 4–12) make with the direction of the central ray, that is, with the direction taken by the two horizontal dashed lines in Figure 4–12. Each of these two dashed lines is perpendicular to the line S_1S_2 joining the slits; consequently, these two lines

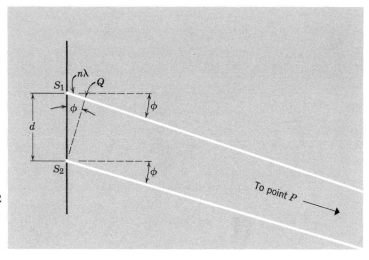

FIGURE 4–12
Diagram illustrating the geometry relating λ, d, and ϕ.

are parallel. But since S_1P and S_2P are nearly parallel, the angles they make with the parallel dashed lines are nearly equal.

The line S_2Q is perpendicular to the line S_1P, and the line S_1S_2 is perpendicular to the central ray; therefore, the angle that S_2Q makes with the line S_1S_2 is also ϕ. This follows from the geometric relationship: two angles are equal if each side of one angle is perpendicular to a side of the other angle. You might convince yourself of this geometric relationship by drawing angles whose sides are so related.

In the triangle S_1S_2Q of Figure 4–12 the angle at Q is a right angle and, therefore, the triangle is a right triangle. The angle at S_2 is ϕ. As you can see, this angle ϕ depends upon the ratio of the distance $n\lambda$ (side S_1Q) to d (side S_1S_2). This ratio is $n\lambda/d$; it is called the sine of ϕ* and is abbreviated sin ϕ:

$$\sin \phi = \frac{n\lambda}{d} \qquad (4\text{-}1)$$

For constructive interference, $n = 0, 1, 2, 3, \ldots$ This is the quantitative relationship we have been seeking. If we know any two of the three quantities, ϕ, λ, or d, we can find the third. For the central ray the difference in the distances from slits S_1 and S_2 is 0, and therefore, constructive interference results for any wavelength λ. For the central ray, $n = 0$.

Can you see how the rays of destructive interference are located? At any region where crests from one slit meet troughs from the other slit, destructive interference (cancellation) occurs. These regions must fall where the difference in distances from the two slits is $(\frac{1}{2})\lambda$, $(\frac{3}{2})\lambda$, $(\frac{5}{2})\lambda$, $\ldots$.

EXPERIMENT 4-2   Measuring the wavelength of light

In Experiment 4–1 you observed the general characteristics of the double-slit interference pattern. In this experi-

---

* The sine of an angle in a right triangle is defined as the length of the side opposite the angle divided by the length of the hypotenuse.

ment you will repeat your earlier work, concentrating this time, however, on the quantitative aspects of the experiment. The same equipment will be used.

Using a red filter, look at the lamp through the double slit on the microscope slide which gives the best pattern. Do not use one of the double slits made with a spacer. Locate the best section of this double slit, and scratch a small window just above or below it. You are going to look through this window toward the lamp. A meter stick clamped in a horizontal position above the lamp will enable you to make measurements.

Hold the microscope slide so that you can look through both the double slit and the window at the same time. As one step in the measurement of the wavelength of light, you are going to count the number of bands in a given length along the meter stick.

From Equation 4–1, we see that the wavelength of light is given by

$$\lambda = \frac{d \sin \phi}{n}$$

First, we must determine d, the distance between the two slits of the double slit. As you can see from Figure 4–13 the distance between the slits (the distance between the cutting edges of the razor blades) can be determined by measuring

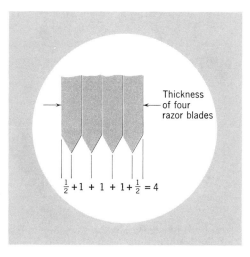

FIGURE 4-13
Diagram illustrating the method of determining the distance between the slits made by the razor blades.

the thickness of a stack of razor blades and dividing that thickness by the number of blades.

We now need to determine sin ϕ for a value of n. In order to find the value of n of any particular band, we must identify the central band from which to count (as from 0 to 1, 2, 3 in Figure 4–10b). As you look at the lamp through the red filter, the band resulting from the central ray is in line with the filament of the lamp. Clamp the meter stick in position directly above this filament. Have one partner use a piece of paper to indicate, on the meter stick, where the other partner sees the farthest distinct band. Now count the number of bands between the filament (zero band) and the farthest band. The number of bands is n. Now the only missing quantity for the determination of λ from Equation 4–1 is sin ϕ.

Sin ϕ can be determined by measuring the distances OP along the meter stick from the zero band to the farthest band, and SP, the distance between that farthest band and the slits on slide S. (See Fig. 4–14.) Sin ϕ is the ratio of OP to SP.

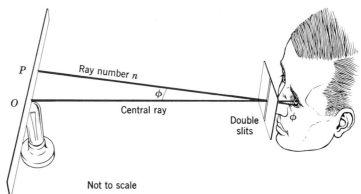

FIGURE 4-14
Schematic diagram of Experiment 4-2.

Change places with your partner and repeat the experiment. Each observer should use his own values of n, d, and sin ϕ to calculate the wavelength of the light that comes through the red filter.

We have seen that the wave model of light is consistent with the results of Experiments 4–1 and 4–2. Would particles flowing through two parallel slits give similar effects?

If we poured sand through two parallel slits of appropriate size, it would fall onto a surface below and pile up as two ridges of sand. Although the two piles might overlap, this simple particle model of light could not explain the distribution into the many bands which you observed.

Apparently, the wave model is the satisfactory one for interpreting diffraction experiments.

EXPERIMENT 4-3 Measuring the separation of wires in a wire mesh (a demonstration)

In Experiment 4–2, you used the known spacing between two slits to measure the wavelength of light. This experiment reverses that procedure. You will use the wavelength of red light that you have already measured to determine the spacing between the wires of three pieces of different wire mesh.

Since the wire mesh has wires in a two-dimensional pattern, it is better to use a point of light rather than the straight-line source used with the double slits (see Fig. 4–15). Our point source is a slide projector holding a slide that is entirely opaque except for a very small pin hole. With the slide in the projector, the pin hole can be focused on the screen. The wire mesh should then be inserted into the beam of light. The whole class can then see the diffraction pattern.

The wire mesh behaves in much the same manner as the double slits, so you can use the same equation to determine the spacing of the wires in the mesh.

$$n\lambda = d \sin \phi$$

But now you must solve it for d.

$$d = \frac{n\lambda}{\sin \phi}$$

If white light is projected through the pin hole in the slide and then through the wire mesh, what type of diffrac-

FIGURE 4-15
Diffraction of light from a distant street lamp by a screen and mesh curtain. (Fundamental Photographs)

tion pattern would you expect to see on the screen? How will the diffraction pattern change when a red filter is placed in the path of the light?

Using the same red filter you used in Experiment 4–2, and making the appropriate measurements, determine the spacing d of the wires in the wire mesh.

What value of n can you use?

Which measurements should be made to determine the value of sin ϕ?

Refer to your notes of Experiment 4–2 to obtain the wavelength of light passing through the red filter.

In Chapter 6 we will be concerned with making measurements of distances between atoms, and we will consider a technique similar to the measurement made in Experiment 4–3. The distances considered in Chapter 6, however, are so small that even the wavelength of visible light is too long. To make measurements of distances between atoms, x rays must be used.

QUESTIONS   **4-13**   Two waves travel along a rope in opposite directions, as shown in Figure 4–16.

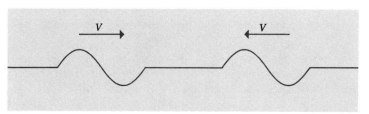

FIGURE 4-16
Two waves approaching each other.

(a) Draw the shape of the rope when the leading half of the left wave coincides with the leading half of the right wave.

(b) Draw the shape of the rope when the two waves coincide completely.

(c) Draw the shape of the rope when the trailing half of the left wave coincides with the trailing half of the right wave.

(d) What point(s) on the rope always remains at rest (never moves up and down)?

4-14   Consider Figure 4–6. Make a tracing of the 3/24 sec and the 7/24 sec frames. In what direction (up or down) is each part of the spring going? Put arrows on your tracings to show the directions.

4-15   Consider Figure 4–9. Trace *every other* circular arc to the right of screen *B*. These circular arcs are twice as far apart as those in the original. (Note: Use a pencil or pen which gives a dark line.)

(a) If these arcs represent wave crests, what is the wavelength of these waves compared to the original? How does their frequency compare with that of the original? (The velocity of all electromagnetic waves, regardless of wavelength, is equal to 3×10^{10} cm/sec in vacuum.)

(b) When you look along the right edge of this new pattern, how does the alternation of light and dark rays compare to the original?

(c) Can you predict qualitatively what the pattern would look like if you used every third circular arc? Twice as many circular arcs as shown in Figure 4–9?

4-16 As in Question 4-15, trace out *every other* arc in Figure 4–9. On another piece of paper trace out the remaining arcs in Figure 4–9. Now you have two figures with the same wavelength, but the waves of one figure have traveled $\frac{1}{2}$ of a wavelength farther than the other. Now look along the rays of both figures. Are the light and dark bands of reinforcement and cancellation at the same angles in the two drawings? What can you conclude from this?

4-17 Using a protractor, construct a right triangle with an angle of 30°. Measure the appropriate sides to determine the sine of 30°. Do the same for an angle of 37°, 45°, 60°. What is the sine of 90°? 0°?

4-18 Using the definition of the sine of an angle in a right triangle as the side opposite the angle divided by the hypotenuse, determine $\sin \phi$ for rays of constructive interference in Figure 4–10b. Compare this with the value obtained by measuring d and λ and by using a form of Equation 4–1, $\sin \phi = n\lambda/d$.

4-19 Measure the angles ϕ of the corresponding rays of reinforcement in Figures 4–10a and 4–10c. Compare the angles in one figure to the corresponding angles in the other figure. Are they equal? Why?

4-20 Draw a diagram similar to Figure 4–13 and indicate the separation distance between two slits made with a third blade used as a spacer. How does this separation distance compare with the separation when no spacer is used?

4-21 Suppose the sharp edge of the razor blade is not halfway between its two sides but is one-fourth of the way from one side to the other. What can you say about the separation of slits made with a pair of such razor blades held together? How could you find out that the razor blades you used were not made this way?

4-22   In Experiment 4–2 in which you measured the wavelength of light, why couldn't you use the slits which you had made with a spacer?

4-23   When you performed the double-slit experiment to measure the wavelength of light, what was the distance (approximately) between your eye and the slits? How many wavelengths of red light were there in this distance? Compare these numbers with the number of wavelengths shown in the photograph, Figure 4–10*b*.

4-24   A source of red light ($\lambda = 6500$ Å) produces interference through two narrow slits separated by a distance $d = 0.01$ cm.

(a) Suppose that the pattern produced was projected onto a screen. How far away would the screen have to be so that the bands near the central band would be 1 cm apart?

(b) If you viewed the source by placing one eye directly behind the slits as you did in Experiment 4–2, how far should the slits be from the source so that the bands near the central band appear to be 1 cm apart?

(c) Why do you think we used the method described in part (b) of this question instead of that in part (a)?

4-25   A knowledge of technical terms is not one of the primary goals of this course, but studying new terms whose meaning has been essential in the development of Chapters 3 and 4 would be a useful way of reviewing the material of those chapters. Make a list of these terms.

4-26   What concepts from Chapter 3 were necessary to the understanding of Chapter 4?

References   1. Baez, A. V., *The New College Physics*, W. H. Freeman & Co., 1967, Chapter 23, Sections 23.1 and 23.2, pages 281 to 285.

Young's experiment, with a brief discussion of the need for a constant phase difference between sources.

2. PSSC, *College Physics,* Raytheon Educ. Co. (D. C. Heath), 1968, Chapter 6, Sections 6–1 through 6–3, pages 87 to 93.

Waves on springs (as in our text), superposition, and interference. See also the color plates between pages 132 and 133.

3. Shamos, M. H., *Great Experiments in Physics,* Henry Holt & Co., 1959, Chapter 7, pages 93 to 107.

A short biography of Thomas Young, including Young's description of his experiment.

Albert Steiner, Photo Researchers

CHAPTER 5

CRYSTALS IN AND OUT OF
THE LABORATORY

5-1 *Crystal growth*

"The beauty of large single crystals is arresting. The flatness of their faces, the sharpness of their angles, the purity of their colors will give you deep satisfaction."[*]

Earlier in this course you grew crystals of salol on a glass slide by melting the salol and letting it solidify again. You could see "the flatness of their faces, the sharpness of their angles" and compare them with those grown by others. By this time you may also have grown crystals in a different way (Experiment 1–2) by letting a solution evaporate slowly. Again "the flatness of their faces, the sharpness of their angles" were evident, as well as "the purity of their colors." In this chapter we want to take a closer look at these crystals —to think about how they grow and what we can learn from our observation of them.

First, you should repeat the experiment of melting and crystallizing salol which you did at the beginning of the course. If you watch closely as a salol crystal grows, you will notice that its diamond-shaped faces increase in area but that the angles between the edges of the diamond remain the same. Two of them are acute angles (less than 90°), and two of them are obtuse angles (more than 90°) as in the diamond shape on playing cards.

Furthermore, no matter who grows the salol crystal, or where (in the laboratory, at home, in Washington, D.C., or in Moscow), the grower will observe exactly the same shape of face forming as the salol crystal grows. You can melt it and grow it again repeatedly, and the angles between the edges will be always the same.

Consider what kind of process must be taking place to

[*] From *Crystals and Crystal Growing* by Alan Holden and Phylis Singer, 1960, Educational Services Incorporated. Reprinted by permission of Doubleday and Company.

achieve such a result. Can you suggest a model for what is going on right at the surface of the crystal as the crystal grows? What happens to the liquid? How does it get used up? What happens when you heat the crystals again? Where does the liquid come from? Does your mental model describe what you observe? Melt and recrystallize the salol repeatedly and use your observations as a basis for answering these questions.

5-2 *Harvesting your home-grown crystals*

Now take a look at the crystals grown from solution in Experiment 1–2. If the crystals have grown reasonably large, pour off the liquid and carefully dry the crystals by pressing a wad of cleaning tissue into the jar. The blue substance found in some of the jars is copper sulfate and is poisonous. Be sure to dispose of the liquid and wash your hands. The crystals found in any of the jars may be stuck to the bottom of the jar; leave them there for the time being.

In what ways is this experiment similar to and different from the salol experiment? When you added water to the powder at the beginning of the experiment, the powder disappeared and the water became uniformly colored. The substance of the powder must have spread throughout every region of the water. Try to imagine yourself able to see what is going on during this process, to see details so small that you could not ordinarily see them even with the most powerful microscope. Can you make a mental model of the matter that was once solid and later permeated the water thoroughly?

What happened to the liquid to cause the growth of crystals? Can you suggest a model for what is going on right at the surface of the crystal as it grows? To repeat the salol experiment you heated the crystals. How would you proceed to repeat Experiment 1–2, using the same crystals?

Review your original observations made before you added water to the powder. All the jars originally contained a mixture of colored powder and white powder. Now compare the crystals you have grown with those that your classmates grew. In all cases a colored solution formed when you added water, but in half of the jars two kinds of crystals have grown, whereas in the other half of the jars a single kind of crystal has grown.

Clearly you could not have predicted that one mixture would produce two kinds of crystals with different shape and color, whereas the other gave crystals which appear all pretty much alike, having a color intermediate between those of the two original powders. Can you extend your mental model to suggest what might be going on, again on a submicroscopic scale, to give the observed results?

Observe the crystals; a good way to examine them is to hold the jar under a bright light and look at the crystals with something white behind them. Turn the jar and notice how the various faces of the crystals reflect light. Record your observations, using sketches if they are helpful. Now try to pry the crystals loose from the bottom of the jar so that you can examine them more closely. Try not to break the jar.

Recall that the angles between the edges of the flat salol crystals are dependably the same—in *all* salol crystals. Are there constant angles between the *edges* of the crystals you have just examined? Are there constant angles between the *faces* of those crystals? You can make an approximate comparison by eye without actually measuring the angles. There will be different angles between different faces on any one crystal, but corresponding pairs of faces will meet at the same angle on each and every crystal of the same substance. The fact that each different substance does have its own characteristic shape is one of the earliest observations made when men started to look closely at crystals.

Compare the interedge angles on different kinds of crystals: the blue, the violet, the colorless. How do these compare with the angles you observed in salol? Compare interfacial angles if you can.

It will be instructive to compare two other kinds of

crystals: common sugar and salt. These can be grown at home in a week or so, according to the instructions given in Experiment 5–1.

EXPERIMENT 5–1 Salt and sugar

Part A Salt

To about one-third of a glass of water add more salt than will dissolve, even when stirred vigorously. The cloudiness of the solution is due not only to undissolved particles of salt but also to the presence of a substance used by the salt producers to prevent the salt grains from sticking together. This substance does not dissolve in water but forms a finely divided material that will remain suspended in water for hours. Let the solution stand for several hours, and the material will settle out. Then the clear solution containing the dissolved salt can be poured into another container. This must be done slowly and carefully so as not to disturb the sediment at the bottom. Such a procedure is known as *decanting* the liquid.

A useful trick for keeping the solution from dribbling down the outside of the glass is to hold a glass rod (or soda straw or teaspoon) in a vertical position against the lip of the glass. The liquid will follow down the vertical surface into the container.

Observe the salt solution at intervals for the next few days and record your observations.

Remember that some undissolved salt as well as an insoluble substance was on the bottom of the first container. Can you predict what would have happened if you had left the solution in that container instead of decanting it off into another one? Perform the experiment to check your prediction. Record the results of the two experiments in your notebook and compare them.

Part B Sugar

Add one-fourth of a glass of hot water to one-half of a glass of sugar. Stir the hot water and sugar until the sugar is dissolved. The mixture will become thick and syrupy. If it is allowed to sit on the shelf for a day or two, crystals may not form in the same manner as they did in the salt-water solution. Can you suggest an explanation for this?

You may recall that in a different situation, when we were crystallizing salol, we added a bit of salol crystal to the molten salol to get the crystallization started. Such a bit is sometimes called a *seed*, and the process of adding it is known as *seeding*. The salol became a super-cooled liquid—one cooled below its crystallizing temperature. In the sugar solution we have a supersaturated solution—one containing more than enough sugar to form crystals. Seeding may result in crystal growth.

Drop one or two grains of sugar from the sugar bowl into the thick solution. If they dissolve, the solution is not supersaturated, and further evaporation must take place before growth will occur on the seed crystal.

Another way of producing a seed is to hang a soft thick string in the solution. The solution soaks up into the fibers and evaporates there so that some tiny crystals may grow, and these may act as seeds. Crystals will form in parts of the solution other than on the string and tend to retard the growth of the crystals you want. To encourage crystals to grow on the string, you can use two glasses; when the unwanted crystals form in one glass, the sugar solution can be poured through a fine sieve into the other glass. However, make sure that the other glass is clean. The string, with its crystals, can then be lowered into the solution without the distracting crystals. Depending on conditions, the string may have to be tranferred from one glass to the other every few days.

You have seen a number of examples in which a solid with a definite shape became fluid and shapeless either by being heated or by being attacked by water. You have seen the solid form again with its regular faces, and you have tried to make a mental model of what is going on. The next experiment is a model one—a real model rather than a mental model. This model illustrates one way that atoms can be arranged in the solid state.

EXPERIMENT 5-2   Packing of spheres

Place 200–300 glass spheres, 2.5 mm in diameter, in a clear plastic box, about 9 × 5 × 1.5 cm in size. Put the lid on the box and do not open it again. Shake the box and then let the spheres run together into one end or one corner of the box, keeping them in a single layer as much as possible. What do you observe? Is there any regular arrangement of spheres? How many spheres does each sphere touch? Could you arrange pennies the same way? Mixed coins?

Could the spheres be arranged in a single layer so that there are more of them on a given area of the floor of the box, or are they as close-packed as possible? Do the spheres always fall into the same arrangement after the box is jiggled? Are there imperfections in the regularity of arrangement—places where the orderliness is interrupted?

When the box is tipped so that the spheres pile up in one corner, they form a three-dimensional arrangement, several layers deep. How does the arrangement in the second layer compare with that in the first? How many spheres does each sphere touch now? Are they close packed, or could you arrange them so as to occupy less space?

Some metal crystals have their atoms arranged in just the same orderly array as the balls in the box. In the same way their three-dimensional array has imperfections. These make it easier for the atoms to move past each other, and therefore make it easier for man to bend metals, roll them into sheets, or draw them into wires.

If you shake the box violently, you destroy the orderly array, and the balls move randomly past each other. How could you destroy the orderly array of atoms in a metal crystal so that its atoms move randomly past each other?

These spheres in the box do not represent a satisfactory

model for the way all crystals are constructed, although the simple close-packed array of this model does result in angles between edges and between faces that are found in some crystals. However, many other crystals have shapes that could not result from such an array of crystal units.

5-3   *Harvesting crystals grown in nature*

With but one exception, rocks are made entirely of crystals. That one exception is the volcanic glass found near some geologically recent volcanoes. Crystals that occur in rocks, sand, silt, and soil are called minerals.

When rocks which have been melted deep in the Earth become cooler, they crystallize, just as the salol did on your glass slide. Neighboring crystals interfere with each other's growth just as the salol crystals did. Such rocks are called *igneous rocks*.

Whenever igneous rocks are exposed at the surface of the Earth, changes begin to occur in them as a result of *weathering* and *erosion*. Weathering processes cause breakdown or disintegration. The greatest amount of weathering results from chemical combination of parts of rocks with oxygen and carbon dioxide in the air, and with water. Simultaneously, erosion (the removal of the disintegrated rock) occurs. When the rock fragments are small enough, wind and water pick them up and carry them away to deposit them in other places. Water-carried fragments become rounded and sorted. The larger pieces drop out first as the slope of the stream bed becomes gentler and the velocity of the water decreases on its downward trek to the ocean. Then the next smaller fragments drop out, and so on, until the stream loses most of its load of materials. Because of this method of deposition, water-laid materials are characterized by rounded shapes and evenly sorted sizes of fragments.

In drier regions, wind picks up small fragments and carries them along before depositing them. Again, sorting occurs, but this time according to the carrying power (speed) of the wind.

In glaciated regions glaciers pick up loose materials, carry them along, much as a giant conveyor belt would do, and dump them in piles, called moraines, of unsorted sediments.

Such rocks, weathered to fragments, are carried away and redeposited as sand, silt, clay, and limey mud. Deposits formed in this way—at the mouths of rivers, in the beds of rivers and lakes, or wherever water drops them—pile up, layer after layer. The weight of added layers on top compresses the lower layers, and fine material seeping in between the grains helps to cement them into a hard mass. They become *sedimentary rocks* such as sandstone, shale, and limestone.

Both igneous and sedimentary rocks can be further pressed, heated, and changed in structure or *metamorphosed*. These are called *metamorphic rocks*. During the process of metamorphism,° crystalline minerals already present are changed into different crystal forms. They are not just squeezed into different shapes, as you would squeeze clay; instead, one kind of crystal disintegrates and another grows in its place. Great pressure can reorder the structure so that limestone becomes marble, sandstone becomes quartzite, and shale becomes slate. In many rocks the grains are too small to see without the aid of a magnifying glass. This is partially true of shale and fine sandstone which are just silt and sand, respectively, stuck together to form rock. Most sand and sandstone is made up largely of grains of minerals tumbled by rushing water or washed back and forth by the waves of the sea. The less durable minerals get broken up mechanically or decomposed chemically so that finally only quartz grains are left. Look at some of them with your magnifying glass. What do they remind you of? Glass? Ice? After looking at grains of quartz sand with your magnifying glass, the following quotation from *Crystals and Crystal*

° The zoologist uses a somewhat different word, *metamorphosis*, for change in form of living organisms.

Growing by Holden and Singer will take on more meaning for you.

Are glaciers made of quartz? Several centuries ago people thought they were. But today you know better. Glaciers are made of ice.

Here is an interesting instance of what a word can do to confuse matters. The word "crystal" comes from Greek roots meaning "clear ice." Who now would deny that quartz is "crystal" and glaciers are "clear ice"?

Even today the word "crystal" is still a confusing one: an unabridged dictionary will show you many different uses of it. Is a crystal a clear ball in which to gaze at the future? A cut-glass punch bowl? A gem set in a ring? Has it plane faces and sharp angles? Can you see through it? In one definition or another the dictionary will probably say yes.

But to these questions a physicist will answer "Maybe; it all depends." He has a definite use of the word "crystal"—one of many uses given in the dictionary—which he shares with the chemist, the metallurgist, and the crystallographer. The shortest definition he could give you would be, "Crystals are solid, and solids are crystals," not a very helpful definition until he expands it. And "solid"? He has a definite use for that word too.

To your question about the punch bowl, he may give you the answer, "No, that isn't crystal, because it isn't solid." It is a surprising answer, perhaps even an annoying one to anybody who has broken a punch bowl and found that the fragments of glass are "solid" enough to cut him. But be patient with the physicist and hear his reasons. You will get new insights into the world of matter; you will learn to see his crystals in the most unexpected places.

One very good place to see the crystals that make up rocks is in the polished pieces of rock that are often used to decorate the street-level walls of buildings. Many of these are worth close examination with a magnifying glass.

As you walk past such polished rocks on a sunny day, note how some of the minerals reflect the sun brightly at certain angles so that the rock surface twinkles and sparkles as you pass. The cause of this is that some crystals have a tendency to break along very smooth planes called *cleavage planes;* and when the polished surface of the rock is nearly parallel to such a plane, the crystal breaks along that plane during the polishing. Then the cleavage plane reflects the sunlight as brightly as a mirror.

5-4 *Breaking crystals*

If you have not already pried loose the crystals you grew in the jar, do so now. When you break them, what does the broken surface look like? Do you think it is a cleavage plane? Not all crystals have cleavage planes; for example, quartz is a mineral that does not. Try to find a piece of quartz to break. Since quartz is the commonest of all minerals, this should be easy. It is white or colorless and looks glassy when broken. Nearly all white pebbles are quartz.

EXPERIMENT 5-3 Cleaving crystals

You will be given some crystals to break. Some of them, perhaps all, will have cleavage planes. You can start by giving the crystal a light tap with a hammer, rock, or other heavy object. Watch carefully when the first break occurs, to see whether the break occurred along a plane surface. (Note that the word *plane* here means perfectly flat and smooth.) If so, you may want to try cleaving the crystal as the diamond specialists cleave diamonds.* Place a sharp knife on its surface, preferably near the middle of the crystal, as though you were going to cut it in two. Then strike the back of the knife a rather hard blow with some blunt weapon. The handle of a screw driver does nicely, but the heel of your shoe will serve almost as well.

When you have cleaved your crystal into many small pieces, you should have made some very important observations.

* When a large raw diamond (which commonly looks much like a large quartz pebble) must be broken into smaller pieces before it is faceted as gem stones, it is always cleaved. To saw it would waste the diamond material that would be worn away by the saw.

Question 5–1   How many differently oriented cleavage planes did the crystal have? (Parallel planes have the same orientation. These constitute one set of parallel cleavage planes but are usually referred to as one direction of cleavage or simply one cleavage plane.) If you were given more than one kind of crystal to break, you should answer this question for each crystal.

Question 5–2   Can you predict how a particular piece will break, that is, what the orientation of the surface of the break will be?

Question 5–3   Do the remaining pieces still have cleavage planes along which they would break? If so, how many?

Question 5–4   If you crush some of the crystal so that it is too small for you to detect its shape, can you predict what it would look like if you examined it with a magnifying glass? Try it.

Question 5–5   How small a bit of the crystal do you think would still be bound by cleavage faces oriented in the same way as those you have been observing?

Your experience in breaking crystals into small bits repeats the experience of the Abbé Haüy (pronounced "Howie"), who lived in the latter half of the eighteenth century. His story is a famous one and has often been told.

When René Just Haüy was teaching at Lemoine College in France, he began to devote all his spare time to the study of botany, but a good friend of his was a mineral collector, and probably because of trips he had taken with his friend, Haüy had a small collection himself. One day his friend showed him a particularly fine specimen of calcite.

While Haüy was examining the treasured specimen, he dropped it. One of the large crystals broke off, and his good friend let him have it for his collection. But Haüy had noticed a surprising thing. The broken surface was a smooth bright face, and he was curious to see whether the same sort of face would result if he broke it again.

"The prism had a single fracture along one of the edges of the base," he wrote later, "by which it had been attached to the rest of the group. Instead of placing it in the collection which

FIGURE 5-1
*Photograph of a topaz
crystal. (From Elizabeth
Wood's* Crystals and Light:
An Introduction to
Optical Crystallography,
*1964, D. Van Nostrand
Company, Inc., Princeton,
New Jersey.)*

I was then making, I tried to divide it in other directions . . ."
He found that there were three directions in the crystal along
which he could break it and get a bright plane surface like the
first one. . . .

When Haüy observed that the angular relations between the
cleavage faces were the same in every fragment, he erroneously
concluded that he could break the crystal into the small pieces
which were the ultimate building blocks of which it was constructed: ". . . and I succeeded after several trials," he wrote,
"in extracting its rhomboid nucleus." In spite of this error, he
did have the concept of a repeat unit, and he showed that he
could reconstruct the original sharply pointed crystals, with the
proper interfacial angles, by stacking up the little cleavage
blocks which he called "molecules integrantes."° [One of his
figures is illustrated in Figure 5–2.]

You have repeatedly had evidence that crystals are made
up of some sort of bits that can be added on, bit by bit, as
the crystal grows, and that these bits have a shape which is
the same for all crystals of one substance but may differ
from one kind of crystal to another. Here is one more piece

° Here quoted from *Crystals and Light, an Introduction to Optical
Crystallography* by Elizabeth A. Wood, Van Nostrand, 1964. Reproduced
by permission of D. Van Nostrand Company, Inc.

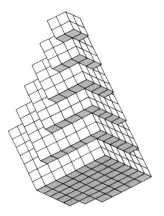

FIGURE 5-2
Calcite units according to Abbe Haüy.

of evidence in support of this hypothesis. The angles that crystal faces make with each other have been found to be related to each other in a special way. In Figure 5–3 you see the outline of a crystal of topaz with the crystal faces represented as lines, just as though you had sliced the crystal in two and were looking at the cut face. In addition to the vertical face on the left side of the crystal, two sloping faces are shown. Now the interesting observation is that the slope of one of these faces is just exactly twice as steep as the slope of the other, as shown in Figure 5–4. In the same horizontal distance, the steeper-sloping face goes up just twice what the gentler-sloping face does. The slopes of crystal faces are always related in this way by small whole numbers.

What does this suggest? To the early crystallographers, those who observed crystals and measured the angles between their faces, it suggested that crystals were made up of some sort of bits with a definite shape, packed in a

FIGURE 5-3
Outline of a topaz crystal.

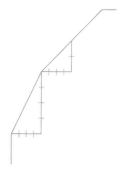

FIGURE 5-4

Comparison of slopes of faces.

regular way, and set back stepwise in a specific way at the boundary of the crystal to form the differently oriented faces, as shown in Figure 5-5.

If this is so, why can't we see these little steps on the face of the crystal? If they exist, they must be very small. How small? An examination of Figure 5-6 will suggest the answer. This figure shows water waves incident on two different surfaces. The first surface (in Figure 5-6a) is a side of a block. That surface is rough, but the irregularities of the surface are smaller than the wavelength of the water waves incident on the surface. The waves reflect from the side of this rough block as if it were perfectly smooth.

Figure 5-6b, on the other hand, shows water waves incident on a surface which has irregularities that are as large as the wavelength of those water waves. It is clear that the waves do not reflect from this very irregular surface in the same manner that they reflect from the rough surface of Figure 5-6a. Their reflection from the surface with big

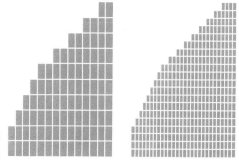

FIGURE 5-5

Units of structure in topaz.

FIGURE 5-6
Waves are approaching an oblique object from the left.

(a) The surface on which the waves impinge and from which they reflect is smooth, that is, it has irregularities which are small compared to the wavelength of the waves. This results in a reflected wave with a straight wavefront.

(b) The reflecting surface has irregularities large compared to the wavelength. This alters the shape of the reflected waves completely. (Fundamental Photographs)

irregularities is itself irregular; their reflection from the rough surface (the one with small irregularities) is regular. Similarly, a piece of glass that has been scratched with sand paper reflects the light irregularly.

The consideration of the two photos of Figure 5–6 leads us to conclude that since crystal faces reflect light with the regularity of a mirror, the step-like irregularities on the crystal faces of Figure 5–5 are smaller than the wavelength of light. In Chapter 3 you measured the wavelength of red light to be about 6.5×10^{-7} meter or 6500 Å. This means that each of the orderly bits of material that make up the crystal has an overall dimension smaller than 6500 Å.

We have developed a model for the structure of crystals made up of a very large number of very, very small units, all similar in shape for any one kind of crystal, repeated hundreds of thousands of times. However, we do not know how to measure the size of the little units of crystal material nor do we know precisely what they are.

This is the unsatisfactory state of knowledge about crystals that vexed crystallographers just before the year 1912. In that year there occurred a very exciting breakthrough to a far better understanding of the structure of crystals. We will deal with this in the next chapter.

QUESTIONS

5-6 In Experiment 1–2 all jars originally contained a mixture of a colored powder and a colorless powder. In some of the jars crystals intermediate in color between the two powders grew. In other jars two different kinds of crystals grew: one blue and one colorless. Suggest a model or models to describe these results.

5-7 You have observed that in certain cases two types of crystals were formed separately in the jars. However, when the original solids were dissolved in water, the solutions were homogeneous. Describe a model that would be consistent with this behavior.

5-8 In the topaz crystal in Figure 5–4 it was shown that the slopes of the crystal faces were related as 1 to 2. Draw a diagram of the same crystal with faces whose slopes are related as 1 to 3.

5-9 Suppose you lower the temperature of a solution of common salt (sodium chloride) until it starts to freeze. Design an experiment to determine whether the first ice crystals formed contain any salt.

5-10 Assume that each of the following figures is the outline of the top view of a solid figure resting on the paper. Furthermore, assume that each of the figures is composed of building blocks smaller than the figures. What shape must these building blocks be, and what arrangement must they take in each figure, if all of the building blocks are identical, and if either of the following is true.

(a) All the building blocks are oriented in the same way.
(b) Any orientation is permissible.

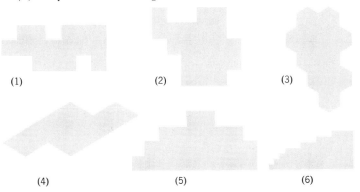

(1)   (2)   (3)

(4)   (5)   (6)

5-11   In Experiment 5–1 you grew some salt and sugar crystals from solution. Do you suppose the solutions were more concentrated near the bottom or near the top? Propose an experiment that might permit you to answer this question were you able to perform it. How would this experiment help you answer the question?

References   1. Baez, A. V., *The New College Physics*, W. H. Freeman & Co., 1967, Chapter 10, Sections 10.1 and 10.2, pages 120 to 127.
Simple examples of evidence for matter existing in "bits."

ALSO OF INTEREST:

2. Holden, Alan and Phylis Singer, *Crystals and Crystal Growing*, Anchor Books, Garden City, New York, 1960, Chapters IV and V.
Chapter IV discusses two methods for growing crystals, and Chapter V gives twelve recipes for growing crystals. The remainder of the book is of value to those who desire to study crystals in detail.
3. Hurlbut, Cornelius S., Jr., *Minerals and How to Study Them*, 3rd Ed., John Wiley and Sons, New York, 1949, pages 63–68.
Directions are given to the person interested in growing crystals.
4. Jensen, David E., *Getting Acquainted with Minerals*, Revised Edition, McGraw-Hill, New York, 1958, Chapter 2.
The entire book is written for the hobbyist. Chapter 2 contains useful hints and information about collecting minerals. Other chapters describe minerals and means of listing them.

Bough, Fundamental Photographs

CHAPTER **6**

WHAT HAPPENED IN 1912

6-1 *The story*

This is the story of an important event in science that took place in the university town of Munich in the year 1912. The University of Munich was one of the best in Germany and a center of experimental research in science. Many graduate students were going beyond their four years of college work to do advanced study. Every afternoon they would go across the street for coffee at the Cafe Lutz and distress the waitresses by drawing diagrams of their problems on the marble table tops.°

One of the three people of interest to us in this story is W. C. Roentgen, a very shy professor who kept to himself as much as possible. He was a very careful experimenter, not one to jump to conclusions without painstaking examination of the evidence. He was working on a problem that puzzled him. Seventeen years earlier he had found a new kind of ray whose nature remained unknown. Since x generally stands for an unknown quantity, he called these new rays "x rays." He discovered that these rays affected photographic film in much the same manner as light. He also discovered that x rays can pass right through matter, for when he placed his hand between the source of x rays and a photographic plate, the plate, when developed, revealed the bones of his hand. His observations indicated to him that if x rays are made of particles, they must be very small particles indeed. On the other hand, x rays might be electromagnetic waves similar to radio waves, light, and ultraviolet radiation. But was it possible for electromagnetic radiation, even of a very short wavelength, to pass right through solid matter, such as a hand? In the years right after the discovery of x rays in 1895, there had not been enough

° From an account by P. P. Ewald in *Fifty Years of X-ray Diffraction*, published in 1962 for the International Union of Crystallography by N.V.A. Oosthoek's Uitgeversmaatschappij, Utrecht. Later parts of this account are from the same source.

observations to substantiate any theory about their nature. X rays remained a puzzling enigma.

The second person of interest to us is Max von Laue,* a young instructor who already had his doctor's degree and was working on an article for an encyclopedia. The article had to do with wave optics.

The third person is Paul Ewald (pronounced "Ayvald"), a young graduate student who was studying for examinations, attempting to decide which of two job offers to accept, and trying to finish writing his thesis for his Ph.D. His thesis had to do with the way light interacts with a regularly arranged set of tiny oscillators of some sort—a theoretical problem, not an experimental one. However, it seemed likely to him that the results might have some application to the study of light in crystals since the evidence seemed to be that crystals are made up of a regular array of *something* (Chapter 5).

These three people were all at the University of Munich. One evening Ewald walked home with von Laue and discussed some problems of his thesis. In the course of their talk, the instructor learned from the student that the crystallographers had reason to believe that a crystal is made up of layer after layer of repeated units—a very regular array but on a very small scale. (Recall the evidence for this in Chapter 5.) Von Laue had been working on his encyclopedia article on light and had probably been writing the part that had to do with diffraction.

So that the rest will be meaningful, we must interrupt the thread of the story to investigate further the nature of diffraction.

6-2   *The nature of diffraction*

We have studied in some detail interference of light coming from two slits. What happens if we just continue to increase the number of slits? Figure 6–1 shows three ripple-tank photographs of just such a sequential increase in the

---

* The *Lau* part is pronounced just as *lou* in the word *loud*. The *e* on the end is pronounced just as the *a* in *about*.

a

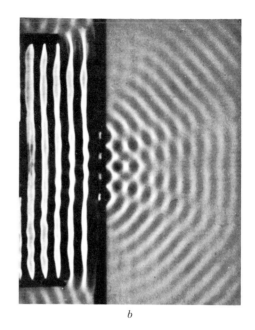

b

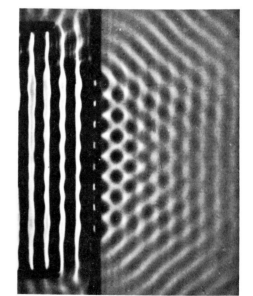

c

FIGURE 6-1 (a) An interference pattern of
straight water waves passing through a double
slit. (b) The same, except with four slits having
the same size and spacing as in a. (c) The
same, except with eight slits having the same
size and spacing as in a. (Courtesy of the
Physical Science Study Committee, Educational Development Center.)

number of equal-sized, equally spaced slits. Figure 6–1a is
a two-slit pattern. (Compare this with Figure 4–9a, which
is the equivalent of a two-slit system except that the sources
are vibrators rather than slits.) Figure 6–1b is a four-slit interference pattern, and Figure 6–1c is an eight-slit pattern.
Note that the outer beams (top-most and bottom-most)

make the same angle with the plane of the slits, independent of the number of slits, so long as their spacing and the wavelength of the waves is the same. Recall the relationship between wavelength, slit spacing, and angle you found in Chapter 4.

We could perform similar experiments with light, provided we could construct a series of equally spaced slits, each with an exceedingly narrow width. Fortunately, these are available to us—they are known as *diffraction gratings*, although they might be more properly called interference gratings. The diffraction grating you will be given is a thin, transparent film which has grooves and ridges, forming thinner and thicker parts acting in the same way slits do. It has been prepared by replication; that is, by pressing a transparent acetate film material against an original metal grating. The original grating was produced by the laborious process of scribing parallel lines very close together on a metal base using a special ruling engine. These gratings have 13,400 lines to the inch. This means that the separation between lines is just under 2×10^{-4} cm, or 20,000 Å, and is equal to about three wavelengths of red light or five wavelengths of blue light. Thus, these transparent replicas satisfy the conditions necessary for a diffraction grating: that the spacing between lines be close to, but greater than that of the wavelength of the radiation to be diffracted.

The reasons the spacing between the grating lines must be close to the wavelength of light may be made clear by considering Equation 4–1, which can be written in the form

$$\sin \phi = \frac{n\lambda}{d}$$

Consider first what the angle ϕ would be if λ were equal to d. For $n = 1$, you can see that $\sin \phi$ would equal 1, and, as you found in your answer to Question 4–17, $\sin 90° = 1$. This would mean that the beam for $n = 1$ would be at 90° to the primary beam (the one proceeding straight ahead) and would lie right along the grating. You can also see from Figure 4–10 that as λ gets larger, ϕ gets larger and when

$\lambda = d$, with $n = 1$, ϕ will equal 90° and the diffracted rays will lie right along the screen carrying the slits. Therefore, no constructively reinforced beams can come from the grating if the wavelength of the electromagnetic radiation is longer than the grating space.

If the wavelength λ is much shorter than the grating space d, then according to Equation 4–1 the various values of n (1, 2, 3 . . .) will give many closely spaced values of ϕ together; and it will be difficult, perhaps impossible, to separate one from the other as you were able to do in Experiment 4–2 when you measured the wavelength of red light.

Question 6–1   Figure 5–5 is a good model of the regular, step-like nature of the face of a crystal. If the faces of the crystals you grew have a regular array of steps and niches, why don't they act as diffraction gratings for visible light?

If the plastic diffraction grating is placed in one end of a long tube, and if a slit, limiting the beam of light, is placed in the other end with its long direction parallel to the lines on the diffraction grating, the spectrum of the entering light can be clearly seen off to one side within the tube. Such an instrument is called a *spectroscope.*

You will be given a spectroscope. With the grating near your eye and the slit at the far end, look through the spectroscope at a bright source of light. Do not look directly toward the Sun, as this might injure your eyes, but do use the spectroscope to look at many different sources of light. Look at fluorescent lights, neon lights, and other colored displays of light. Compare their spectra with that of a bright white light bulb. A regular light bulb is an *incandescent* source, that is, one in which the light is produced by a hot wire. Later in this course we will find out more about these spectra, and at that time you will use your spectroscope again.

Compare each of the spectra from the above sources with the spectrum you obtained with the prism in Chapter 3. The color separation in each instance is obtained because of a range of wavelengths present in the light. Each color passing through the grating has rays of reinforcement and cancellation appropriate to its particular wavelength and to the spacing of the lines of the grating.

6-3 *The idea*

Now we can return to the story of what happened in 1912. From your knowledge of interference effects and work with a diffraction grating, you can see that a measurement of the angle at which a diffracted ray leaves a diffraction grating can tell you the spacing between the slits or ridges of the grating—provided you know the wavelength of the diffracted light. Max von Laue had probably been trying to describe just this in his encyclopedia article. He had also perhaps been wondering whether to include Professor Roentgen's new x rays in his encyclopedia article on waves. Even Professor Roentgen himself wasn't sure that they were wave-like. He suspected, however, that if they were waves they must have a very short wavelength, since he knew they did not behave like any of the longer wavelength radiations with which he was familiar.

Now the grating spacing of a diffraction grating has to be pretty close in length to the wavelength of the light being diffracted. Max von Laue knew that diffraction gratings of the type used for light would be too coarse for x rays. One would need a diffraction grating with a very tiny spacing to diffract Professor Roentgen's x rays if they were electromagnetic radiation with a very tiny wavelength. With all this in his mind, he learned from Paul Ewald that crystals have some pattern which regularly repeats with a very tiny repeat distance that no one had ever measured. Could it be that a crystal would act as a diffraction grating for x rays? Max von Laue became so lost in thought about this, that he stopped listening to Ewald, who finally gave up trying to discuss his thesis and went home.

6-4 *The experiment*

Very soon afterward, Dr. von Laue asked two young graduate students, Friedrich and Knipping, to place a crystal in a beam of x rays from a Roentgen x-ray generator to see whether the beam was diffracted by the crystal. If it was diffracted, then, instead of there being one beam after it left the crystal, there would be several separate beams, just as there are when visible light is diffracted by a diffraction grating. If the crystal would act as a diffraction grating for much shorter wavelength electromagnetic radiation, and if x rays were in fact just like light but with much shorter wavelength, then diffraction would occur.

Friedrich and Knipping placed a crystal in position to receive the x-ray beam. From Professor Roentgen's experiments with objects scattering x-ray beams that fell on them, they decided that the most favorable place to put the photographic plate was either between the source of x rays and the crystal or beside the crystal, with the plate parallel to the x-ray beam. They arranged the plates in these ways and protected them from visible light which would expose them. Then they turned on the x rays. After some time they turned off the x rays and took the plates into the darkroom to develop them.

It was a very exciting moment. Would the plates show separated spots from diffracted beams? When the plates were developed and they could turn on the red light in the darkroom to have their first look, they found no separate spots—only a general darkening of the plates.

It has been said that people doing scientific research spend much of their time being either elated or dejected. This must have been a moment of dejection for Friedrich and Knipping. But they did not give up. They decided to try it again. This time they placed photographic plates in various positions around the crystal, however unpromising they seemed. And this time, when they developed the plate that was placed on the far side of the crystal from the x-ray source, they found a very important result: separate, distinct spots, made by separate, distinct x-ray beams diffracted by the crystal (see Fig. 6–2).

This historic observation indicated two very important things. First, it indicated that x rays are wavelike in nature and that they have a very short wavelength. This work

FIGURE 6-2

Photograph of the x-ray diffraction pattern of a crystal. This photograph, made with modern equipment, is much better than the first one taken by von Laue. (From Elizabeth Wood's Crystals and Light: An Introduction to Optical Crystallography, *1964, D. Van Nostrand Company, Inc., Princeton, New Jersey.)*

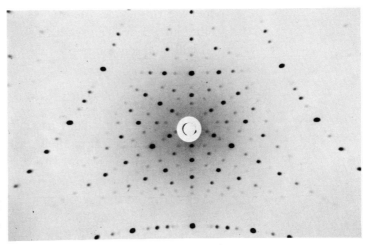

showed that x rays are part of the electromagnetic spectrum and have shorter wavelengths than ultraviolet radiation. Second, this observation showed that the crystal had acted like a three-dimensional diffraction grating; the wavelike x rays were diffracted by the crystal on passing through its fine-scale repetitive structure. Two years later, Max von Laue was awarded a Nobel prize in physics for achieving this major advance in our knowledge of the world around us.

Probably the most remarkable thing about Max von Laue's idea was that he had so little on which to base it. He didn't know that x rays were wave-like in their behavior; he knew even less about crystals, except that Paul Ewald had told him that they were made up of something that was repeated on a very fine scale. Was his idea a "stroke of genius"? It has been said that "genius favors the prepared mind." Max von Laue's mind had just been occupied with the task of describing how diffraction takes place. It was prepared to see the possibility of diffraction taking place where others would not have thought of it.

After the success of the experiment, there were still problems. A diffraction grating is an array of slits or light-scattering grooves lying in a plane. A crystal is a three-dimensional array of some sort of object, and the x rays penetrate this array. The objects at depth must contribute to the diffraction. The special nature of this sort of diffraction will be discussed in the following section.

6-5   *Diffraction from a regular array of objects*

We have seen that light passing through a row of regularly spaced slits (diffraction grating) will produce a diffraction pattern consistent with the wave model diffraction pattern. This pattern results from the interference of the waves scattered by the posts, and it is called either a diffraction pattern or an interference pattern. In Figure 6–3 the interference pattern formed when water waves move through a regular array of posts is shown. The incoming waves are traveling from the upper left toward the posts. There is also a set of outgoing waves which appears in the upper right as a set of light and dark bands. Note that these begin deep in among the little posts. If this were a movie, we would see that these are moving up and to the right. In effect, each little post acts as a point source of water waves just as each slit acted as a point source of light waves. The first row of posts scatters only a portion of the incoming waves. The rest of the waves pass through and can be

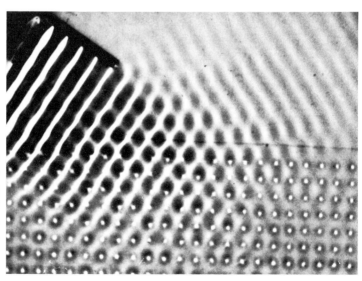

FIGURE 6-3
Interference pattern of water waves diffracted from a periodic array of objects. (Courtesy of the Physical Science Study Committee, Educational Development Center.)

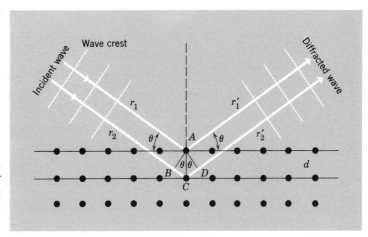

FIGURE 6-4
*Schematic line drawing of
the experimental
arrangement shown in
Figure 6–3.*

scattered by the second row of posts, the third row, etc. If there are many rows of posts in a regular array, we obtain an interference pattern, like that shown in Figure 6–3, under certain conditions which are discussed below.

The direction of travel of the incoming waves is perpendicular to the parallel white lines representing the crests of the incoming waves. As can be seen in Figure 6–4, the direction of travel of these crests (indicated by the arrows) makes an angle which we call θ (theta) with the parallel rows of posts (lines have been drawn through these rows for convenience). In this example, θ is about 35 degrees. Note that the sequence of waves leaving the array and traveling upward to the right also makes an angle of about 35 degrees with the same row of posts. Experimentation shows that waves will travel outward from such an array only for certain angles θ of the incoming waves. Let us see what determines these angles.

The geometry of the arrangement in Figure 6–3 lends itself to a simple quantitative analysis. A line drawing, representing this arrangement, is shown in Figure 6–4. What is the relationship between the angle θ, the vertical spacing d between the adjacent rows of posts marked by lines, and the wavelength λ of the incoming waves? The argument that answers this question is similar to the one used in the analysis of double-slit interference.

You may have found in the double-slit analysis that following a mathematical development is different from other types of reading. For a better understanding follow these developments step by step, referring to the figure so that you understand each step before proceeding to the next. You will find it helpful to take pencil and paper and write out the steps yourself. Only by such a systematic procedure will you acquire understanding and thus gain confidence in the conclusions. The conclusion of the mathematical development that follows is of great importance in the study of solids. It is the key to the use of the discovery made in 1912.

The waves traveling in ray directions r_1 and r_2 (Fig. 6–4) traverse the same distance from their source to line *AB;* the crests and troughs are exactly aligned at right angles to the direction of travel until line *AB* is reached. Line *AB* is perpendicular to the incident ray. At this point, ray r_2 must travel an additional distance, $BC + CD$, before reaching the line *AD*. The line *AD* is perpendicular to the direction of the diffracted ray. The rays r'_1 and r'_2 then traverse equal distances before reaching the observation point. If $BC + CD$ equals a whole number of wavelengths, the waves traveling along r'_1 and r'_2 will also have their crests and troughs coinciding from *AD* outward to the observation point. Constructive interference will result. If $BC + CD$ equals an integral number of wavelengths plus one-half of a wavelength, the wave leaving point *D* will always have its crest coinciding with the trough of the wave leaving point *A*. The result will be total cancellation—destructive interference (see text just preceding Experiment 4–2). Quantitatively, the condition that determines whether a wave will be diffracted may be expressed by

$$BC + CD = n\lambda,$$

where $n = 1$ or 2 or any whole number

Now we want to express the lengths BC and CD in terms of the angle θ and the separation distance d between the rows of posts. This can be done by referring to the right triangles *ABC* and *ADC* in Figure 6–4. Since *AB* is perpendicular to the incident ray, and *AC* is perpendicular to the rows of posts, the angle between the lines *AB* and *AC* is equal to θ. (Recall that two angles are equal if each side of one angle is perpendicular to a side of the other angle.) By the same argument the angle between the lines AD and AC is also θ.

In the right triangles ABC and ADC we can use the sine relationship defined in Section 4–3. For triangle ABC,

$$\sin \theta = \frac{BC}{AC}$$

therefore

$$BC = (AC) \sin \theta$$

For the triangle ADC,

$$\sin \theta = \frac{CD}{AC}$$

or

$$CD = (AC) \sin \theta$$

Consequently

$$BC + CD = 2(AC) \sin \theta$$

But AC is the distance d between the rows of posts so that

$$BC + CD = 2d \sin \theta$$

For constructive interference to occur, $BC + CD = n\lambda$. Therefore, we can express the condition necessary for constructive inference in Figures 6–3 and 6–4 as

$$n\lambda = 2d \sin \theta \qquad (6\text{--}1)$$

This equation is similar to Equation 4–1 for double-slit interference, although it describes a different physical situation. Notice for example, that the angles are defined differently. How do these definitions differ? Equation 6–1 was first proposed for x-ray diffraction from crystals by Lawrence Bragg, who worked with his father, William, in England in 1913. Later, father and son shared a Nobel prize for their work in x-ray diffraction analysis of crystal structures.

Although we have derived the Bragg equation by using a model of rows of posts in the regular two-dimensional array most conveniently shown on our two-dimensional piece of paper, the same equation holds for regular three-dimensional arrays of objects as well. If we know the wavelength of the radiation incident on the three-dimensional periodic array and measure the angles at which diffraction occurs, we can deduce the separation between the scattering objects in all directions, even though the objects themselves are far too small for direct visual observation.

The Bragg equation is a powerful relationship among the spacing d, wavelength λ, and diffraction angle θ. It is central to the science of crystallography. Light waves passing through two narrow slits in Experiment 4–1 were found to behave in a manner analogous to water waves passing through two openings. Similarly, x rays diffracted from the orderly array of atoms in a crystal behave in a way analogous to the water waves diffracted from the posts in Figure 6–3. X-ray diffraction is to the crystallographer what the microscope is to the biologist or the telescope to the astronomer. Every model showing the structure (the arrangement of atoms) of crystals is the result of x-ray diffraction studies.

An example of how the Bragg equation can be used to find the distance between rows of units making up a crystal can be obtained by applying the equation first of all to the water waves of Figure 6–3. These waves are redrawn in Figure 6–4, and we ask how far apart the rows of posts are. The angle θ is equal to $35°$; $\sin 35° = 0.57$. The distance $BC + CD$ is equal to $n\lambda$; assume that this distance is 1.2 cm. If we assume that $n = 1$, then $\lambda = 1.2$ cm. We can now solve Equation 6–1 for d, the quantity we wish to obtain:

$$d = \frac{n\lambda}{2 \sin \theta}$$

$$d = \frac{(1)(1.2 \text{ cm})}{(2)(0.57)}$$

$$d = 1.05 \text{ or about } 1.0 \text{ cm}$$

Question 6–2   Suppose that a beam of monochromatic x rays of wavelength $\lambda = 3.4 \times 10^{-10}$ meter is used to determine the spacing d between the rows of regularly spaced units in a crystal. Those x rays are observed to be diffracted when they are incident on the crystal at an angle of $37°$ with one of the crystal's cleavage planes; $\sin 37° = 0.60$. The x rays are not diffracted for any angle less than $37°$, so n must equal 1. Calculate the spacing *d* between the parallel rows of units that make up this hypothetical crystal.

Since 1912 the study of the orderly arrangement of x-ray beams diffracted from crystals has given us a rich yield of information about the orderly arrangement of atoms in crystals. The repeat units of structure, for which the crystallographers already had evidence, could now be measured. Together with earlier chemical and physical data, the data from x-ray diffraction enabled the scientists to deduce the nature of the structural units of which crystals are made. In some crystals they found that clusters of atoms occurred, separated from other clusters to which they were not as strongly bound. Such clusters are called *molecules.* Other crystals were found to be made, not of molecules but of continuous arrays of atoms, each as close to its neighbor on the left as to its neighbor on the right. Nevertheless, one could still distinguish a repetitive unit of arrangement of atoms, just as one can distinguish a unit of arrangement in a continuous sheet of wallpaper.

One such array of atoms is represented by the model in Figure 6–5. You may recognize it as the sort of pattern which

FIGURE 6–5
Hexagonal close-packed structure typical of some metals.

was assumed by the glass balls in the plastic box when they were shaken down into one corner. Several metals are made up of atoms arranged in just this way. Cobalt, magnesium, zinc, and zirconium are examples. X rays diffracted by such a regular array of atoms act in essentially the same way as water waves do when they are diffracted by a regular array of posts, as in Figure 6–3. The same equation, $n\lambda = 2d\sin\theta$, is applicable.

QUESTIONS   **6-3**   With lights out in your room at night, look at a distant bright light through a window screen, a thin woven curtain, or a tightly stretched linen handkerchief. Describe the pattern that you see, and draw a sketch of it. Interpret it in terms of the wave model of light.

6-4   (a) In an experiment such as that shown in Figure 6–3, suppose the spacing d between rows of posts is 2 cm and the wavelength of the water waves is also 2 cm. For $n = 1$, what angle do the incident ray and the diffracted ray make with the rows of posts?

(b) What is the longest wavelength for which a diffracted ray could occur from rows of posts with $d = 2$ cm ($n = 1$)?

References   1. Baez, A. V., *The New College Physics*, W. H. Freeman & Co., 1967, Chapter 23, Section 23.3, pages 285 to 288.
   The diffraction grating. Also, on page 596, a von Laue photograph.
2. Christiansen, G. S., and P. H. Garrett, *Structure and Change*, W. H. Freeman & Co., 1960, Chapter 25, Section 25–8, pages 435 to 438.
   X rays, von Laue's experiment, the Bragg equation.
3. Shamos, M. H., *Great Experiments in Physics*, Henry Holt & Co., 1959, Chapter 14, pages 198 to 209.
   Wilhelm Roentgen and the discovery of x rays.

ALSO OF INTEREST

4. Heathrate, Niels H. de V., *Nobel Prize Winners in Physics, 1901–1950*, Henry Schuman, New York, 1953.
   (a) 1914—Max Theodore Felix von Laue, pages 118 to

124. This is a matter-of-fact account of the circumstances leading to the discovery of x-ray diffraction by crystals. Von Laue's Nobel prize acceptance lecture is included.

(b) 1915—William Henry and William Lawrence Bragg, pages 125 to 138. Biographical sketches of father and son are given as is an account of the discovery that x-ray diffraction patterns can be explained by using an analysis based upon diffraction of waves from parallel planes.

CHAPTER 7

MATTER: A CLOSER
LOOK AT DIFFERENCES

7-1 The need to question

Have you ever been kept waiting for an appointment, read all of the current magazines, and then found yourself restlessly inventing ways to while away the time? If so, your choice of pastime was probably influenced by your immediate surroundings. You may have idly counted the number of squares in the ceiling, or listed all the colors in a pair of draperies, or figured out the repeating unit in the wallpaper. Suppose that you decided to list all solid materials in the room and to associate with each a characteristic property. For example, a rubber band lying on the receptionist's desk might remind you that rubber is elastic, or a quick glance through the window to check on the weather that glass is transparent. Take a minute or two to list a half-dozen materials in your present surroundings and a distinguishing property of each.

The variety of solids in our everyday world is so great that listing six different materials is easy. But you may be surprised to find it is quite a bit more difficult to associate a characteristic property with each one. For one thing you may find it difficult to list a property rather than a particular use (what is a characteristic property of leather? or a window shade?). More often than not, we make quite passive observations of our environment, and passive observations are not very informative.

To illustrate the limitations of passive observations consider a rubber band. You know that a rubber band is elastic because you have had previous experience, active experience, with rubber bands. But suppose for a moment that you saw a rubber band for the very first time and you were asked to state or describe its most characteristic or distinguishing property. By just looking at it you could describe only its size, shape, and color; its elastic property would remain unnoticed. However, if you picked up the band, you would immediately recognize its lack of rigidity. If you played with it a little, you would observe it actively and no doubt would soon discover that it stretched when you pulled on it and returned to is original shape when you stopped pulling. But think about this discovery process. In the random process of examining the rubber band you may have consciously or unconsciously asked the question, "What happens when I pull on it?" Having posed the question, you devised a method of finding an answer, that is, you pulled on it and discovered its *stretchability*. A second question might be, "Does it return to its original shape when I stop pulling?" You then devised methods of answering the second question and discovered its *elasticity*—its property of regaining original shape and size after deformation. By this definition of elasticity a steel ball, for example, though not very stretchable, is very elastic. Although it is difficult to deform, once deformed, it does regain its shape and size very well. Some quantitative questions about the rubber band would be more difficult to answer, such as "How much does it stretch?" or "How hard must one pull to stretch it?" or "How well does it regain its shape and size?" Further experience with other rubber bands would convince you that the properties of stretchability and elasticity are common to all rubber bands, and, in the absence of conflicting evidence, you would conclude that these are characteristic properties.

More often than not, answers to questions suggest other questions. The simple observation of the stretchability and elasticity of a rubber band might raise many questions of which the following are examples. Can you suggest answers for these?

Question 7-1 Do you think that a quantitative definition of stretchability is possible? What factors would have to be taken into account? How would you define quantitatively the stretchability of a rubber band? Several equally useful definitions are possible.

Question 7-2 Suppose you were given a rubber band, a nylon thread, a cotton thread, and a leather thong. Can you devise a simple experiment that would enable you to arrange these materials in order of decreasing stretchability? Do you think the order would be the same if they were ranked according to elasticity?

Question 7–3   One might wonder why or how a rubber band can be made to stretch. Suppose research on the structure of a hypothetical rubber band revealed that an acceptable model of it would be composed of discrete balls held together by springs. A section of such a model might be represented as in Figure 7–1. Using this ball and spring model, can you suggest what might happen when you stretch a rubber band?

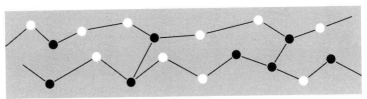

FIGURE 7-1
Section of a hypothetical rubber band.

We have considered differences in the behavior of various substances when they are stretched. Now let us investigate differences in the behavior of various substances in another way—by heating them.

7-2   *Another way to explore differences*

Wood is a very familiar substance and the burning of wood a common occurrence, so ordinary that one scarcely pays it any conscious attention. As you well know, it takes some time to start a wood fire because the wood must reach a fairly high temperature before it bursts into flame. You may have wondered how hot the wood must get before it begins to burn. What happens when wood burns? Gases, including carbon dioxide, are given off; smoke, which must consist of some kind of particles, is given off; ashes are left. We could make an analysis of the products of burning wood, but you already have a fairly good idea of what they are. What else can we do to learn more about wood in general and the burning process in particular? Clearly, what we find depends on what we ask, how we design our experiments, and how carefully we observe. Suppose you were to heat wood in the absence of air. How would this differ from heating and burning wood in plenty of air? The following experiment, in which you will heat some wood, is intended to answer this question, to give you practice in observation, and to acquaint you with some common manipulations frequently encountered in the laboratory.

EXPERIMENT 7-1   The distillation of wood*

What will happen if you heat some wood splints in a closed test tube? This experiment is more complicated than any you have performed thus far.

Question 7-4   Before you read on, try to predict on the basis of your experience, what will happen to the wood.

Pack a pyrex test tube with wood splints, clamp it to a pegboard, and then connect the glass tubing, another test tube, and the stoppers as shown in Figure 7–2. Make sure that the open end of the glass tubing is narrowed like a medicine dropper.

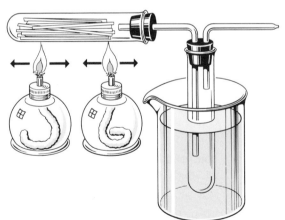

FIGURE 7-2
Apparatus used in distilling wood.

Heat the tube strongly with two alcohol burners, moving them back and forth along the test tube to heat all parts of

* Taken from *Introductory Physical Science*, Prelim. Ed., pp. 3–5 (prepared by the PSSC).

the wood. What do you observe happening when the wood gets hot? If wood will burn in plenty of air, will the gas that is given off burn as it comes out of the narrow opening of the tube? Try lighting the gas but avoid deliberately breathing it; it is slightly poisonous.

Now attach a piece of rubber tubing to the apparatus and collect the gas in a bottle by displacement of water (Fig. 7–3). Do not remove the alcohol burners while the open

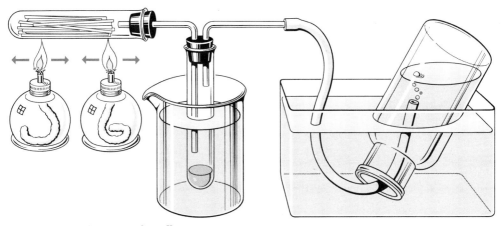

FIGURE 7-3   *Apparatus for collecting gas.*

end of the rubber tubing is under water. The cooling of the gases in the test tube will draw water up through the rubber tubing. When the bottle is nearly full of gas, disconnect the rubber tubing. Leave the bottle with the gas trapped above the water, and relight the gas still coming from the splints. Continue to heat the splints and keep the gas lit; it may come out in puffs, so you may have to relight it frequently. You probably will have to tilt and move the burners to heat all of the wood. However, try not to disturb the upright test tube. Stop heating the splints when you can no longer keep the gas lit.

Question 7–5   Look at the gas you have collected. (a) How does its volume compare to that of the splints you used? (b) What is its color? (c) Is it soluble in water?

If you wanted to determine its composition, you would

need to perform several tests. However, for the present we are concerned with other things. Turn the bottle upright, set it in a well-ventilated place for a few minutes to get rid of the gas, and then pour out any residual water.

Question 7–6   Without shaking the liquid in the upright test tube, examine it carefully. (a) Is it one liquid or more than one? Disconnect the tube containing the condensed liquid and put a few pieces of broken porcelain in it. Connect the apparatus as shown in Figure 7–4 and gently heat it. Once it boils, the

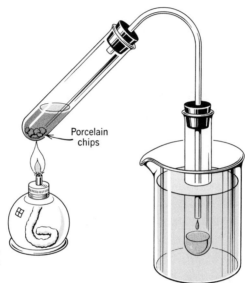

Porcelain chips

FIGURE 7–4
Apparatus for redistillation.

porcelain chips will keep it boiling more evenly. Continue to boil it gently until about half the liquid has evaporated. (b) What happens? (c) Is the liquid that condensed in the right-hand test tube the same as that in the left-hand one? (d) How do the two liquids compare? (e) What happens if you mix them together?

Question 7–7   After the test tube containing the wood has cooled, examine the remains. (a) What is the appearance of the solid remains of the wood splints? (b) Take one piece and hold it in a flame. Does it burst into flame? Does it burn at all? Does it leave any ash? Does it resemble any substance you know?

Question 7-8   Refer to your predictions in Question 7–4. (a) Did you predict that all these gases and liquids would be obtained from
the wood? (b) Can you get the wood back by mixing all the
substances you have collected? (c) Were these substances
actually present in the wood or were they formed by heating? Have you any evidence for your answer?

The behavior of heated wood is much too complicated to
describe in simple terms, and the experiment you have just
performed was included to show that results are not always
simple, nor are they immediately obvious. This experiment
was intended as an exercise in experimental technique and
careful observation. Although it did not enable you to understand the phenomenon you observed, you certainly became
aware that wood is a much more complex substance than
the crystals you heated in Experiments 2–6 and 2–7.

By asking questions, by devising experiments, and by
making careful observations we can learn about unfamiliar
substances as well as familiar ones. Remember, our goal is
to learn about the structure of solid matter and this includes
both crystal and wood. However, since the crystals are
simpler, we will learn more about them than about the
wood.

EXPERIMENT 7-2   Effect of heat on various solids

Part A   General Observations

You will be given or asked to obtain small amounts of the
following substances: zinc, tin, magnesium ribbon, zinc
chloride, sodium chloride, naphthalene, iodine, sugar, and
para-dichlorobenzene. It will be convenient to record your
observations in the form of a table with column headings:
Substance, Appearance, Effect of Heat. On the basis of appearance, which you should list now for each substance,
could you describe for each a characteristic property? For
example, if you didn't know beforehand, could you tell, by

just looking, which substance is suitable for sweetening coffee?

It is clear that to learn something about these substances we must *do* something to them. Let's see how they behave when heated.

Put a small piece of zinc in a test tube, place the tube in a test-tube holder, and heat it carefully and gently over an alcohol flame. Gentle heating is best accomplished by slanting the test tube (approximately 45 degrees from the vertical) and passing it carefully back and forth through the flame so that just a second or two of direct heat is applied during each pass.

If no observable change occurs after a few minutes of gentle heating, heat the tube more strongly by placing it directly in the flame and holding it there (be careful to point the open end of the tube away from you and from any others near you). Note carefully whether melting or any other change occurs, and record your observations. (If a solid melts during any stage of the heating process, do not heat it any further.) Repeat this procedure with each of the substances. Be sure to use only one or two small crystals of iodine, about the size of a pinhead. (Do not touch iodine with your hands; use forceps, a spatula, or a slip of paper.) The magnesium ribbon is the only substance which you need not heat in a test tube. Hold a tiny (5 mm) strip of magnesium ribbon with a pair of forceps, and heat it gently in the alcohol flame. Be sure to observe and record your results carefully.

Question 7–9 Sugar and salt (sodium chloride) look the same. Did they behave similarly upon heating?

Question 7–10 A solid is said to sublime if, when heated, it passes directly (without liquefying) from the solid to the gaseous state and, on cooling, condenses directly to the solid state again. Did any of the solids you heated sublime? Can you think of any other solids that do?

Question 7–11 You have seen that the nine solids tested demonstrated marked differences in behavior when they were heated. You can now answer such questions as: "Do all the solids melt when heated in an alcohol flame?" "What happens when you heat magnesium?" What questions have been raised by your observations? Make a list of questions for possible future study.

Part B   The Melting Point of a Mixture

In Part A you observed that some of the solids melted when heated, whereas others did not; but each substance was fairly pure, none was a mixture. In Experiments 2–6 and 2–7 you actually made a measurement of the freezing temperatures of naphthalene and *para*-dichlorobenzene. Those measurements, however, were not as accurate as they might have been. For example, are you sure that each substance was well mixed so that its temperature was uniform throughout?

You will now perform an experiment from which you will obtain more reliable data. The purpose of this experiment is twofold: (1) to make a more accurate measurement of the melting temperature of those same substances when pure, (2) to determine the melting points of two different mixtures of those same two substances.

The class should be divided into four teams: One team will be given a melting-point capillary tube with a small amount of pure naphthalene; the second, a capillary tube with pure *para*-dichlorobenzene; the third, a capillary tube with a mixture of roughly five parts of naphthalene to one part of *para*-dichlorobenzene; and the fourth, a capillary tube with a mixture of roughly one part naphthalene to five parts *para*-dichlorobenzene. Each team should make at least three determinations of the melting point of their particular material to establish the best value they can. Use the following procedure, and then compare the results of the four teams.

Fasten the capillary to a thermometer with a small rubber band cut from a piece of rubber tubing so that the sample is adjacent to the bulb of the thermometer (Fig. 7–5). Clamp a test tube about two-thirds full of water to the peg board in an upright position at a height convenient for heating with an alcohol burner. Carefully immerse the ther-

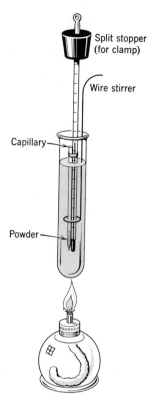

Split stopper
(for clamp)

Wire stirrer

Capillary

Powder

FIGURE 7-5
Melting-point apparatus.

mometer and capillary in the water as shown in Figure 7–5. Be sure that the powder is below the water level, but do not let water into the capillary. Clamp the capillary and thermometer in place. Use an 8- to 10-inch length of copper wire, coiled on the lower end, for stirring the water as you heat it with the alcohol burner.

It is often a good idea to heat the sample rapidly the first time to determine the approximate range in which it melts. Then repeat the procedure making a more careful determination. When you get about 10° below the melting point on the second heating, try to regulate the heat so that the temperature rises about 2° per minute. Record the temperature at which the sample first begins to melt (soften) and the temperature at which melting is complete (you will get a clear liquid). Record in your notebook the melting-point range of the sample. If your range is greater than 2°, you probably heated the water bath too rapidly near the melting point and will need to repeat the melting-point determination.

Cool the water bath (either by running cold water over the tube or by replacing the hot water with cold), and repeat the process to obtain three good values of the melting point. The average of these three values is your best result. Record that result and the range.

Question 7–12   How precisely do you know the melting point of each sample? For each sample specify a range of temperatures within which you are confident the melting point lies.

Question 7–13   Compare the melting temperatures of the pure substances and the mixtures. Did the melting temperatures of the mixtures follow your expectations? Can you form a model that describes the differences in the melting temperatures between the pure substances and the mixtures? The melting process is not simple, but our investigations will eventually lead you to a deeper understanding of this experiment.

Question 7–14   How would you need to modify the apparatus you used to be able to determine melting points greater than 100°C?

7–3   *Comparison of solids, liquids, and gases*

Now that you have heated a number of solid materials you know that even at the temperatures you achieved with only an alcohol burner, a variety of results are possible: melting, decomposition, sublimation, and no change. Why do solids differ markedly in their behavior toward heat? Of course, with what you have found so far you can't really answer this question. First you need to know more about the particles that make up a solid. In particular, you need to know something of how they interact with each other and what happens to them when heat is applied.

What would you say is the most characteristic property of a solid? Most people choose to describe its rigidity or its ability to hold a shape without being placed in a container.

From our earlier examination of the properties of solids, we now picture a crystalline solid as an orderly array of particles. Although we have not yet considered any evidence relating to the nature of these particles, our description up to this point has implied two basic concepts. First, crystalline solids are composed of discrete particles. And second, there is some kind of force which holds these particles in ordered positions. It is the second of these two concepts which will explain why solids are rigid.

In contrast to solids, what is the most characteristic property of liquids? You might say that it is their ability to flow: particles in a liquid are able to move more freely than those in a solid. However, the particles in a liquid must have some attraction for each other since, as we know, a liquid does not come apart into small pieces. We can state this differently by saying that the volume of a liquid is fixed, but its shape is not. With gases, neither the volume nor the shape is fixed; that is, a gas always expands or contracts (within limits) so that it completely fills the container it is in.

We could use what we know about the structure and properties of solids to make reasonable predictions about the structure and properties of liquids and gases. Conversely, a study of both gases and liquids should help us understand solids. We propose to take this latter approach and study gases in some detail.

You may think that studying gases will be more complex. It is true that the properties of gases under everyday conditions are less readily observed than the properties of liquids and solids. However, as we look more closely at the situation, we find that the characteristics of gases are more easily measured and described than it might at first appear. For instance, we know that a gas has flow properties like a liquid. With fans it can be made to flow in ducts and tubes, and differences in pressure and temperature make it flow across the surface of the Earth. Since both gases and liquids do flow, exactly how does a gas differ from a liquid? Let us start by observing what happens when a liquid is converted into a gas.

EXPERIMENT 7–3   The vaporization of a liquid

Put about 5 ml of water and a small porcelain chip in a test tube; mark the water level with a grease pencil. Take a small plastic bag (such as a "Baggie"), and squeeze the air out of it as completely as you can. Fasten the empty bag tightly around the mouth of the test tube using a rubber band. Attach a test-tube clamp and heat the water in the test tube using an alcohol burner. Water boils at 100°C; at that temperature the liquid will boil, and some of it will become gas. Continue to boil the water in the test tube until the plastic bag is extended completely. Compare the volume of the water as a gas with the volume it occupied as a liquid.

Use cold tap water to cool the test tube. What happens? If you wish, you can heat the system again and repeat the experiment.

Would the gaseous volume be the same if you had used a larger plastic bag? Did the plastic bag serve as an effective container for the gas? Don't merely give "yes" or "no" answers here; instead, defend your answer by using evidence gathered during the experiment.

Up to this point we have seen that solids have the property of rigidity, the ability to stand independent of any container, and we picture a crystalline solid as a regular, orderly array of particles. We have characterized liquids by their ability to flow; we pictured them as being composed of particles that move relatively freely but are attracted to each other. Finally, we have just seen that a large increase in volume takes place when a liquid changes to a gas. Judging by the way a gas expands, the particles must exert relatively smaller forces on each other. In view of the lesser effect of forces between particles in a gas, it appears that it will be easiest and most profitable to study gases first, rather than solids, and then apply what we learn about gases to liquids and solids.

7-4   *Alternative models of a gas*

You saw in Experiment 7–3 that there is a large increase in volume when a substance changes from a liquid to a gas. Can you think of an explanation for this? There are two fairly obvious possibilities. One is that the particles of the substance may be moved farther apart so that there is empty space separating them. Using this model, we would predict that the particles in the gas are, on the average, much farther apart than in the liquid. A second possible explanation is that the particles themselves become much larger. Using this model, we would predict that the particles in the gas are much larger (and therefore less dense) than in the liquid.

Which of these two alternative hypotheses or models is more consistent with experimental results? To decide this we will need to ask further questions concerning the models and to do additional experiments. We must examine each model carefully to determine what predictions we can make from it concerning properties of matter, other than those from which the model was originally generated. A satisfactory model must account not only for the large expansion when a liquid becomes a gas, but also for other properties of gases. We will not have enough time to do a large number of experiments. Some observations may be made very simply in the classroom or laboratory. In other experiments, we will just describe the results. Try to decide for yourself which of the two models is preferable for each of the following observations.

1. A water solution of ammonia is poured into a dish on the lecture table. How long does it take for gaseous ammonia to diffuse through the air at a distance of 10 ft? 15 ft? 20 ft? How do you know?

2. Let us take a glass tube with cotton plugs at both ends. Soak one plug in a water solution of ammonia and the other in a water solution of hydrogen chloride. Both ammonia and hydrogen chloride are gases at room temperature, and both escape readily from their water solutions. Ammonia and hydrogen chloride react with each other to form a white solid, ammonium chloride. This reaction takes place

readily within the tube. The result is that a ring of white "fog" appears where the two gases come into contact.

3. Is diffusion peculiar to gases? What happens if you put a drop of blue ink in still water and don't stir it? Try it. Do liquids diffuse into liquids?

4. Solids also diffuse into liquids. We saw in Experiment 1–3 that when potassium permanganate crystals were placed in water and allowed to stand undisturbed, the water became uniformly purple after several hours. Compare the rate at which diffusion of ammonia through air occurred with the rate at which potassium permanganate diffused through water. It can be shown that solids diffuse into solids, but the rate of diffusion is extremely slow.

5. The following demonstration of gaseous diffusion is a very useful one in deciding between the alternative gas models. A small sealed glass vial of liquid bromine is put in a long glass tube. The tube is then evacuated (that is to say, nearly all of the air particles are pumped out). When the tube is inverted, the vial of bromine breaks. Almost instantly, the tube is filled with the red-brown color of bromine gas from one end of the tube to the other. It is significant to observe what happens when this experiment is repeated without removing the air particles. This time, on breaking the vial, the reddish color of bromine gas first appears near the vial, then spreads slowly for several hours until it is uniformly distributed throughout the tube. Air is still present in the tube, so the bromine particles must have moved among the air particles. This process can be shown to accompany any mixing of gases, whether or not one starts with a liquid that expands as it is converted to a gas.

Question 7–15   Which model of a gas is consistent with the experiment showing the diffusion of bromine when air has been pumped out of the tube? When air has not been removed? Does the same model explain both situations? Are both models reasonable?

7-5   *Which model shall we choose?*

A careful review of the experimental results in Section 7–4 suggests that the particles in gases have space between them.

If this were not so, particles of one gas couldn't move among particles of a different gas. The fact that particles diffuse more rapidly through a gas than through a liquid or a solid indicates that there is more space between particles of a gas than between particles of a liquid or solid. If the particles of a substance increased in size when the substance changed from a liquid to a gas, there would be little space between the gas particles to allow the diffusion of other particles into the gas.

The individual, widely spaced particles of a gas are called *molecules*. When the gas condenses to a liquid, the particles come much closer together, but each particle behaves as a unit which we will continue to call a molecule.

The diffusion of one gas into another and the vaporization of a liquid to form a gas is good evidence for the idea that even in liquids, the molecules are not at rest but are in constant motion. In fact, the interaction of molecules in motion with larger particles in their vicinity was first observed in liquids. In 1827 the botanist, Robert Brown, examined some grains of pollen with the aid of a microscope (some pollen grains are too small to be observed without magnification). He discovered that when the pollen grains were suspended in water, they moved continuously in erratic fashion. This phenomenon, *Brownian motion,* is explainable in terms of the incessant movement of the water molecules which strike the pollen grains from all directions. When the number of collisions with water molecules is greater on one side of the pollen grain than on another, the net effect is the movement of the grain away from the side with the most hits. Here is evidence that liquid molecules are in motion. Smoke particles suspended in air also show Brownian motion, indicating that gas molecules are also in incessant random motion.

Incidentally, Brownian motion is easy to see, even when only a microscope with limited magnification is available. Just place a drop of some latex or water-base paint on a microscope slide. It may help to dilute the paint first with a small amount of water. Cover the drop with a cover slide, and observe it under a magnification of about one hundred times.

We have made some important observations on gases. These have led us to a preference for a model in which gas molecules are separated by relatively large distances and are in constant random motion. Now we shall pursue our study of gas behavior in somewhat greater detail and see whether our model of a gas is consistent with some additional observations. Our first step will be to examine how a gas behaves when its pressure or its temperature changes.

7-6 *Volume changes of a gas*

The following experiment is designed to enable you to determine quantitatively how the volume of a gas varies with changes in temperature.

EXPERIMENT 7-4 Effect on a gas of a change in temperature

You will be given a glass tube containing a drop of oil that keeps a given amount of air trapped in the tube. The mass of the air trapped in this gas tube will remain constant throughout the course of the measurements you make even though its volume may change. Use care when measuring or manipulating the gas tube.

In making your measurements, you will need to assume that the volume of trapped air is proportional to the length of the trapped air column. That is, if the length of the air column doubles, the volume also doubles. If the length of the air column triples, the volume also triples. One quantity is said to be "proportional" to another when they are related in this way.

Our assumption that the volume of the trapped air is proportional to the length of the trapped air column is a reasonable one, since the diameter of the gas tube is uniform (constant), and thus the cross-sectional area is also uniform. You will recall that to find the volume of a cylinder you multiply the cross-sectional area by the height of the cylinder; doubling the height would double the volume. Thus, your measurement of the change in height of the trapped air column in the gas tube is also a measure of the change in volume of the trapped gas.

We will perform this experiment at a constant pressure, so we will not be concerned with measuring the pressure. Set up the apparatus shown in Figure 7–6. Use a large test

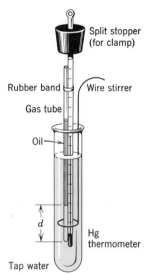

FIGURE 7-6
Apparatus for volume-temperature measurements.

tube, and clamp it on a pegboard or ring stand at a height convenient for heating with a burner.

Secure the thermometer to the glass tube with a rubber band. Support the thermometer and gas tube inside the test tube by means of a split cork around the combination, and attach the thermometer and tube to the pegboard by a clamp. Fill the test tube about two-thirds full of ice water, being sure that the length d can double and still be entirely in the water.

To determine the dependence of the volume of trapped air in the gas tube on temperature, you will need to measure the height of the trapped air column (d in Fig. 7–6) at various temperatures. Can you measure d conveniently when the gas tube is immersed in the water? If not, devise a method to determine d without direct measurement. Describe your method in your notebook.

Record the temperature of the water; calculate and record the length d.

Gently heat the test tube until the temperature rises about 5°. Stop heating, wait 10–15 seconds, read and record the temperature and determine the length d of the trapped air column. Repeat the measurements at approximately 5° intervals until the water in the test tube is close to boiling (stop at about 90°C).

Record all of the temperature-volume data in table form. You will need a table with five columns. It will be convenient to head the first two columns: (1) Temperature °C, (2) Length d of Trapped Air Column. The other three will be labeled as the need arises.

A graphical presentation of data will be far more informative than the "raw" data in your table above. First, plot Celsius temperature on the x (horizontal) axis and plot the height d (which is proportional to the volume) on the y (vertical) axis. It is convenient to have the two axes meet at a temperature of 0°C and a height (volume) of zero. When this is done, any point actually on the y axis represents a temperature of 0°C; any point actually on the x axis represents zero volume. You will need a temperature range from about -350°C to 100°C along the x axis. Plot your experimental points, one value of d for the particular temperature reading at which that d was measured. When all of your points are plotted, draw the best straight line through the experimental points. The best line will usually have as many points on one side of the line as on the other. However, it is not just the number of points on each side of the line that determines the best line, but the sum of the distances of those points from the line. For example, one point at a given distance from the line will balance two points closer to and on the other side of the line.

Extend the straight line so that it intersects the temperature axis. At that temperature, the height of the air column d, and thus the volume of the gas, would presumably be zero. Record that temperature. The results of experiments using equipment capable of more precise results than are possible with your gas tube show this value to be -273.15°C. This is often referred to as the *absolute zero* of temperature. In fact, a temperature scale may be defined such that $-273.15°C$ is $0°$. This is the Kelvin scale. A degree is the same size on both the Kelvin and Celsius scales; to convert from Celsius to Kelvin one need merely add $273°$ ($273.15°$ for more precise measurements). Thus $20°C$, which is approximately room temperature, is changed to Kelvin degree as follows:

$$20°C = (20 + 273)°K = 293°K$$

In the same manner, $-70°C$ is changed to Kelvin degrees as follows:

$$-70°C = (-70 + 273)°K = 203°K$$

With this in mind, you can now complete your five-column table. Column 3, appropriately labeled, should contain all of your temperature readings converted to the Kelvin scale. Column 4 should contain your values of each length d divided by the corresponding temperature in degrees Kelvin. Column 5 should contain each length d divided by the corresponding temperature in degrees Celsius. You will want to compare the numbers in these two columns, and that comparison is more easily made if those numbers are calculated in decimal notation. Which of the two columns—4 or 5—has values which are more nearly constant? Which of the two temperature scales appears the more significant?

Question 7–16 What do your results show about the relationship between the volume of a gas and its temperature on the Kelvin scale if the pressure and mass are constant?

Question 7–17 Does your graph show that the volume of the confined gas doubles if its temperature doubles?

Question 7–18 From your own graph predict what the volume of the trapped gas should be at $-273°C$. Why do you suppose you were not asked to measure the volume of the gas at temperatures near this value?

Question 7–19 Why were you advised to wait 10–15 seconds after you stopped heating before taking the temperature reading?

In Experiment 7–4 you discovered that the volume of a gas is dependent on temperature (when the pressure and mass of the gas remain constant). Before we consider whether these results are consistent with our gas model, let us investigate gas behavior a little further. Specifically, let us see whether the volume of a gas is dependent upon the pressure when its temperature and mass remain constant.

To investigate this we will need some type of apparatus in which we can keep a fixed mass of gas confined and vary its pressure in a known or measurable manner. You can use the same gas tube as in the last experiment, or you can use a syringe containing a fixed mass of gas. Whichever is chosen, you must have some means to measure pressure. But what do we mean by pressure? How is it measured? Perhaps you have heard pressure defined as force per unit area. What do we mean by force? How are forces measured? What is the physical significance of the pressure of a gas? Do we mean pressure *of* the gas or pressure exerted *on* the gas? These are questions we must be able to answer before we can proceed intelligently with our study of gases.

We began studying gases to learn more about their behavior. We hoped that this would eventually give us a better understanding of the structure of matter, particularly solid matter, and the relationship of structure to properties. Now we need to find out exactly what we mean by force and pressure in order to pursue our study of gases. We will do this in the next chapter.

QUESTIONS

7-20   How did you decide whether the gas evolved in Experiment 7–1 was soluble in water?

7-21   In Experiment 7–2 was there any substance for which you could say there was more than one effect due to heat? If so, write out a plan for a way to isolate one effect from the other.

7-22   Consider the two alternative models of a gas as stated in Section 7–4. List all the evidence in some organized fashion to show the superiority of one model over the other.

7-23 Summarize in tabular form all of the properties of solids, liquids, and gases which are: (1) similar to and (2) different from each other.

7-24 You test a hot iron by wetting your finger and touching it to the bottom. Suppose you hear a hissing sound when you do this? Give all the information you can about the temperature of the iron.

References 1. Greenstone, A. W., F. X. Sutman, and L. B. Hollingworth, *Concepts in Chemistry,* Harcourt, Brace & World, New York, 1966.

Page 195, a discussion of water of hydration; pages 391 to 392, fractional distillation; pages 138 to 139, determination of freezing points; page 194, distillation; pages 127 to 128, gases and temperature changes.
2. Lehrman, R. L., and C. Swartz, *Foundations of Physics,* Holt, Rinehart & Winston, New York, 1965, pages 245 to 253.

Calorimetry, change in phase, and calculations involving heat.

James Drake for Sports Illustrated © Time, Inc.

CHAPTER 8

MATTER IN MOTION

In Chapter 7 we posed questions about the effects of pressure on the volume of a gas. But to understand pressure we must understand force, because the concept of pressure depends on the concept of force. In this chapter we develop the relationship between the force acting on an object and the change in motion of that object. Only after we have done this will we be able to investigate pressure more thoroughly.

Later in the course, we present observations and ideas which lead us to believe that atoms are continuously in motion. Because they are always moving, atoms collide frequently with one another. More often than not, the atoms rebound after collision. By way of analogy, billiard balls bounce apart when they collide but presumably, they would stick together if they were coated with a very sticky substance. Similarly, atoms stick together upon collision if they exert an attractive force on one another. What kinds of forces are these? When atoms fly apart after a collision, what determines which way they go? To understand more about this, we will combine our study of forces with the study of motion. Since atoms are so small that they are invisible, we will first consider the motion of and forces on ordinary-sized things.

8-1 *Some simple kinds of motion*

In the world around us, there are many objects: baseballs and gloves, birds and trees, automobiles and people, stars and grains of sand. Objects are classified as such because they all have properties which remain constant while we observe them, and they all have a specific location in space at any particular time. Some are at rest, and some are in motion; some are speeding up, and some are slowing down; some are going uphill, and some are going downhill; some travel in straight lines, but most travel in curved paths. In this chapter we will consider simple motions of objects which travel in straight lines.

EXPERIMENT 8-1   Motion of objects on a horizontal surface

For our observations on objects moving on a horizontal surface we will use a wooden block with wheels (see Fig. 8–1a).

(a)

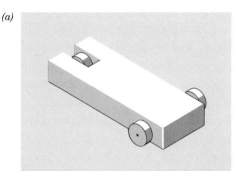

(b)

FIGURE 8-1   *(a) Wooden cart. (b) Air table with pucks. (Courtesy of The Ealing Corporation, Cambridge, Mass.)*

1. Place the wooden cart upside down on a table with a smooth, clean surface. Push the cart, and when you get it moving, let go. What happens after you stop pushing?

2. Now turn the cart over so that it rolls on its wheels, and give it about the same push as before. What happens after you stop pushing? How do the distances traveled by the cart in the two different situations compare?

3. Can you imagine conditions under which an object once set in motion would continue moving indefinitely?

As you observed in Experiment 8–1, the tendency of a moving object to slow down varies considerably when certain conditions are changed. What would happen if an object had absolutely no outside influences on its behavior? We cannot actually achieve such a situation, but it is possible to reduce the outside influences by various means. For example, we can make a smooth table with small holes in it through which air can be forced by a blower. When a small puck (see Fig. 8–1b) with a smooth flat bottom is placed on such a table, it will float on a layer of air. This layer of air is an excellent lubricant; the puck moves almost without friction. If given a little push the puck slows down only very gradually. Perhaps it is reasonable to suppose that given a push, a body completely isolated from interaction with other objects—a body out in space between two galaxies—would continue to move at the same speed indefinitely.

Galileo Galilei (1546–1642), the great Italian natural philosopher, was the first to propose that a moving object, if it were free of what he called "resistances and external impediments," would continue moving without slowing down and stopping. This very important step in the study of motion of objects was later expanded by Isaac Newton, who was born in the year of Galileo's death.

Newton stated, "Every body perseveres in its state of rest, or of uniform motion in a straight line, unless it is compelled to change that state by forces impressed thereon." This is known as *Newton's First Law of Motion*. Newton recognized that a moving body will continue to travel with undiminished speed *and* in a straight line unless some other body interacts with it. This property of maintaining a state of rest or of uniform motion in a straight line is called *inertia.* It is characteristic of all objects.

The state of motion of the cart or of the air puck—or perhaps we should say the change in the state of motion—depends on the interaction of the moving body with other bodies. We need, therefore, to investigate the relationships between the change in the state of motion of a body and its interactions with other bodies. In order to establish these relationships quantitatively, we will need to define: (1) terms which describe motion and (2) terms which describe interactions. We will begin with motion.

Figure 8–2 illustrates an air puck which has been set in

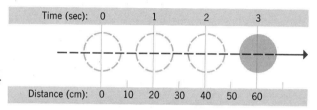

FIGURE 8-2
Successive position of a puck floating on a film of air moving to the right.

motion to the right along a horizontal surface. (The successive positions of the air puck could have been determined by examining successive frames of movie camera film, as discussed in Chapter 4 on waves.)

As we can see from Figure 8–2, the distance traveled by the puck during each 1.0-second time interval is the same, namely 20 centimeters. However, it is very unusual for the distance covered in equal time intervals to remain the same over an extended period of time. Because of this, it is convenient to use *average speed.* The average speed of an object is the total distance covered by the object during a given time interval divided by that time interval. Or we may express it as

$$v_{av} = \frac{\text{distance covered}}{\text{time interval}} = \frac{\Delta s}{\Delta t} \qquad (8\text{--}1)$$

where v_{av} is the average speed, and Δs is the distance covered in the time interval Δt. Δ is the Greek letter for our D and is commonly used to represent "difference in" or "change in." Suppose an object moves forward along the *x* axis and occupies a position x_1 at time t_1, and position x_2 at time t_2. The distance covered Δs then becomes Δx

$$\Delta x = x_2 - x_1$$

and the interval of time Δt becomes

$$\Delta t = t_2 - t_1$$

Therefore, the average speed v_{av} becomes

$$v_{av} = \frac{x_2 - x_1}{t_2 - t_1}$$

If the puck in Figure 8–2 was at $x_1 = 0$ at time $t_1 = 0$ (and we can establish the *x* axis so that its position the moment we start our stop watch is $x_1 = 0$) then the average speed during the time interval is

$$v_{av} = \frac{20 \text{ cm} - 0}{1.0 \text{ sec} - 0} = 20 \text{ cm/sec}$$

Question 8–1   For the puck whose successive positions are indicated in Figure 8–2, calculate the average speed for the time interval between 1.0 second and 2.0 seconds, and for the interval between 1.0 second and 3.0 seconds.

Question 8–2   When the average speed of an object has the same value for all time intervals, we say the speed of the object is constant or uniform. Is the speed of the air puck whose positions are illustrated in Figure 8–2 constant over the time interval shown in the figure?

Question 8–3   What can you say about the average speed of the puck over the time interval from 0 to 5.0 seconds?

From Equation 8–1 you can see that the units for speed are those of distance divided by time. If the distance is measured in centimeters and the time in seconds, the unit of speed is centimeters divided by seconds, or cm/sec. If the distance is measured in miles and the time in hours, the unit is miles per hour, or mile/hr. The combination of any other pair of distance and time units is possible, as for example, feet per second, or ft/sec.

To the physicist, the words speed and velocity have

different and distinct meanings. *Speed* refers to *how fast* an object travels; *velocity* refers to *how fast and in what direction* an object travels. The speed of the puck in Figure 8–2 is 20 cm/sec; its velocity is 20 cm/sec *to the right*. If this puck were moving to the left at the same speed of 20 cm/sec, its velocity would be 20 cm/sec to the left—quite different from the velocity it had when moving to the right. If we had chosen to call the direction to the right our positive direction, then all pucks moving to the right would have a positive velocity (e.g., +20 cm/sec) while all pucks moving to the left would have a negative velocity (e.g., −20 cm/sec).

In this section we have found that bodies free of interactions with other bodies move at a constant velocity. Very few objects in the world around you move with constant velocity. Have you ever thought about what the Universe would be like if all bodies moved without interacting with each other? Happily, objects do influence one another. In the next section we will see how these interactions account for changes in velocities.

8–2   *Forces*

The interactions which account for changes in the velocity of objects are called *forces*. Consider an object at rest hanging from a string. Obviously, there are interactions between this object and its surroundings. Yet these interactions are not causing any change in its velocity, which is zero. The string is under tension and, therefore, it must be pulling upward on the object. If we cut the string, so that it can no longer pull on the object, the object acquires a downward velocity. From this change in its velocity we can conclude that there is a downward force on it. Since this force must also have been acting before the string was cut, it must have been balanced by an equal and opposite force exerted upward by the string. In short, before the string was cut, the object had two forces acting on it, an upward force by the string and a downward force, and these two forces combined in such a way that the effect was the same as if no force at all acted on it.

Let us return to the puck in Figure 8–2. Suppose this air puck is at rest on a surface with no unbalanced forces acting on it. What will happen? If the puck is then given a horizontal push (or pull), what happens? Is there a simple relationship between the push on the puck and its change in motion? If we give it a slight push, the change in velocity is small—a hard push produces a large change. When the puck is started from rest, in what direction does it move when pushed? The motion of the puck is dependent on both the direction and the magnitude of the push.

If the puck is already moving, the effect of a push is more complicated. Figure 8–3a represents a puck moving with an

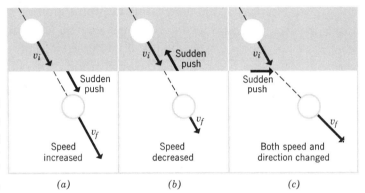

FIGURE 8-3
Positions of moving objects upon which forces—acting in different directions—operate. Initial velocity v_i*, final velocity* v_f.

(a)   (b)   (c)

initial velocity v_i. It is then given a push in the direction of its motion, which increases the speed of the puck but does not change its direction of travel. Here we indicate the velocity of the object by an arrow whose length (magnitude) represents the speed of the object and whose direction indicates the direction of motion. In Figure 8–3b the puck is moving with the same initial velocity v_i. It is given a push of the same magnitude as before, but the push is in the direction opposite to that in which the puck is moving. This push decreases the puck's speed. Applying a push which is not along the line of motion of the puck can cause both a change in speed and a change in direction of travel, as shown in Figure 8–3c.

Recall that Newton's first law refers only to situations in which no unbalanced forces are acting. We know that when an unbalanced force acts on a body, as in Figure 8–3, the velocity of the body will change. Let us see if we can determine a quantitative relationship between the change of velocity of a body and the unbalanced force acting upon it. Before we can determine this relationship, we will need to learn how to measure the total of all the external forces on the object. We will also need to have a simple quantitative measure of change of velocity.

How can we measure a force? One method is to use the elastic property of a rubber band. A stretched rubber band exerts a force. We can arbitrarily call the force exerted by a particular rubber band stretched to a particular length one unit of force. Then by attaching this same rubber band to an object and stretching it to this particular length, we can exert one unit of force on that object. However, to be useful as a standard of force, the rubber band must reproduce the same force every time it is stretched to the same length. We could test this by applying the unit force repeatedly to the same object. If the object always changes its velocity in the same way, we can be sure the unit force is constant and that the rubber band is a good standard for measuring force.

By connecting such a standard rubber band to an object and pulling horizontally on the other end, you can exert a constant horizontal force on that object. To keep the force constant, you will need to pull so that the rubber band remains stretched the same amount (Fig. 8–4). If we call the initial (unstretched) length of the rubber band L_0, and the length when stretched L, then we can represent the amount

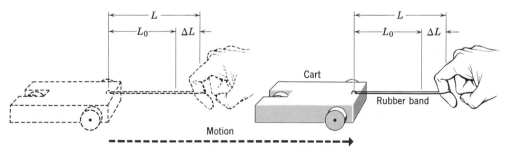

FIGURE 8-4 *A cart pulled by a rubber band.*

it is stretched by $\Delta L = L - L_0$. Because ΔL, the change in length of the rubber band, remains constant, the rubber band exerts a constant force on the object.

Suppose we exerted a constant force on a puck floating on a thin layer of air. Furthermore, suppose we took photographs of the puck every second as it scooted over the film of air. From this photographic record we could then study the motion of the puck during the time the force is being applied and during the time the force is not applied. A diagrammatic representation of such a record might look like Figure 8–5. We can measure the distance traveled

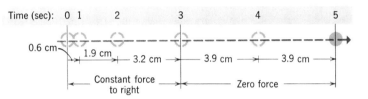

FIGURE 8-5

Successive positions of a puck moving to the right.

during each successive second, and from this we can determine the average velocity. Measurements from this figure are listed in Table 8–1; see whether you can verify the dis-

Table 8–1

	d	t	v
1st second	0.6 cm	1.0 sec	0.6 cm/sec
2nd second	1.9	1.0	1.9
3rd second	3.2	1.0	3.2
4th second	3.9	1.0	3.9
5th second	3.9	1.0	3.9

tance intervals by actual measurements with a meter stick. From the table we see that the puck moved 0.6 cm during the first second; therefore, the average speed during the first second is

$$v_{av} = \frac{\text{distance covered}}{\text{time interval}} = \frac{\Delta s}{\Delta t}$$

$$v_{av} = \frac{0.6 \text{ cm}}{1.0 \text{ sec}}$$

$$v_{av} = 0.6 \text{ cm/sec}$$

During the second second, the puck moved 1.9 centimeters, and its average velocity was 1.9 cm/sec, etc. Further measurements yield average velocities for succeeding one-second time intervals: these are listed in Table 8–1. This table tells us the distance the puck traveled and its average velocity during each successive second, but this is not a complete picture of the puck's motion. Perhaps we can get a better idea of this motion if we show the changes in velocity on a graph.

In Figure 8–6a we have plotted the average velocities during each of the successive seconds. This graph contains the information obtained by measurements from Figure 8–5. The process of going from the figure to the graph involves: (a) measuring the distances traveled during successive seconds, (b) forming the table by using Equation 8–1 to calculate the average velocities, and (c) plotting the graph.

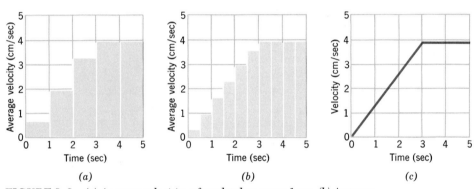

FIGURE 8-6   (a) *Average velocities of puck taken every 1 sec. (b) Average velocities of puck taken every ½ sec. (c) Instantaneous velocities of puck.*

Had we been given only the graph, however, we could reverse the process and calculate the puck's position at any of the one-second time intervals. For example, during the first second of travel, its average velocity was $v_{av} = 0.6$ cm/sec. During that one second it must have traveled a distance

$$\Delta s = v_{av}(\Delta t) \qquad (8\text{–}2)$$

This equation is derived from Equation 8–1 by multiplying both sides of the equation by (Δt) and then writing it as solved for Δs. The distance traveled by the puck during the first second must be

$$\Delta s = (0.6 \text{ cm/sec})(1.0 \text{ sec})$$
$$\Delta s = 0.6 \text{ cm}$$

Question 8–4   Calculate how far the puck traveled during the first two seconds of travel.

The information in Table 8–1 and in Figure 8–6a is not, however, very complete. We know neither the puck's velocity nor how far it traveled after, say, 1.50 seconds. In short, we cannot use either that table or that figure to make a reliable general statement about the force acting on the puck and the puck's resulting change in motion. We could, however, obtain more information concerning the puck's motion if we had made photographs every 0.50 second. We could then compute the average velocity over the time intervals 0–0.5 sec, 0.50–1.00 sec, 1.00–1.50 sec, etc. These average velocities could then be plotted against time as in Figure 8–6b; this would give us a better idea of the puck's position and velocity. Using these more refined data, it is not difficult to calculate how far the puck traveled during the first one-half second or during the first 1.50 seconds.

Question 8–5   Using the information obtained from Figure 8–6b find: (a) the average velocity during each of the one-half second intervals and (b) the distance the puck traveled by the end of the first 0.5 second; by the end of the first 1.00 second; by the end of the first 1.50 seconds.

Question 8–6   Using the information now available to you, can you find out how far the puck traveled during the first 1.20 seconds? In order to make this calculation, you could assume that it traveled with the average velocity for the first second indicated in Table 8–1 (and in Figure 8–6a) and calculate the distance traveled during that first second. Then you could assume that it traveled with the average velocity indicated for the time interval 1.00–2.00 seconds for the remaining 0.20 second of travel. The two distances added together would approximate the total distance traveled during the first 1.20 seconds. Calculate the total distance.

Question 8–7   Repeat the calculations you made in Question 8–6 for the distance traveled by the puck, but this time use the intervals and average velocities obtained in Figure 8–6*b*, and again estimate how far it traveled during the first 1.20 seconds.

Question 8–8   Of the two answers to Question 8–6 and 8–7, which do you think is closer to the "actual" distance traveled by the "real" puck during the 1.20 seconds?

Apparently, if we measure positions at more frequent time intervals, we are able to describe the motion of the puck more accurately. By observing the positions of the puck over shorter and shorter intervals of time, you can finally, at least in your mind, idealize our representation of the puck's motion by considering the average velocities over such exceedingly short time intervals that the "flight-of-stairs" appearance of the graph gives way to a "ramp" appearance (Fig. 8–6*c*).

For any motion, a plot of average velocities over sufficiently small time intervals will have the appearance of a smooth curve. The velocity read from this curve corresponding to any particular time is called the *instantaneous velocity* or simply the *velocity*. The smooth straight lines in Figure 8-6*c* show that the instantaneous velocity of the puck changed continuously and regularly as long as the constant force acted. After the force ceased to act, the velocity remained constant.

During the first three seconds, the velocity of the puck changed (Fig. 8–6*c*). If in this same time interval the velocity had changed more, the sloping line in the graph would be steeper. The slope (steepness) of this line indicates how rapidly the velocity changed with time. This slope is called the *acceleration*. It is defined as the ratio of the change in velocity to the time interval during which the change took place:

$$\text{acceleration} = \frac{\text{change in velocity}}{\text{time interval}} = \frac{\Delta v}{\Delta t} \qquad (8\text{–}3)$$

As with instantaneous velocity, we can think of the *instantaneous acceleration* by considering exceedingly short intervals of time Δt in Equation 8–3.

Not all kinds of motion can be represented with straight lines on velocity-time graphs. Sometimes the instantaneous acceleration varies, and for such motions the slope of the velocity-time curve must change. When the acceleration is constant, however, the average acceleration equals the instantaneous acceleration. We will consider only cases where the acceleration is constant. Therefore, we need not distinguish between average and instantaneous acceleration.

To calculate the acceleration of our puck we must use two velocities at different times to find the change in velocity over an interval of time. For example, from the graph of instantaneous velocities (Fig. 8–6c) the velocity at time $t_1 = 1.0$ sec is $v_1 = 1.3$ cm/sec; the velocity at time $t_2 = 2.0$ sec is $v_2 = 2.6$ cm/sec. Since

$$a = \frac{\Delta v}{\Delta t} = \frac{v_2 - v_1}{t_2 - t_1}$$

then $$a = \frac{(2.6 - 1.3)\ \text{cm/sec}}{(2.0 - 1.0)\ \text{sec}}$$

$$a = \frac{1.3\ \text{cm/sec}}{1.0\ \text{sec}} = 1.3\ \text{cm/sec/sec} = 1.3\ \text{cm/sec}^2$$

where a denotes acceleration.

Question 8–9   Calculate the acceleration during each second of the 5 seconds of travel of the puck in Figure 8–5. Then make a graph of the acceleration versus the time. Does the acceleration change at all during the first 3 seconds? During the last 2 seconds? What can you say about the acceleration at time $t = 3.0$ sec? How does the acceleration compare with the force applied? What conclusions can you draw?

Question 8–10   Propose, by means of a velocity-time graph, a description of the motion of the puck—without the film of air to float on—under the following conditions. The puck is first accelerated by a constant force, and then after 3.0 seconds the force ceases to operate. The puck, however, continues to move for 2.0 seconds before coming to rest. Assume that the force of friction is constant throughout the motion.

EXPERIMENT 8-2 Acceleration of falling objects

To determine the acceleration of a falling object in the laboratory, you will measure the object's change in velocity over small time intervals. In order to measure intervals of time smaller than a second, you will use a simple doorbell whose hammer vibrates up and down at a nearly constant rate. Under the hammer place a piece of carbon paper; under that, in turn, a strip of paper will pass. Each time the hammer comes down it hits the piece of carbon paper, making a dot on the paper tape. As the tape is pulled along, the hammer makes a series of dots on the tape (see Fig. 8–7). If we assume that the time between successive dots is

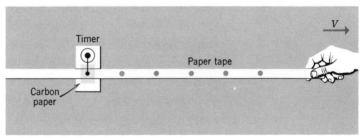

FIGURE 8-7
Drawing showing the position-time record of a tape moving at constant velocity v.

constant, then we can define an arbitrary time interval as that interval during which some particular number of dots has been made on the tape. It is possible to establish the relationship between this selected time interval and the standard second of time.

With the doorbell timer secured to a vertical support by means of a C-clamp or masking tape, turn the timer on and pull a long strip of paper tape through it for an interval of 5 seconds. The tape should be pulled fast enough so that the timer makes separate dots, but not so fast that an excessive amount of tape is used. You may want to practice once or twice with short lengths of tape. From the number of ticks on your tape determine the relationship between the tick and the second for your timer. Remember that the first dot you use is the zero of your time interval; the second dot ends the first interval of time we call a tick. For counting purposes it is convenient to mark the tape every 5 dots.

Question 8–11   For your timer, how many ticks are there in 5 seconds? in 1 second?

Question 8–12   How much time (in seconds) is one tick?

Now attach a heavy object to a piece of paper tape about 1 meter long so that the object hangs freely. Put the tape through the timer and under the carbon paper, holding the tape so that the object is just below the timer. Start the timer and release the tape. Recover the tape from the floor, and identify your tape by writing your name and pertinent information on it. Your partner should also perform the experiment so that an average result can be made. When this has been done, attach a second object (identical in mass to the one used above) to the first object and collect tapes on this falling double mass.

The last 20–30 centimeters of your paper tape may have fluttered as the tape passed through the timer; this leads to excessive friction. Therefore, you might restrict your study to the first 70–80 cm of tape.

How much information about the motion of the falling object can we extract from these tapes? Let us indicate what information is available by asking a few questions and restricting our attention to the tapes.

Question 8–13   From the record of dots on the tapes, how can we tell whether the velocity of the falling object is increasing or decreasing?

Question 8–14   How can we tell whether or not the velocity of the object changed uniformly with time?

It will be helpful to record your data in the form of a table. The measurements you will make will be the distance traveled during successive selected intervals of time. If using every tick of the vibrating door bell involves too much work, you may want to let 2 ticks (dots) be your basic unit of time. We call this arbitrarily chosen unit of time the *twotick*. If you have 27 dots on your paper, the interval of time for the entire run would be 13 twoticks (the first dot is zero). The average velocity is simply the distance traversed during that twotick divided by 1 twotick.

Choose the first clearly defined dot on your tape, circle it, and call that dot zero. Now count out and mark every second dot on your tape, marking the tape in units of two-ticks and recording the distance traveled (in centimeters) during each twotick. Head this column of numbers Δs, for distance interval. The next column can be the average velocity during each twotick, which, since our interval of time is one twotick, is numerically equal to the distance but is in units of centimeters per twotick.

Once you have arrived at a series of values of the average velocity, it is instructive to make an average velocity-time graph. This can be done either by drawing a graph from the table or by simply cutting the paper tape at each dot previously marked to designate twoticks, and lining up these short strips of paper in a stepwise fashion to resemble the graph in Figure 8–6a. Since the length of each strip of paper is the distance traversed by the object during that particular twotick, these lengths represent the average velocities of the object over twotick intervals.

From your graph of the average velocity versus time, draw the best straight line to indicate the instantaneous velocity-time curve. Figure 8–6 can serve as a guide. Lay the straight edge of a paper on Figure 8–6a so that the edge goes through the point $t = 0$ and $v = 0$, and so that the edge has the appropriate slope. Repeat this exercise with Figure 8–6b. You will see that the edge of the paper will go through the midpoints of each of the steps.

Therefore, assuming constant acceleration, the best velocity time curve for your falling object will be that straight line which most nearly passes through the center of each of the steps of your graph. You may want to identify the center of each step by a pencil point with a circle around it. Remember that your first dot on the tape was the first clearly defined dot; it was made after the object had already fallen a bit, so your straight line will not pass through zero velocity at zero time.

The slope of this line is the acceleration of the falling object. How can you determine what it is? Since each of your measurements of average velocity may be in error for any number of reasons, you cannot use them individually to determine the acceleration. You can, however, use the straight line you have drawn, for by drawing the best possible fit to the velocity-time graph, you have averaged out some of the small errors of the individual measurements. All you need do, therefore, is to determine the slope of that line as was done on page 166 for Figure 8–6c. Select two points near the extremities of your line; two points close together will not give as good a result as two points spaced some distance apart.

What is the acceleration of the single object in cm/(two-tick)²?

To convert the units of time from twoticks into seconds, you will need the results of your determination of the number of ticks per second. Let us assume that your timer produces 50 ticks per second, that is, 25 twoticks per second. Furthermore, let us assume that the acceleration of your object was 2 cm/(twotick)². To convert your units, you must multiply the acceleration in cm/(twotick)² by (twotick)²/(sec)²:

$$a = 2\frac{\text{cm}}{(\text{twotick})^2} \cdot \frac{(25 \text{ twotick})^2}{(\text{sec})^2}$$

$$a = 2(625)\frac{\text{cm}}{(\text{sec})^2}$$

$$a = 1250 \text{ cm/sec}^2$$

Question 8–15   What was the acceleration of the single object in cm/sec²? in meter/sec²?

Question 8–16   What was the acceleration of the double object in meter/sec²? How does it compare with the acceleration of the single object?

8–3   *Force, mass, and acceleration*

Now that you know how to measure the acceleration, that is, the change in velocity, of any object, it will be easier to relate the instantaneous velocity-time graph of the puck in Figure 8–5 to its change in motion. Consider again the motion of our puck under the influence of a single, constant, horizontal force as illustrated in Figure 8–6.

A constant force produced the constant acceleration of the puck as shown in Figure 8–6c. What acceleration would result if twice that force were applied to the same puck? three times the force?

In order to discover the answers to these questions we must be able to produce two or three times the original force. You could, for example, stretch the rubber band more and produce a larger force. But doubling the stretch of a rubber band will not produce twice the force. (How could you verify this accurately with the equipment you have?) However, if two different rubber bands produce the same force, then presumably, together, they will produce twice the force on an object if they both pull on the object in the same direction.

As defined in Section 8–2, two forces are considered equal in magnitude if they simultaneously pull in opposite directions on the same movable object, and that object does not move. Whenever forces act on a movable object, and that object does not move in response, those forces and that object are said to be in *equilibrium*. We can use the condition of equilibrium to test our rubber bands; if two rubber bands pulling in opposite directions on an object leave that object in equilibrium, then those two rubber bands are exerting equal forces. You can use a paper clip as the object. If the rubber bands are identical, they will both be stretched the same amount to produce this force. If your rubber bands are not identical, try others until you find two that are.

We could now connect these same two rubber bands to the paper clip so that they pull in the same direction (as shown in Fig. 8–8b). If each is stretched the same amount as before (Fig. 8–8a), together they will produce twice the force that either would alone. If a third identical rubber band is added to the paper clip so that it pulls in the same direction and is stretched the same amount, the force exerted by these three parallel rubber bands will be three times that of any one.

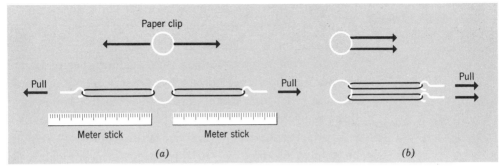

FIGURE 8-8 *(a) Two rubber bands stretched to produce equal forces.*
(b) Rubber bands pulling in the same direction.

Note again that in Figure 8–8*a*, even though there are horizontal forces acting on the paper clip, it remains at rest because the two forces are equal in magnitude and act in opposite directions. These two forces balance each other, and the effect is the same as if there were no horizontal forces. When this situation occurs, the object is in equilibrium.

EXPERIMENT 8-3 Force and motion

If you see a cart standing still, you know that there are no unbalanced forces acting on it. Furthermore, if you see a cart moving along a horizontal straight line at a constant speed, you know that there are no unbalanced forces acting on it. Why? In this experiment you will discover what happens when you apply constant *unbalanced forces* to a wooden cart when it is at rest. Some of you will do the experiment by keeping the load constant and applying several different unbalanced, yet constant, forces to the cart. Others of you will use a single constant force throughout but vary the amount of the load. Using all the data gathered, we will then attempt to establish a relationship among: (1) the forces applied, (2) the resulting acceleration, and (3) the amounts of inertia of the cart with different loads.

By changing the cart's load, you will vary its mass. If you maintain a constant accelerating force, you will expect the resulting acceleration to be constant for *any one* loading;

but will the acceleration be the same for *every* loading? If you vary the force applied to the cart by changing the number of rubber bands exerting forces and hold the mass constant, will the acceleration be different for each different force?

To answer these questions intelligently we will need to determine a relationship among force, mass, and acceleration. Therefore, we will need a method of measuring acceleration. You know that the acceleration of the cart can be calculated by determining the cart's change in velocity over some convenient interval of time. To obtain accelerations over short time intervals we can use the same timing device as was used in Experiment 8–2. See Figure 8–9 for a picture of the experimental arrangement.

When you wish to apply constant forces which differ in magnitude, it will be convenient to select three rubber bands which are identical. This will enable you to exert forces of one unit, two units, and three units on the cart.

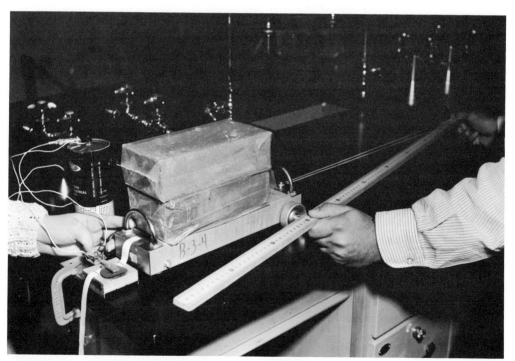

FIGURE 8-9 *Accelerating-cart experiment.*

To select three identical rubber bands, use a fourth one as a standard or measuring instrument. Attach one end of this rubber-band standard to a fixed object and a paper clip to the other end. Use a meter stick and select the distance you will stretch the standard rubber band. Call this the standard stretch. For this experiment the standard stretch is about 10–15 cm. Other rubber bands will be identical to the standard if, when connected to the standard by means of the paper clip as shown in Figure 8–8a, the stretch of the band is the same as the stretch of the standard. Both of these stretches should be approximately equal to the standard stretch.

To conduct this experiment divide up into teams of three so one member can be the timer, another the operator, and the third the catcher. The timer will start the door bell buzzing and release the cart; the operator will pull the cart with a constant force; the catcher will catch the cart with its load before it runs off the table onto the floor.

Attach one of the three identical rubber bands to the front of the cart, and hook the other end over the end of a meter stick. While the timer holds the cart as shown in Figure 8–9, the operator should stretch the rubber band by the standard stretch. Then the timer should release the cart. The rubber band will exert a force on the cart causing it to accelerate. You will remember that we want to study the motion of the cart while it is influenced by a *constant force* throughout the run. How can you make the force exerted by the rubber band remain constant? To acquire the technique of walking along in front of the cart so you can keep the stretch in the rubber band the same, it is a good idea to make several practice runs before attaching the paper tape. The cart, during its acceleration, should move a distance of about 1 meter or more. Each member of your team should learn this technique of applying constant force.

When you have mastered this technique, you will be ready to make measurements. Begin by placing two bricks or other objects with about 4 kg mass on the cart. The timer should hold the cart stationary with one hand while the operator stretches the rubber band by the standard stretch.

When you are ready, the timer should turn on the doorbell timer and then release the cart. The operator must move along with the cart maintaining a constant force (constant stretch). Let the cart accelerate at least 1 meter before releasing the rubber band and allowing the cart to proceed with its load to the catcher. Obtain tapes for three different forces or three different masses, depending on which quantity you are varying. Mark on each tape the force and mass used in making that run.

Make a table with two columns in order to organize your data. You will need measurements of the distances traveled during successive selected intervals of time. From these you will be able to determine the average velocity. Using every other tick of the vibrating doorbell would involve a lot of work, so let ten ticks (dots) be your basic unit of time; let us call this arbitrarily chosen unit of time the tentick. Thus, if you have 121 dots on your paper, the interval of time for the entire run would be 120 ticks or 12 tenticks.

Choose the first clearly defined dot on your tape, circle it, and call it zero. Now count out and mark every tenth dot on your tape so that the tape is divided into units of tenticks. Label a column of the table in your notebook Δs (for distance interval), and record the distance traveled in centimeters during each tentick. Label the next column v_{av} (for average velocity) and insert the value for each tentick. In this instance, the average velocity is numerically equal to the distance, since the interval of time is one tentick. Because the distance is in centimeters and the time in tenticks, the unit of velocity is centimeters per tentick. At this time you may want to determine the relationship between the tentick and the second for your timer (it may be different from what you determined in Experiment 8–2, since you may have different batteries and a different timer). From this relationship you can determine the average velocity of the cart in cm/sec during each successive tentick.

Once the average velocities are computed, you can obtain a better idea of how the velocity changes by making an average velocity-time graph similar to the one in Figure 8–6*b*.

Question 8–17   Does it appear from your average velocity-time graph that the velocity of the cart changed uniformly with time?

If you feel that the velocity did change uniformly with time, you are justified in following the procedure of Experiment 8–2 to draw the instantaneous velocity-time graph. The slope of this straight line is the acceleration of the cart.

Question 8–18   Examine how well the straight line of the instantaneous velocity-time graph fits the points of the average velocity-time graph. Was the acceleration of the cart constant during the application of the constant force?

Now that you can determine the acceleration of the cart, we can investigate the relationship among force, mass, and acceleration. It will save time if one-half of the class will determine the relationship between mass and acceleration for a constant force, while the other half determines the relationship between force and acceleration for a constant mass. We shall discuss these one at a time.

A   *Hold the force constant and vary the mass*

It is easy to hold the force constant. You already know how to do this from the last experiment. The mass of the cart can be varied by changing the load. First try doubling the mass to see what happens to the acceleration when the same force is applied. Don't forget that you can't just double the number of bricks—you will need to add another cart as well. (Share equipment with another team if necessary.) How does the acceleration of the cart with double mass compare with the acceleration of the original cart? Take another measurement for a third mass.

In analyzing the relationship between mass and acceleration when there is a constant force, we could designate the mass of the single load (cart plus brick) as m and the double load (2 carts plus 2 bricks) $2m$. But what if we wanted to use a single cart and three bricks? For situations such as this it will be much easier if we know the actual mass, in kilograms, of the cart and one brick. Determine this mass with a beam balance or in any other way that you can.

Make a graph of mass versus acceleration using three

values of mass in kilograms and the corresponding values of acceleration that were obtained with an equal force for each mass.

B *Hold the mass constant and vary the force*

To determine the relationship between force and acceleration while keeping the mass of the cart and bricks constant, the force should be varied. This can be done by using the three identical rubber bands. You have already accelerated the cart with two bricks on it by pulling it with one rubber band. Now you should pull that same cart with two and then with three rubber bands, always stretching the bands by the standard stretch. Determine the acceleration for each run and make a graph of force versus acceleration with mass held constant.

Question 8–19 Those students who applied a constant force and varied the mass should connect the three points of the mass-acceleration graph with a smooth curve. Try to predict the general shape of that curve for increasingly larger and then decreasingly smaller masses. You may find it helpful to determine the product of the mass and acceleration for each of your three values.

Question 8–20 Those students who applied different forces to a constant mass should connect the three points on the force-acceleration graph with a smooth curve. Try to predict the general shape of the curve for increasingly larger and then decreasingly smaller forces. When you use a constant mass, how does the acceleration depend upon the force?

Question 8–21 According to your force-acceleration curve, what acceleration will result when a zero force is applied?

By combining all the data collected, we can now discover the relationship among force, mass, and acceleration. One group found that the relationship between force and acceleration is a simple one. If the force acting on the object is doubled, the acceleration of the object is doubled; one-third the force produces one-third the acceleration. In other words, you found that an object is accelerated in the same direction as the unbalanced force, and that the magnitude of the acceleration is directly proportional to the magnitude of the force. This statement can be represented much more concisely by the mathematical expression $F \propto a$ (where $\propto$ means "is proportional to").

The other group found that the relationship between the mass of the objects and the acceleration is also a simple one. If the mass of the object is doubled, the acceleration is one-half as great; if the mass is one-third of the original, the same force will produce three times the acceleration. In other words, they also found that an object is accelerated in the same direction as the unbalanced force; in addition, they found that the magnitude of the acceleration is inversely proportional to the mass of the object. This can be symbolized mathematically as $m \propto 1/a$.

We can combine these two proprotionalities into a single algebraic expression

$$\text{Force} = \text{mass} \times \text{acceleration}$$
$$F = ma \qquad\qquad (8\text{--}4)$$

This relationship is Newton's second law. Like all physical laws, it is a general statement describing a certain group of observations. If the mass m is held constant, doubling the force will double the acceleration; if the force F is held constant, doubling the mass will result in one-half the acceleration.

What are the units of force in the equation $F = ma$? In Chapter 2 we established the unit of mass as the kilogram, the unit of length as the meter, the unit of time as the second. This system of units is used frequently and is called the *mks system.* But what about the unit of force? According to Equation 8–4, the unit of force must be kg-meter/sec²; this combination of fundamental units is called the *newton* in the mks system. One newton is defined as the unbalanced force required to give an acceleration of 1 meter/sec² to an object with a mass of one kilogram.

Sir Issac Newton first published this law in a somewhat different form in 1687. This law, as expressed by the simple Equation 8–4, has been used since the middle of the eighteenth century to describe the motion of almost every object in the Universe. Not until the beginning of the twentieth century did Albert Einstein show that Newton's second law of motion must be modified for objects moving with speeds close to that of light (3×10^8 meter/sec). Scientists of our century have also had to evolve additional and different laws to describe the nature of atoms and molecules. However, except for these situations, Newton's second law of motion is still valid. It applies to the motions of all objects moving at any speed that is small compared to the speed of light; it applies to the motions of the stars in a galaxy and to the motions of myriads of molecules forming a gas. The range of applicability of this law is enormous, and within this range it is as valid today as it was the day it was proposed.

We can now apply Newton's second law of motion (Equation 8–4) to several relatively simple situations. First, let us use some of your results from the experiment with the acceleration of the cart. As you have found, the unit of time, the tentick, is about $\frac{1}{4}$ sec. So let us assume that 1 sec = 4 tenticks. You can now convert your units of acceleration, cm/tentick^2 into meter/sec^2. Suppose, for example that one of your accelerations was 0.7 cm/tentick^2. How many meter/sec^2 is this? (Recall that 1 meter = 100 centimeters.)

$$a = \frac{0.7 \text{ cm}}{\text{tentick}^2} \cdot \frac{1 \text{ meter}}{100 \text{ cm}} \cdot \frac{(4 \text{ tentick})^2}{(1 \text{ sec})^2}$$

$$a = 0.007 \cdot 1 \text{ meter} \cdot \frac{16}{(1 \text{ sec})^2}$$

$$a = 0.11 \text{ meter/sec}^2$$

Using this value for a, you can calculate the force with which the cart was pulled. If, for example, the total mass of the cart plus two bricks was 5 kg, we would have, by Equation 8–4,

$$F = ma$$
$$F = (5 \text{ kg}) (0.11 \text{ meter/sec}^2)$$
$$F = 0.55 \text{ newton}$$

Question 8-22   Referring to your data in Experiment 8–3, calculate the force exerted by one rubber band when stretched to your standard stretch.

Question 8-23   How great a force is required to accelerate a car of mass $m = 2000$ kg (this car would weigh 4400 lb) at a rate of 5 meter/sec^2?

8-4   *The motion of a falling object*

We have arrived at Newton's second law by generalizing from motions of objects on a horizontal surface (Experiment 8–3). As an example of its general applicability, let us analyze the motion of the freely falling object in Experiment 8–2.

As you found in that experiment, the acceleration of a freely falling object is constant as the object falls. Applying Newton's second law to the falling object, we conclude that there was a force which caused the acceleration, and that this force was constant as the object fell. The force is clearly an attraction between the Earth and the object. It is called the force of gravity, the gravitational force, or the weight of the object.

A gravitational force is always present between the Earth and any object. An object which is falling in a vacuum will fall freely and accelerate downward at a rate consistent with Newton's second law of motion

$$F = ma$$

The force F is the weight of the object, which we shall call w. You discovered in Experiment 8–2 that the accelerations of all freely falling objects are the same. We give this special acceleration the symbol g and call it the acceleration due to gravity. Newton's equation for a freely falling object then becomes

$$w = mg \qquad (8\text{--}5)$$

The acceleration of a freely falling object near the surface of the Earth is about 10 meter/sec^2 (actually closer to 9.8 meter/sec^2) or 32 ft/sec^2.

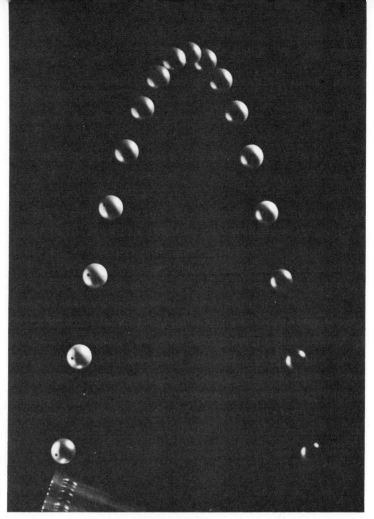

Since g was found to be the same for all objects at
a given location, Equation 8–5 implies that the weight, that
is, the gravitational attraction of the Earth for the object, is
proportional to the mass of the object. Notice that we have
not shown that this ratio, $w/m = g$, is the same at all locations, and in fact it is not; but this need not bother us here.

Figure 8–10 shows a ball that is being acted on by the
force of gravitational attraction between it and the Earth.
It was projected upward, but with a slight horizontal motion
so that the stroboscopic pictures would not overlap. The
horizontal motion is uniform—there being no appreciable
horizontal force—and so it need not concern us in our discussion of this freely falling object. *Is* it freely falling?

During its ascent, the ball's vertical velocity decreases,

until it comes, momentarily, to a position at the very top when its vertical velocity is zero. It then starts to descend with ever increasing vertical velocity.

For the sake of clarity let us call the downward direction positive, since that is, after all, the direction in which the force of gravity acts, and it is a little simpler to discuss a positive force than a negative one.

During its ascent the ball's velocity is negative; that is, it is traveling in the negative direction. At the very top, its vertical velocity is zero. During its descent, the ball's velocity is positive; that is, it is moving in the positive direction.

But acceleration is more important to our discussions of motion than is velocity, so let us ask some questions about the acceleration of this ball.

Question 8-24 During its ascent, does the ball have a positive, a negative, or zero acceleration?

Question 8-25 During its descent, does the ball have a positive, a negative, or zero acceleration?

Question 8-26 At the very top of its flight, when its vertical velocity is zero, does the ball have a positive, a negative, or zero acceleration?

Does a force act on the ball at all times? Is this force counterbalanced by any other force? If only one force acts on the ball, then according to Newton's second law of motion, the ball must accelerate; $F = ma$. The direction of the acceleration must be the same as the direction of the force. We stipulated that the force of gravity acts in a positive direction; therefore, as long as this force—and this force alone—acts, the ball must accelerate in a positive direction. Does this conclusion agree with your answer to Question 8-25? It should.

Question 8-27 Sketch a velocity-time graph for the motion of the ball. Remember that its velocity is initially negative and then becomes positive. Negative velocities appear below the time axis, positive velocities above it. The acceleration of the ball is the slope of the velocity-time curve. Is that curve a straight line? Is the acceleration constant? Is the force acting on the ball constant? Is its mass constant?

At the very top of its flight the vertical velocity of the ball is zero; if its acceleration were also zero, its velocity would not change. But if its velocity does not change, it will remain zero and the ball will remain suspended at the top.

QUESTIONS

8-28 A man of mass 70 kg hangs from a rope tied around his waist. What forces act on him? Calculate in newtons the magnitude and specify the direction of each force.

8-29 Compare the average speeds of a good 100-yard-dash runner and a good mile runner. If you do not know the times in which these distances are run, now is the time to consult your athletic-minded friends. Calculate the ratio of the speed of the dash runner to the speed of the miler.

8-30 If the initial speed of the puck in Figure 8–3c was 10 cm/sec, estimate the final speed v_f and the change in direction of the velocity which results from the push.

8-31 A road up a hill is 5 miles long. If you drive up the hill at a steady speed of 30 mph, and then drive the 5 miles down the same hill at a steady speed of 50 mph, what is your average speed for the whole trip? Note: this is a famous problem since it is a bit tricky. Be careful. The answer is *not* 40 mph.

8-32 A hot rod enthusiast accelerates his car at a rate of 30 ft/sec², starting from rest. Fortunately, however, he runs out of gas 5 seconds later, and thus he decelerates as he coasts to a stop. His rate of deceleration is one-half that of his acceleration. Plot the velocity-time curve for his motion. What is the distance he travels during his period of acceleration? How far did he travel altogether?

8-33 Describe Newton's first law in terms which would be understandable to an untutored layman. (Try it on a friend.)

8-34   A tennis ball, initially moving to the right at 30 meter/sec, strikes a wall and rebounds with the same speed of 30 meter/sec.

(a) What is the velocity after rebounding?

(b) What is its change in velocity Δv?

(c) If it made contact with the wall for a time interval of 0.01 sec, what is its average acceleration during this time?

(d) What is the average force of the wall on the tennis ball during this time? Assume the mass of the ball to be 0.02 kg.

8-35   If your moving car should collide with a bridge abutment, the car is likely to be wrecked. Analyze this in terms of Newton's second law.

8-36   A body is moving 10 ft/sec at the start of an interval. Five seconds later, it is traveling 75 ft/sec in the same direction.

(a) If the acceleration is constant during this period, what is the average velocity?

(b) What is the acceleration during this interval?

References   1. Feynman, R. P., R. B. Leighton, and M. Sands, *The Feynman Lectures on Physics*, Vol. 1, Addison-Wesley Publ. Co., 1963, Chapter 8, pages 8-1 to 8-4.

A beginning discussion of motion, with some of the difficulties encountered illustrated in an interesting way.

2. Lehrman, R. L., and Clifford Swartz, *Foundations of Physics*, Holt, Rinehart & Winston, New York, 1965, pages 133 to 145.

Presents experimental evidence relating to Newton's laws of motion.

3. PSSC, *College Physics*, Raytheon Educ. Co. (D. C. Heath), 1968, Chapter 11, pages 169 to 189, and Chapter 13, Sections 13.1 through 13.4, pages 218 to 225.

A treatment similar to ours, although more detailed.

4. Shamos, M. H., *Great Experiments in Physics*, Henry Holt & Co., 1959, pages 75 to 92.

The original paper by Henry Cavendish, who was the first to measure the density of the Earth.

NASA

CHAPTER **9**

ENERGY

In *Experiment 8–3 when you pulled a wooden* cart on wheels, you found that it accelerated. To cut down on the damage to the cart, one member of your team caught it when it reached the end of the table. It should be clear, at least to those who did the catching, that a moving cart is different from a cart at rest. Certainly it is different in that it has a velocity. But there must be something more than simply a difference in velocity to distinguish between the two carts, because it would have been easier to stop an empty cart than a cart loaded with bricks *even if both were moving at the same velocity*. Clearly this fact is dependent, in part, on the difference in mass. Apparently, then, a moving cart has a property that depends not only on its *velocity* but also on its *mass*.

Four bricks on a table are different from a single brick on a table. As you well know, one brick dropped from the table on your toe would not be as painful as four bricks dropped from the same height. In this situation too, the difference is dependent, at least in part, on mass. In addition, the effect of a single brick dropped on your toe depends upon the height from which it is dropped. Apparently the brick has a property which depends on its *position* as well as its *mass*.

A compressed spring can propel a toy car, or lift an object. Chemical reactions in a battery can run an electric motor that will drive a toy car or lift an object. The combustion of gasoline in an engine can be used to set an automobile in motion, or to lift an object. The electric current generated by a battery heats up a flashlight bulb and causes it to give off light. The motion of your hands when you rub them together produces heat.

Is there something in common among all of these things: chemical reactions, an electric current, heat, moving objects, compressed springs, and lifted objects? We will, in fact, find it convenient to introduce a new concept that will enable us to describe, in a simple way, a great variety of phenomena like those mentioned above. We call this concept *energy*, and we devote considerable time to its study because it will prove so useful to us later on in the course. But before we can understand energy we must first develop the closely related concept of work.

9-1 *Work*

In Chapter 8 we discussed a thought experiment in which a constant horizontal force was exerted on an air puck, which was initially at rest on a flat surface. The force acted for 3 seconds and was then removed. As you can see from Figure 8–5, while the force acted, the puck moved through a certain distance; from the graph in Figure 8–6c we see that while the force acted, the velocity changed. In order to better understand how the force acting on the puck is related to the puck's energy, it is useful to define a concept called *work*. Work is the product of the unbalanced force exerted on the puck and the distance through which the puck moves while the force is exerted on it. We will limit this discussion to forces that act in the direction of the object's motion. We can define work mathematically by

$$W = F(\Delta s) \tag{9-1}$$

where W stands for the work done on the puck, F is the constant force exerted on it, and Δs is the distance through which the puck moved while the force acted. In the mks system of units, force is measured in newtons and Δs in meters; therefore a unit of work is a newton-meter. This unit is given another name: the joule (rhymes with *pool*). (In the 1840's James Joule, an English physicist, proposed the concepts of energy and suggested that energy is conserved.) Figure 9–1 exhibits this relationship diagramatically.

FIGURE 9-1

The 1-newton force does 1 joule of work on the object when it moves that object through a distance of 1 meter.

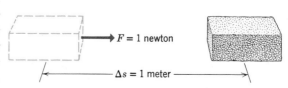

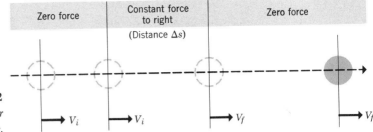

FIGURE 9-2

A constant force acts over a distance Δs.

9-2   *Kinetic energy*

Let us see if we can relate the work done on the body by the unbalanced force to the energy of the body. To do this, consider the simple situation shown in Figure 9–2. A puck with mass m is moving with an initial velocity v_i (the subscript i refers to "initial"). It will continue to move at a constant speed, in a straight line to the right, as long as no unbalanced forces act on it. However, if a constant force F is exerted on the puck, *in the same direction it is moving,* and continues to act while the puck moves a distance Δs, the puck will accelerate to a velocity v_f (the subscript refers to "final"). If the force is then removed, the velocity of the puck will remain constant. The work done is $W_{i \to f}$ (the subscript $i \to f$ means *from* the initial moment *to* the final moment of the action of the force), so that

$$W_{i \to f} = F \, \Delta s \qquad (9\text{-}2)$$

But we know from Newton's second law that the unbalanced force acting on the body is equal to the mass of the body times its acceleration:

$$F = ma$$

This force is the same as that in the definition of work given in Equation 9–2. Furthermore, you may recall that acceleration is the change in velocity divided by the change in time, $a = (v_f - v_i)/\Delta t$. If we substitute this into Newton's second law, we have

$$F = \frac{m(v_f - v_i)}{\Delta t}$$

This can then be substituted for F in Equation 9–2:

$$W_{i \to f} = F \, \Delta s$$

$$W_{i \to f} = \frac{m(v_f - v_i) \, \Delta s}{\Delta t}$$

$$W_{i \to f} = m(v_f - v_i) \frac{\Delta s}{\Delta t} \qquad (9\text{--}3)$$

Does the quantity $\Delta s / \Delta t$ appearing on the right-hand side of this equation look familiar? It is, in fact, the definition of average speed introduced in Chapter 8. Here it represents the average velocity of the puck during the time it is accelerated from v_i to v_f. But how can we use the average velocity? What is it equal to?

If the acceleration is constant, the average velocity is the velocity at the half-way point in time between the initial and the final velocities. For example, if a car accelerates uniformly from 20 mph to 40 mph, its average velocity is 30 mph. A car traveling with a constant velocity of 30 mph will travel the same distance in the same interval of time as a car accelerating uniformly from 20 mph to 40 mph (see Question 8–36). Consequently, we can write the average velocity as

$$v_{av} = \frac{\Delta s}{\Delta t} = \frac{v_f + v_i}{2}$$

In our example: $(20 \text{ mph} + 40 \text{ mph})/2 = 60 \text{ mph}/2 = 30$ mph.

Substituting the expression for average velocity into Equation 9–3 gives

$$W_{i \to f} = \frac{m(v_f - v_i)(v_f + v_i)}{2}$$

Since $\qquad (v_f - v_i)(v_f + v_i) = v_f^2 - v_i^{2\,*}$

* For those of you who don't remember this from algebra, the product of the sum and difference of two terms can be obtained as follows:

$$
\begin{array}{r}
v_f - v_i \\
v_f + v_i \\
\hline
+ v_f v_i - v_i^2 \\
+ v_f^2 - v_f v_i \\
\hline
+ v_f^2 \qquad - v_i^2
\end{array}
$$

we get

$$W_{i \to f} = \frac{m(v_f{}^2 - v_i{}^2)}{2} = (\tfrac{1}{2})m(v_f{}^2 - v_i{}^2)$$

$$W_{i \to f} = (\tfrac{1}{2})mv_f{}^2 - (\tfrac{1}{2})mv_i{}^2 \qquad (9\text{-}4)$$

Let's see what Equation 9–4 tells us about the motion of the puck. The left side of the equation is the work done by the *unbalanced* external force F acting on the puck. That is, it is a measure of the external influences on the puck. The right side, on the other hand, involves only the mass and the initial and final speeds of the puck. The combination of these properties in the form $(\tfrac{1}{2})\,mv^2$ is given the name *kinetic energy* and is abbreviated *KE*. With this new term, Equation 9–4 can be written in the form

$$W_{i \to f} = KE_f - KE_i$$

or

$$W_{i \to f} = \Delta KE \qquad (9\text{-}5)$$

In other words, the work done on the body by the unbalanced force $(F\,\Delta s)$ turns out to be equal to the increase in its kinetic energy. Equation 9–5 is sometimes called the *work-energy theorem*.

In Equation 9–5 you can see that work, on the left side, is equated to energy, on the right. Both quantities may be expressed in joules. Let's see how much kinetic energy a familiar moving object has. Your wooden cart in Experiment 8–3, loaded with four bricks, had a mass of nearly 7 kg. If that cart moved at a speed of 1.0 meter/sec, its kinetic energy would be

$$KE = (\tfrac{1}{2})mv^2$$
$$= (\tfrac{1}{2})(7.0 \text{ kg})(1.0 \text{ meter/sec})^2$$
$$KE = 3.5 \text{ joules}$$

Question 9–1 Refer back to Experiment 8–3. Assume you exerted a force of 1.5 newton with your three rubber bands on the cart loaded with two bricks (a total mass of 4.0 kg), and the cart moved for 2 meters while the force acted. How much work was done on the cart?

Question 9–2 If the cart in Question 9–1 started from rest, what was its kinetic energy when the force ceased to act?

Question 9–3 If the cart in Question 9–1 started from rest, what was its velocity at the moment the force ceased to act?

9–3 *The principle of conservation of energy*

We have introduced the concept of work and we have shown that if we define work as the product of the unbalanced force exerted on an object and the distance it moves while that force is acting, then the work done on an object turns out to be equal to the increase in a quantity $(\frac{1}{2})mv^2$ which we call the kinetic energy.

Much of the usefulness of the energy concept arises from the fact that, when properly defined, the total energy in the Universe is constant. This statement is so fundamental to man's model of the world about him that it has become known as *the principle of conservation of energy*. We will discover that if this principle is to apply to all physical interactions and to all situations whatsoever, it will be necessary to invent a number of different kinds of energy. We will also see the need to invent rules for assigning numerical values to the various forms of energy we invent. These rules will tell us how to compute the values of each kind of energy in terms of the properties of objects under any situation imaginable.

The rule for determining this number has already been set for kinetic energy:

$$KE = (\tfrac{1}{2})mv^2$$

But we must establish other kinds of energy and give the rules for determining a given amount of each kind if we are going to be able to develop the principle of conservation of energy. That principle is the accounting system of the energies as objects interact with one another. If we assume that energy is conserved in each such interaction, then if objects do interact in such a way that the energy of one decreases, the energy of some other must increase. It is possible, of course, for the energies of each of the interacting objects to decrease, but only if they lose energy to the surroundings. Objects can also gain energy from the surroundings.

By this reasoning, we are at liberty to invent as many energies as we wish and in any way we see fit in order to accomplish our objective of producing an energy conservation principle. The principle will only be useful, however, if the number of separate rules necessary to specify all kinds of energy is reasonably small. Generating the law would not be worth the effort if every new phenomenon investigated required a new kind of energy calculated by a new rule. If, however, we can describe all interactions in terms of the exchange of only a few kinds of energy, then the energy conservation principle would provide a new relationship between properties of interacting bodies. Fortunately, it is possible, by identifying only six or eight distinct kinds of energy, to arrive at a principle which turns out to be enormously powerful as an aid in thinking about how bodies interact. Once we are persuaded we have established a principle that is valid for all events, we can use it with confidence to predict what results we shall obtain from experiments we have not yet performed.

9-4 *Potential energy*

Consider a phenomenon which will permit us to propose new kinds of energy in order to develop the principle of conservation of energy. Figure 9–3 is a stroboscopic

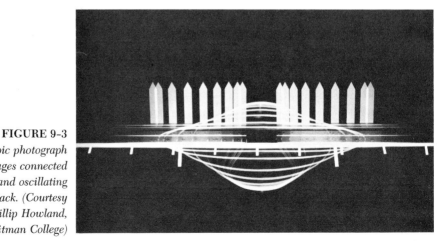

FIGURE 9-3
Stroboscopic photograph of two carriages connected by a spring and oscillating on an air track. (Courtesy of L. Phillip Howland, Whitman College)

photograph of two identical carriages that ride on a thin film of air which was forced through tiny holes in the track. These carriages are connected by a spring made of a material like a watch spring so it can be easily bent from its normal circular shape. When the spring is bent, it exerts a force in the direction that would return it to its natural shape. So that the motion of the carriages could be recorded, the air track was placed in a darkened room and the camera shutter held open for a time exposure. The film was exposed by the flashing light of a stroboscope, which emits flashes at precisely equal time intervals. Thus the faster the carriages went, the farther they traveled between any two consecutive flashes of the stroboscope. The pointed markers in the photograph record the various positions of the carriages during the time the camera shutter was open.

Before the first flash, the two carriages were held close together. Neither carriage was moving, so each had zero kinetic energy. After the carriages were released, the photograph reveals that they were forced apart by the spring. Initially, the velocity of each carriage increased, then each was slowed down by that same spring; and eventually each came to rest at their greatest distance apart.

Once the carriages came to rest at their greatest distance apart, the spring brought them closer again. The carriages, the air track, and the springs have been so designed and constructed that the motion repeats itself over and over again, with little or no appreciable difference from one oscillation to the next. Of particular interest to us is the fact that the carriages will each travel the same distance each trip; and since the carriages are identical, their speeds at any instant will be equal even if they always move in opposite directions. Furthermore, we observed that as each carriage passes through a particular position its speed is always the same.

The total kinetic energy of the system is the sum of the kinetic energies of the two carriages and the spring; but since the mass of the spring is much smaller than that of the carriages, we will disregard its small kinetic energy. Therefore, the total kinetic energy will be divided equally between the two identical carriages. At those places where the velocity is the greatest, the kinetic energy will be at a maximum which we call KE_{max}; at those places where the carriages are at rest, the kinetic energy will be zero. Figure 9–4 is a drawing taken from Figure 9–3 and shows the carriages at five positions.

Question 9–4   At which position is the velocity of each carriage at a maximum? At which position is the velocity at a minimum?

Question 9–5   At which position is the kinetic energy at a maximum? at a minimum?

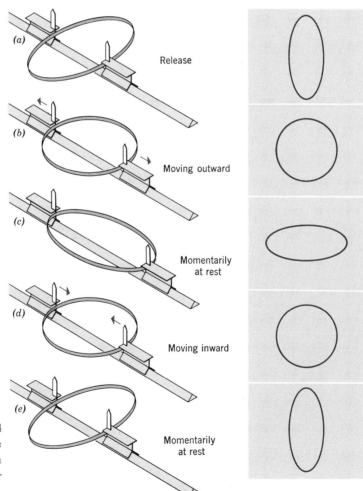

(a)   Release

(b)   Moving outward

(c)   Momentarily at rest

(d)   Moving inward

(e)   Momentarily at rest

FIGURE 9–4
Drawing of the carriages shown in Figure 9–3 in certain positions of their oscillation.

It is obvious that the kinetic energy of the carriages is always changing, and since it is continually changing, we certainly do not have a conservation principle if kinetic energy is the only kind of energy we consider. In order to develop a principle of energy conservation, therefore, we must invent a new kind of energy, an energy that increases as the kinetic energy decreases. It is reasonable to guess that this new kind of energy is somehow related to the spring, so it may be well for us to consider how the kinetic energy is related to the spring.

As the carriages oscillate back and forth, no work is done from the outside. The change in kinetic energy must come from *within* the spring-carriage system—that is, it must be associated with internal forces between the spring and the carriages. When the carriages are in position *a* of Figure 9–4, the spring is bent inward from its normal shape, so it exerts a force outward on the carriages. In response to this force, the carriages accelerate outward until position *b* is reached. In position *b*, the spring is in its natural shape; it is bent neither inward nor outward. As such, it does not exert a force on either carriage. Consequently, the carriages are no longer being accelerated; this is the equilibrium position.

The inertia of the carriages, however, carries them outward from position *b*, and in moving outward they bend the spring outward from its natural position, so it exerts an inward force on the carriages. This inward force slows the carriages down. Consequently, the carriages have their maximum speed, and thus their maximum kinetic energy KE_{max} at position *b*.

As the carriages continue outward, the spring continues to exert a force on them and slows them down until they are brought to rest at position *c*. But since the spring still continues to exert an inward force on the carriages, they do not remain at rest; they accelerate inward. When they reach position *d*, the spring is again in its normal shape and the carriages reach their maximum speed and kinetic energy. Their inertia carries them inward, bending the spring, until they are brought to rest at position *e*. Position *e* is the same as position *a*, so the oscillation starts all over again.

In order to say something about the new kind of energy we must invent (if we are to have a conservation of energy principle), we might start by establishing the total amount of energy of the carriage-spring system. We can then assume that the total amount of energy is conserved and that it must be divided up among the two kinds of energy: the kinetic energy and the new kind of energy. Since the kinetic energy of the carriages changes because the spring exerts a force on them and accelerates them and, furthermore, since this new kind of energy must increase as the kinetic energy decreases, we can relate this new energy to the bending of the spring.

When the spring is in its natural unbent position, the kinetic energy of the carriages is at a maximum. If we specify that the new energy is zero when the spring is unbent, then the maximum kinetic energy is equal to the total energy of the system, which we can call E.

We can now construct our conservation law as follows: for each position of the spring, we can measure the speed of the carriages. Using the definition of $KE = (\frac{1}{2})mv^2$, we can compute the kinetic energy of both the carriages, and thus of the system. If we then subtract this kinetic energy from the total energy E of the system, we are left with the amount of this new kind of energy

$$E - (\tfrac{1}{2})mv^2$$

This new kind of energy was arrived at by considering only the position of the carriages, that is, the amount the spring was bent. This new energy increases as the spring is increasingly bent (until the spring breaks, of course), and it decreases as the spring is permitted to return to its normal unbent position. In short, this new kind of energy depends completely on how much the spring is bent. The new energy is nonzero whenever the spring is bent inward or outward from its natural shape. It is present, for example, if the spring is bent and the carriages are held apart by your hands—even if the carriages are not moving. However, as soon as you release the carriages, they begin to move; the new form of energy decreases as the kinetic energy increases.

You might say that the new energy is stored in the bent spring ready to be transferred to another form. This characteristic has led to the name *potential energy* for this form of energy.

The principle of conservation of energy for our carriage-spring system can therefore be stated mathematically by assigning a symbol *PE*—and later a numerical value—to the potential energy:

$$PE = E - (\tfrac{1}{2})mv^2$$

The total energy *E* is equal to the sum of the kinetic and potential energies:

$$E = KE + PE$$

You may feel that the procedure for assigning potential energy to the system is so trivial that it cannot possibly contribute anything to our understanding of the interactions between different objects, and especially of more complex systems than our simple carriage-spring system. Indeed, if we were to confine our attention to the particular oscillation studied so far, it would be trivial. If, however, the numerical assignments of potential energy of the carriages —as determined by the particular amount of bending of the spring—should always contribute to the total energy of the system so as to yield a constant total energy, even under different conditions of oscillation, then we would have achieved a significant result. Those different conditions might include starting the oscillations with the carriages at different distances from each other or changing the masses of the carriages.

By knowing how the force varies with the bending of the spring, we can calculate the work done by the spring on the carriage. But since the force changes as the bending of the spring changes, the calculation is complicated, so we will not perform it here. The result of the calculation, however, shows that the potential energy of the spring depends *only* on how much the spring is bent. For a given carriage-spring system, then, the amount of potential energy of the system depends only on the location of the carriages with respect to the equilibrium position—their position when the spring is in its natural shape; at that point the potential energy is generally taken to be zero. Consequently, we are able to assign specific values of potential energy to the system, and those values depend only on how much the spring is bent. During oscillation, the potential energy of the system when the spring is bent a maximum amount does equal the kinetic energy when the spring is unbent. The total energy is constant for each position of the spring.

9-5 *Thermal energy*

A real carriage-spring system would not oscillate forever as we assumed in our previous discussion. Even if the air film on which the carriage floats were completely free of friction, we would find that eventually the carriages would come to rest with the spring in its natural unbent shape. Taking this more realistic view of our carriage-spring system, we find something missing. Our rule for computing kinetic energy says—no motion, no kinetic energy; our rule for computing potential energy says—no bending of the spring, no potential energy. But when our carriage-spring system was oscillating, it did have energy; now we say that it will gradually slow down and stop. Are we saying that the total energy, kinetic and potential, is *not* conserved?

Is it possible for us to salvage the principle of conservation of energy? We invented a new form of energy to establish the conservation principle. Perhaps we can invent still another form of energy to enable us to slip out of this predicament. Let's try.

Consider the carriages and the spring once again, but this time imagine that we place a rigid rod on top of the carriages and above the springs, so that it rests on and rubs against the carriages as they oscillate underneath it. If everything is well balanced, the rod will remain stationary as the carriages oscillate. Under these circumstances we would find that each time the carriages reach the equilibrium position, their speed, and thus the kinetic energy, would be measurably less than the previous time. But the kinetic energy at equilibrium is equal to the total energy of the system; and we are saying that it decreases. If we are to preserve the conservation of energy principle, therefore, our still newer kind of energy must increase with each oscillation.

We were led to the adoption of potential energy by considering the action of the spring on the carriages; the potential energy of the system increases as the spring is increasingly bent. Now what property of the system—slightly altered to include the rod—would increase continuously during the oscillations?

Have you ever rubbed your hands together hard and fast? Rub them again, *hard and fast*. What happens? They get hot! Your hands were rubbing together in a manner not unlike the rod rubbing against the carriages; the temperature of your hands increased. Do you suppose the temperature of the rod and the carriages increased?

Accurate measurements of the temperature of the rod and the carriages would answer our question, and, comfortingly enough, would help us out of our predicament. This increase in temperature can be brought into our conservation of energy principle by inventing a new form of energy, *thermal energy* (often called heat energy). To complete this aspect of the principle of conservation of energy we would have to suggest a rule relating the thermal energy to the temperature of the system. Since the temperature of our carriage-rod-spring system increases as the kinetic and potential energies decrease, the thermal energy must increase as the temperature increases. If we can do this, the total energy—the kinetic plus the potential plus the thermal energy—would remain constant. Energy would then be conserved and we would be safely out of our predicament, but only if our rule held for every set of oscillations carried out under any imaginable conditions. This does, in fact, turn out to be possible. A carriage-spring system without the rod would still cease oscillating. The kinetic and potential energy would be transformed into thermal energy which would increase the temperature of the spring and also of the air film between the carriage and the track.

The kinetic and potential energy of an artificial satellite must be converted into some other form of energy before a satellite can be brought back down to the surface of the Earth. Its energy while in a free orbit is transformed into thermal energy during re-entry of the atmosphere (see the photograph that opens this chapter).

We have now reached a position in which we have convinced ourselves (and we hope you) that it is possible to assign to objects kinetic energies that depend on their speeds, potential energies that depend on their positions, and thermal energies that depend on their temperatures, in such a way that the total of these three forms of energy for all of the objects within our system is constant for all times. At least we have done this for our carriage-rod-spring system. But before we proceed too far, we should ask ourselves whether we really have done what we claim to have done. We have proposed a principle; now we must see if that principle is applicable. Is it really applicable for all times? Imagine a carriage-spring system oscillating. We have accounted for the decrease in its kinetic-potential energy by the transformation into thermal energy, but where did that kinetic-potential energy come from in the first place?

For the moment, let us ignore the thermal energy and in so doing call our carriage-spring system an *ideal* system. All of the energy is in a kinetic and potential form. For convenience, these two kinds of energy are often called *mechanical energy*. The two carriages are at rest, the spring joining them is in its natural shape; the system has neither kinetic nor potential energy. To set the carriages in oscillation, they must be given energy. Where does that energy come from?

In order to bend the spring out of shape it is necessary to apply an outside force on the system—the spring will not spontaneously bend itself out of its natural shape. While exerting this outside force, let us say with your hands, you move the carriages through a certain distance. You do work on the carriages and in so doing, you give them energy. Initially, the energy which is given the system is in the form of potential energy, stored in the spring; but if you release the carriages, that energy is converted into kinetic energy as the spring accelerates them.

We can calculate the amount of work you do on the spring by bending it. That work depends not only on how far you move the carriages, but also on how much the spring is bent when you move them. So, the potential energy given the spring by your doing work on the system depends only on the location of the carriages relative to their position when the spring is unbent. The energy given the system by the outside agent—your hands—is sometimes called the work done on the system by an *external agent*.

9-6   *In support of the principle of conservation of energy*

Our assumption that energy is not lost, is, of course, not an original one. In the 1840's the English physicist James Joule, among others, proposed that energy can be neither created nor destroyed, although it can be transformed from one form to another. Since then, new experiments have occasionally revealed the need to introduce new kinds of energy; but the number of distinct kinds is small, and the rules connecting the numerical value of each kind of energy to measurable properties of any system are simple. Hence, we now accept the *principle of conservation of energy* as a fundamental description of nature. Energy itself is a creation of the human mind. Our confidence that we can find a small number of simple rules for assigning energies so that the sum of all kinds is constant has established the conservation principle as a basic premise of science. The applicability of the principle to a wide variety of phenomena has been experimentally demonstrated.

The concept of energy and its conservation is one of the simplest and most successful approaches in describing the many and varied changes occurring about and within us. Energy is continuously being converted from one form to another. For example, the steady conversion of stored nuclear energy into thermal and radiation energy within the interior of the Sun provides us with heat and light. Other processes represent smaller rates of energy exchange. For example, solar radiation from the Sun is converted into chemical energy within a green plant when the sunlight is absorbed by the plant initiating and maintaining the process of photosynthesis. If the trees and plants subsequently become coal and oil, this stored chemical energy is retained in the fossil fuels. Later, if the fuel is ignited, part of the stored chemical energy can quickly be converted into many other energy forms: kinetic energy, thermal energy, electromagnetic energy, and sound energy.

We established the principle of conservation of energy on the basis of two carriages floating on a film of air. These two carriages were joined by a spring and oscillated. Now we want to try out this principle, to test it so to speak, by extending it to other systems. First, we should select a system that is isolated from the rest of the Universe so that no energy can enter or leave. Second, we have to know all the various forms of energy possessed by our isolated system—indeed, we need quantitative expressions for each of these forms. With this information, it is just a problem of bookkeeping. At any instant of time we can tally the various amounts of energy present. If the conservation-of-energy principle is a good description of nature, the total amount of energy present will always be the same. In fact, we performed a simple tally of energy on the carriage-spring system and found that we needed the concepts of potential and thermal energy in addition to kinetic energy if our tally was to balance at all times.

In this test of our conservation principle let us apply it to another two-body system: the Earth and an object close to its surface. Mentally isolate this system from the rest of the Universe, and assume that the only force acting is the force of gravitational attraction between the object and the Earth. Our first problem is this: In the spring-carriage system, we described a potential energy that related to the bending of the spring; can we now describe a potential energy that relates to the gravitational force? If we can, it will be simpler than for the carriage-spring system, for the gravitational force is constant near the surface of the Earth. In fact, we might even be able to express our relationship quantitatively. Recall from Chapter 8 that the gravitational force acting on an object is its weight w, and that

$$w = mg$$

If you lift a brick from the floor to a table top, has the potential energy of the Earth-brick system changed? Has it increased or decreased? Would there have been a larger change in potential energy if we had lifted a pencil, rather than a brick, from the floor to the table?

The potential energy of the carriage-spring system increased whether the spring was bent inward or outward because work was done on the spring. The gravitational potential energy, however, increases only if the object is lifted, that is, if the distance between the object and the Earth is increased. We can use the work-energy theorem to arrive at a quantitative expression for the gravitational potential energy between an object and the Earth.

Because of the large mass of the Earth, we can safely assume that it stays essentially at rest during our thought experiment. For the moment, confine your attention to an object as it falls freely from some height h to the surface of the Earth. A force $w = mg$ acts on that object as it falls through a distance $\Delta s = h$. Consequently, that force must do work on the object since the object accelerates at the rate of 9.8 meter/sec or about 10 meter/sec. Work done on the object increases its kinetic energy KE:

$$KE = F(\Delta s)$$
$$KE = wh$$
$$KE = mgh$$

In order to determine the kinetic energy of the object an instant before it hits the ground, we need to know only its weight mg and the height through which it has fallen h. An object of mass 4 kilograms weighs about 40 newtons,

$$mg = (4 \text{ kg})(10 \text{ meter/sec})$$
$$mg = 40 \text{ newtons}$$

If that object should fall through a height of 2 meters, what would be its kinetic energy an instant before it struck the ground?

$$KE = mgh$$
$$KE = (4 \text{ kg})(10 \text{ meter/sec})(2 \text{ meter})$$
$$KE = 80 \text{ joule}$$

Having calculated the amount of kinetic energy acquired by a falling object, we now need to relate this to its gravitational potential energy at any given height. If we are to relate kinetic and potential energy through the principle of conservation of energy as we did with the carriage-spring system, we should propose that the total energy of the falling object remains constant. As it falls, its kinetic energy increases so its potential energy must decrease and by the same amount. If the object is released from rest at an elevation Y above the surface (see Fig. 9–5), it has zero velocity and hence zero kinetic energy. We have shown that after falling a distance h (where $h < Y$), it will have gained mgh kinetic energy. This increase in kinetic energy means a decrease in potential energy of the same amount, mgh. Consequently, just before it strikes the Earth, its kinetic energy will be mgY. It is convenient to assign a potential energy of zero to the object when it is at the surface of the Earth—then the kinetic plus potential energies of the object just before striking the Earth are mgY. If this

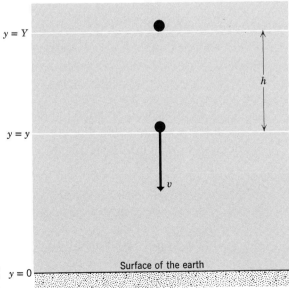

FIGURE 9-5
Positions of a falling object.

sum is to be constant, then when the object is first released, the potential energy must be mgY, since the kinetic energy at that instant is zero. When the object has fallen a distance h, but is still a distance $y = Y - h$ above the surface, the kinetic energy is mgh, so the potential energy is $mgY - mgh = mg(Y - h) = mgy$. If we use the symbol PE for potential energy, we have shown that

$$PE = mgy \qquad (9\text{–}6)$$

for an object of mass m at a distance y above the surface of the Earth.

Question 9–6 How much work must you do on a 1.0-kg block to lift it from the floor onto a table top 1.0 meter high? How much has the gravitational potential energy of the Earth-block system increased? Has the energy of the Universe changed? Why?

Since the gravitational potential energy depends on the height of an object above the surface of the Earth, and since we are assuming that near the surface of the Earth the gravitational force on any given object is constant, the gravitational potential energy is constant on any plane above any small region of the Earth's surface, such as your room. The potential energy of any object on a horizontal table top is constant no matter where it is placed on that table top. Furthermore, if the object is then moved from the table top to the floor, it loses potential energy; its loss of potential energy will be the same no matter where it had been on the table top and no matter where it is placed on the floor, so long as the floor is also horizontal. The path it took in moving from the table to the floor is of no consequence.

If the object falls, its potential energy is transformed into kinetic energy. To illustrate the point that its loss of potential energy is dependent only on the height through which it falls, and not upon the path it takes in falling, see Figure 9–6.

This figure shows a pendulum with a long string. In part *a* the pendulum is lifted so that the bob is at a height h above the bottom of its swing. We can, for convenience, as-

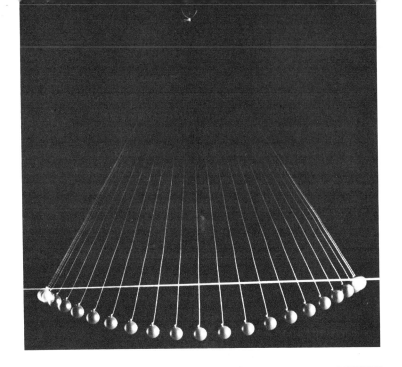

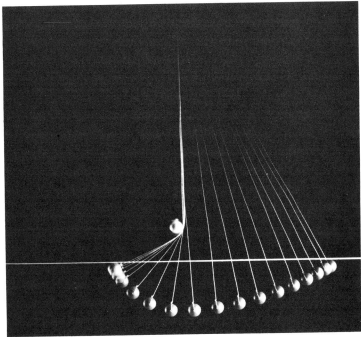

sume that its potential energy is zero at the bottom of the swing. As the pendulum in *a* is freed, it swings along the sweeping arc to the bottom of its swing where all of its potential energy has been transformed into kinetic energy.

At the bottom of the swing, the energy of the pendulum

206

bob is all kinetic. Since energy is conserved, it cannot lose that kinetic energy without transforming it into another form of energy. The pendulum is practically free of friction so thermal energy may be neglected. Consequently, the bob continues moving, and its kinetic energy is converted into potential energy. As a verification of the conservation of energy principle, you see that the pendulum arrives at the same height from which it fell. The kinetic energy gained in falling along the sweeping arc from the left to the bottom is enough to carry it back up to that same height again.

Now if you direct your attention to Figure 9–6*b* you will see that the pendulum falls through the same height *h*, but because of a peg in the way of the string, its path is different from its path in Figure 9–6*a*. Yet in falling along this different path it acquires a kinetic energy that is enough to carry it back up to the original height again. The loss of potential energy of the pendulum bob does not depend on which of the two different paths it falls along—even if one path is actually longer than the other. The important thing is that in falling along each path it fell through the same vertical height.

So far, our discussion of energy has been centered on either thought experiments or photographs of carriages and pendulums. It is time to perform a simple laboratory experiment to demonstrate the principle of conservation of energy of an Earth-object system.

EXPERIMENT 9–1 Conservation of mechanical energy

The principle of conservation of energy states that energy can neither be created nor destroyed. However, this principle is not obvious, and its manifestations are not easily observed. This experiment will give you an opportunity to measure changes in mechanical energy in a relatively uncomplicated situation.

Set up a simple pendulum as in Figure 9–7 by hanging an object that has a mass of about 200 grams from a string about 1.0 meter long. Three points, *A*, *B*, and *C*, have been labeled as reference points.

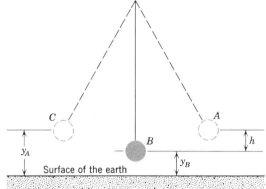

FIGURE 9-7
Swinging pendulum.

Start the pendulum from rest at point *A*, let it swing, and observe it.

Question 9-7   Where does the pendulum appear to have the greatest speed?

Question 9-8   Notice the dimension *h*. How can you change *h*?

Question 9-9   Start the pendulum in a number of different positions so that *h* is different each time. How does the speed of the pendulum as it passes point *B* appear to vary as *h* is varied?

Question 9-10   We have defined kinetic energy as $(\frac{1}{2})mv^2$. Where in its swing do you think the pendulum has the greatest kinetic energy?

Question 9-11   Can you detect a relationship between the maximum kinetic energy of the pendulum and the height *h*? (Don't try to be quantitative—you haven't made any measurements yet.)

In order to convince yourself that the principle of conservation of energy is valid, you may need very precise measurements of both the height through which an object falls and its velocity when all of its initial potential energy is converted into kinetic energy. The equipment you will use in this experiment may not yield such precise measurements, however, it will be able to demonstrate the transformation of energy from one form into another.

The hypothesis we wish to test is: as the pendulum swings from *A* to *B* (see Fig. 9-7), the increase in kinetic energy equals the decrease in potential energy. If energy is conserved, then the sum of the kinetic and the potential energies at point *A* must equal the sum of the kinetic and the potential energies at point *B*. Mathematically,

$$KE_A + PE_A = KE_B + PE_B$$

The kinetic energy at point *A* is zero, for the pendulum starts from rest, so $KE_A = 0$. The potential energy at point *A* is $PE_A = mgy_A$. At point *B* the kinetic energy is $KE_B = (\frac{1}{2})mv^2$; the potential energy is $PE_B = mgy_B$. The equation for the principle of conservation of energy becomes:

$$KE_A + PE_A = KE_B + PE_B$$
$$0 + mgy_A = (\tfrac{1}{2})mv^2 + mgy_B$$

This equation will make more sense if we solve it for the kinetic energy term and see how the kinetic energy is related to the potential energy.

$$(\tfrac{1}{2})mv^2 = mgy_A - mgy_B$$
$$= mg(y_A - y_B)$$
$$(\tfrac{1}{2})mv^2 = mgh$$

Not a really surprising result; but does this, in fact, hold true for your pendulum?

You can easily determine the change in the gravitational potential energy. To measure the change in height *h* all you need to do is measure the height of the *center of mass* of the bob above the table top in both positions *A* and *B*. The difference in these heights is *h*.

It is a little more difficult, however, to determine the kinetic energy at point *B*. To measure the velocity at point *B*, set up your doorbell timer as shown in Figure 9–8. Tie the timer tape to the pendulum bob, and thread the tape through the timer. The bob will not, of course, push the tape through the timer, so you will obtain results from only one swing of the pendulum. The timer should be at nearly the same height as the bottom of the swing of the bob, and this should not be too far above the floor (or table top). There may be a tendency for the tape to sag a bit and thus give a spurious velocity, but this can be minimized by letting the tape run along the floor (table top).

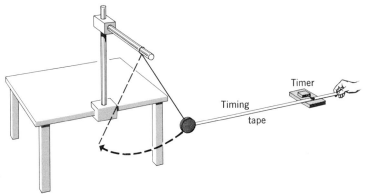

FIGURE 9-8
Measuring the speed of a pendulum bob.

Clamp the timer firmly to the table. Thread the free end of the tape through the timer, and use the tape to pull the pendulum bob back to position *A*. Measure the height from the floor (table top) to the center of the mass of the bob; make sure that the tape will run freely through the timer; start the timer; and then release the bob. After the run, measure the height of the center of mass of the bob in position *B*. The difference in these two heights is the distance through which the bob fell.

Those dots on the tape that were made at the beginning and the end of the swing are obvious. Can you identify the dots which were made when the pendulum was closest to its maximum velocity? After you have located those dots, you will have to determine the maximum velocity, which is needed to calculate the maximum kinetic energy.

However, to calculate the velocity you will first have to calibrate your timer as you did in Experiment 8–3. The mks system of units does not include our arbitrary unit of time, the tentick, so you must determine the interval of time between successive ticks. The calibration of the timer for Experiment 8–3 may not be valid for this experiment. You may have a different timer, or the batteries may have deteriorated since you used them last, making the timer beat with a different frequency.

When you have recalibrated your timer (and recorded it), you can calculate (in meters/second) the maximum velocity of the pendulum bob. This calculation is most easily done by first calculating the average velocity for each of the successive intervals of time and drawing an average velocity-time graph. This graph will have a step-like appearance as the velocity first increases then decreases.

To find the maximum velocity, however, you should draw a smooth instantaneous velocity-time curve that passes close to the middle of each of the steps—with the exception of the highest step. The smooth curve will more than likely go higher than the top of the highest step. The very top of the smooth instantaneous velocity-time curve is the maximum velocity achieved by your pendulum, so draw that curve carefully.

To obtain greater accuracy, you should run this experiment three times in all. It will help to use the same value of h each time. If the same value of h is used each time, the average of the three maximum velocities can be used which should be more accurate than any one of the three individual values.

You must now check to see if our hypothesis is correct:

$$mgh = (\tfrac{1}{2})mv^2$$

Use either 9.8 meter/sec^2 or 980 cm/sec^2 for g.

QUESTIONS 9-12 If you used the same h each time, how did the three maximum velocities compare? The difference between the highest and lowest speed measured is what percent of the average speed? Why is this of interest?

9-13 Do your measurements convince you that, in this experiment, mechanical energy is conserved? (They might not.)

If your results show that the increase in kinetic energy was less than the decrease in potential energy, compare your results with those of your classmates to see if they obtained the same results. Perhaps there is some other form of energy involved in addition to the potential and kinetic energy. Or perhaps the discrepancy is an indication of the accuracy of your measurements. What possible sources of error can you identify in this experiment? How would you expect these errors to affect the results?

9-14   A superball rolls off the top step of the basement stairs and bounces down the stairs to the basement floor. If the basement is 9 ft below the first floor, how fast will the ball be going when it hits the basement floor? The acceleration of gravity g is 32 ft/sec². A superball is very elastic and loses very little energy when it bounces. Hint: Consider the potential energy with which the ball begins and ends its trip. Consider its kinetic energy at the same points in time.

References   1. Feynman, R. P., R. B. Leighton, and M. Sands, *The Feynman Lectures on Physics,* Vol. 1, Addison-Wesley Publ. Co., 1963, Chapter 4, pages 4-1 to 4-8.

An ingenious discussion of the principle of conservation of energy from a point of view that was adopted by the PSNS authors in this chapter. Also, Chapter 52, pages 52-1 to 52-4. Symmetry of various forms, real and abstract. Symmetry and its relation to conservation laws.

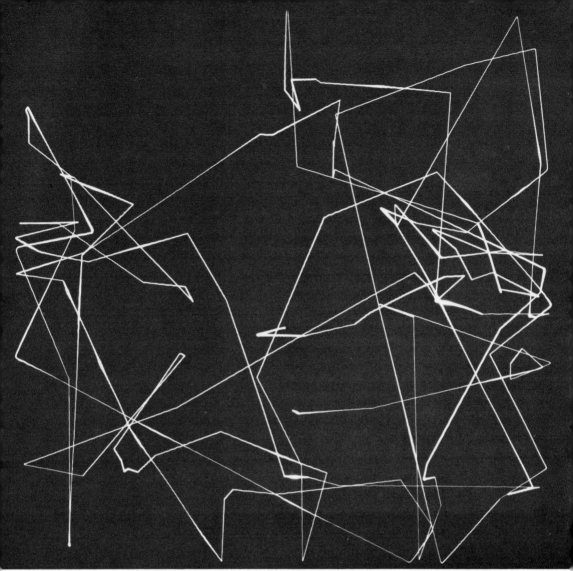

CHAPTER **10**

THE KINETIC THEORY
OF GASES

In *Chapter 7 we developed a model of a gas as being com-*
posed of a large number of small particles called molecules
which dash about every which way in a lot of open space.
That model was suggested in part by the results of Experiment 7–4, in which we trapped some air under an oil drop
in a gas tube. When that trapped air was heated, it expanded
to a larger volume. But when heat was neither added nor
removed, the drop of oil did not accelerate—it didn't even
move with constant speed. Since the oil did not accelerate,
we know from Newton's second law of motion that the
unbalanced force of the oil was zero. If there was no
unbalanced force on the drop of oil, were there any *balanced*
forces? To answer this question let us suggest another
thought experiment.

10-1 *Molecules and gas pressure*

What would happen if we were to pump air out of a
metal container, a container with a close-fitting lid? Could
you take the lid off easily? What is the origin of the force
that holds the lid on?

With the air inside the container we would not detect
any force holding the lid on, but with the air removed from
the interior we would detect a force. We might suspect
that any air in contact with the lid pushes against it. When
there is air on both sides, the forces on the inside and outside balance one another; if air on one side is removed, the
lid is pushed hard against the container by the air on the
other side. Upon what does this force depend? Does it, for
example, depend upon the size of the lid?

If we took several containers with openings of different
sizes, each with a close-fitting lid, pumped all the air out of
each container, and measured the force required to lift
each lid off, what would you expect to find? Which lid
would be the most difficult to remove? (Neglect the weight
of the lid.) If we actually performed this experiment, we

would find that the force would increase with the area of the lid, but if we were to divide the force exerted on each lid by the area of each respective lid, we would find that the force per unit area of the lid would be constant for all containers. So it is the ratio of the force on the lid to the area of that lid which is important. Furthermore, since this ratio is constant for lids of different size, it is reasonable to suppose that this constant ratio is a property of the exterior air.

It is convenient to give a special name to a property when that property helps us describe the object of our studies. The ratio of the force to the area is defined as *pressure*—in Experiment 7-4 the pressure of the air against the bubble of oil. The pressure of the air is exerted on any surface in contact with the air and acts in a direction perpendicular to the surface, no matter how that surface is oriented. This does not mean the pressure can't ever change; but if conditions influencing the gas are not changing rapidly, then the pressure throughout a container of moderate size will be everywhere the same and will push in any direction—but always perpendicular to the walls of the container.

A familiar example of these ideas is offered by the pressure of our atmosphere. This pressure is frequently expressed in millimeters of mercury; a pressure of 760 mm of mercury is said to be 1 atmosphere of pressure and is regarded as a standard of comparison for scientific work. When the pressure is 760 mm of mercury, the force the gas would exert on a surface 1 cm^2 in area is equal to the weight of a column of mercury 760 mm high and 1 cm^2 in cross-sectional area. The pressure of the atmosphere, however, is not constant; it does change.

At any given time, measurements will reveal that the pressure is very nearly the same at any point in a container the size of a room and, therefore, the air pushes equally on equal areas of all surfaces in the room: down on the floor, up on the ceiling, and horizontally on the walls. However, if you measure the pressure of the atmosphere on the top of a mountain or in an unpressurized airplane at 10,000 ft, you will find that it is less than at sea level. Clearly then, atmospheric pressure has something to do with gravity. In fact, the force holding down each lid on an evacuated container in our thought experiment results from the weight of the column of air as high as the atmosphere and with a cross-sectional area equal to the area of each lid. The weight of a column of air as high as the atmosphere and with an area of one square inch is 14.7 pounds.

But clearly, this very static picture does not tell the whole story. It does not, for instance, explain the upward force on the ceiling since forces of gravity are always down. And what about the gases trapped in a closed space capsule as it leaves the vicinity of the Earth? Does the pressure of this gas go down as the capsule reaches outer space where the force of gravity is much smaller? An astronaut will tell you it does not. Consequently, gravity is not the only cause of air pressure; we must propose a better model of how gas exerts pressure than to think of it simply as one substance pressing against another in a static relationship. Let us see if we can do this.

Consider again the idea suggested earlier that a gas consists of many tiny particles moving at random. These particles will frequently bump into one another and into the walls of the container. This must be so, for if they did not experience such collisions, they would never feel any forces and would thus move forever in one direction at a constant speed. Since the gas does not leave a tightly sealed container, the molecules must occasionally have their directions of motion changed, and this requires a force.

The force which alters the motion of the molecules is a force exerted *by* the walls of the container *on* the molecules. But the pressure we are trying to understand involves a force *on* the walls. Are these forces related?

Suppose that a man is standing still but acting as a backstop to basketballs thrown at him from behind. If those basketballs were thrown hard and frequently enough, he might have to brace himself or he would fall frontwards. Whether he falls frontwards or stands steady, he knows those balls exert a force on his back. Since the balls accelerate, that is, they do not continue traveling in a straight line with a constant speed, we know that he exerts a force on the balls (recall that when the speed of an object decreases, the object still accelerates, but the acceleration is negative). How are these two forces related—the force exerted by the basketballs on the man and by the man on the basketballs? Newton himself gave the answer to this question in his third law of motion, sometimes stated as follows: *For every action there is an equal and opposite reaction.* Here "action" and "reaction" should be interpreted as being forces. It is important to note that the action and reaction are on different bodies. Another way to say this is, "If body *A* exerts a force on body *B*, body *B* exerts an equal but oppositely directed force on body *A.*"

Suppose further that the man being pelted with basketballs braces himself against a spring balance. That spring balance will be deflected each time a basketball strikes his back, but if they come frequently enough, the balance will not have time to recover from one blow before the next basketball hits. The spring balance will indicate a sort of average force exerted by the balls striking his back.

Correspondingly, we can think of the molecules of gas bouncing off the walls of a container as exerting an average force on those walls. Since the number of molecules striking one very small area at one instant of time may be different from the number of molecules striking another very small but equal area in the same instant of time, we must think of the average force exerted by all those molecules over the entire surface. This leads to the concept of *average force per unit area,* which is a better definition for pressure than the one we gave on page 215.

10-2 *Molecules, temperature, and pressure*

Let us now see how the pressure on the inside walls of a sealed container of definite, unchanging volume varies as the temperature of the enclosed gas is increased.

EXPERIMENT 10-1 The relation between the temperature and pressure of a gas (a demonstration)

According to our model of gas pressure, air molecules are in constant motion; molecules collide with other molecules and with the walls of the container. Notice, however, that so far the model tells us nothing about the temperature, color, odor, or other properties of the gas. In this experiment you will observe the behavior of a fixed volume of a gas at different temperatures.

We will use a large, hollow copper ball which is attached to one end of a short length of copper tubing. A pressure gage is attached to the other end of the tubing.

To start the experiment, the air or gas is removed from the ball and tubing. What does the pressure gage read?

Air is now introduced into the apparatus at room pressure and temperature. Record the pressure on the gage and the temperature, and close the valve on the ball. Now submerge the ball first in freezing water and then in boiling water. Record the proper gage reading with its corresponding temperature.

A fourth pressure reading can be obtained if dry ice and alcohol are available. The temperature of a slushy mixture of these two substances will be close to $-78°C$.

Plot these sets of pressure-temperature points, and draw the best smooth curve through them.

Question 10–1   Based on your experimental results, explain why there are more tire blowouts on the superhighways in July than in December.

Question 10–2   When the apparatus was completely evacuated, we observed a zero pressure reading. After studying your graph, can you suggest a temperature at which a zero pressure reading might be expected when there is gas in the tube?

Question 10–3   How is the temperature referred to in Question 10–2 related to the temperature found in Experiment 7–4?

From the plot of pressure versus temperature in Experiment 10–1, we see that when the volume remains constant, equal increases in temperature cause equal increases in pressure. If our model of how a gas exerts pressure is a good one, it should help us to understand this experimental result. First we need to understand, according to our model, what happens to the particles when we heat the gas.

We learned in Chapter 9 that when the temperature of a body increases, a form of energy that we call thermal energy increases for that body. Since total energy is conserved, either this thermal energy increased at the expense of some other form of energy the body possessed, or the energy was transferred from some other body. However, the container of gas with the oil bubble did not move about as you heated the water surrounding it; its *overall* kinetic and potential energy did not change. Therefore, since the thermal energy of the gas did increase, it must have absorbed energy from the surrounding water.

Consequently, we find no difficulty in satisfying the conservation of energy principle, at least in a qualitative way. But we would like to do more than that by using our model of a gas. Much of the value of a model is the insight it gives about the nature of such things as thermal energy. We would feel we understood thermal energy better if we could relate it to the properties of the particles of our model. How can we do this?

To begin, let us consider a thought experiment. Picture an air table with several pucks at rest scattered at widely spaced positions (see Fig. 8–1*b*). The rails around the table are wires under tension that do not heat as the pucks collide with and rebound from them. Now an additional puck is projected onto the table with considerable kinetic energy. Initially, that one puck has all the kinetic energy; the others are all at rest. As we watch, however, the moving puck collides with others; those it strikes begin to move, and they strike still others. If we waited long enough and made measurements from a motion picture of the experiment, we would find that all of the pucks share the available kinetic energy (see Fig. 10–1). While the division of energy changes after each collision, we would find that the kinetic energy of each puck, averaged over time, is the same. Perhaps this is what happens when we add energy to our gas molecules.

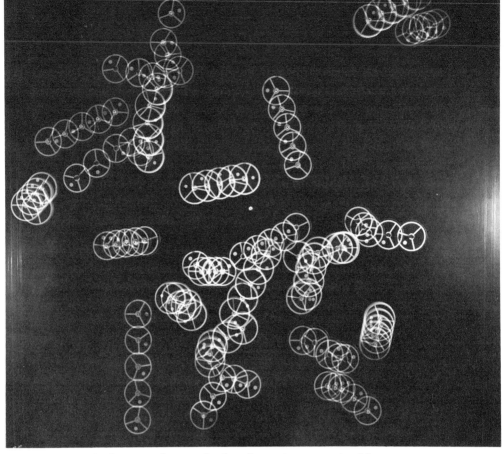

FIGURE 10-1 *Stroboscopic photograph of pucks moving on an air table.*
(Courtesy of The Ealing Corporation, Cambridge, Massachusetts)

If this model of pucks moving about on an air table is a good one, it ought to apply to the air trapped under a drop of oil in Experiment 7–4. What is it we actually did to heat the air? First, we raised the temperature of the glass container by immersing it in a hot water bath. Had we removed the glass tube from the hot water bath after the glass was heated but before the trapped air inside was heated and, furthermore, had we then carefully measured the temperature of both the glass and the air several times, what would we have found? The temperature of the container would drop, the temperature of the air would go up, and these changes would continue until the glass and the air reached the same temperature.

How was the thermal energy transferred from the glass container to the air trapped inside? Perhaps the particles of

220

the hotter glass had a larger kinetic energy than the particles of the cooler gas. As collisions between the air molecules and the particles of the glass continued, kinetic energy was transfered from the solid to the gas until all of the particles had the same kinetic energy. Of course, they would not all have the same kinetic energy all of the time, but on the average they would.

By comparing the pucks of our thought experiment with air molecules trapped beneath a drop of oil in Experiment 7–4, we are led to conclude that *the temperature of a large collection of particles can be identified with the kinetic energies of the individual particles.* If this connection, suggested by our model, is reasonable, it should help us explain why, in Experiment 10–1, the pressure increased with the temperature. Let us analyze this experiment.

According to our model, the temperature of a material increases as we add energy to it; the energy is shared equally by the particles of the substance in the form of kinetic energy. To relate the kinetic energy of the molecules of a gas to the pressure of the gas, we can ask the question: Would we expect the pressure of a gas to increase if we increase the average kinetic energy of molecules of that gas? Your performance of Experiments 7–4 and 10–1 would permit you to conclude that the pressure of a gas would increase if the average kinetic energy of the molecules of that gas were increased. But, would the pressure *double* if the average kinetic energy of molecules were *doubled?* This second question is quantitative and not so easy to answer as the first question, which is qualitative in nature. To discuss pressure, we must rely on our previous discussions of force, because pressure is an average force per unit area.

The force exerted by any particle on the wall of a container is, according to Newton's second law of motion, the mass of the particle times its average acceleration. And a particle rebounding from the wall of the container is accelerating. A molecule rebounding from the wall is like the ball of Figure 8–10 that was tossed into the air; each suffers a change in its direction of motion. Each is first slowed down, then brought to a stop, and finally starts off in a direction opposite to its initial direction. This change from negative velocity, to zero velocity, to positive velocity constitutes an acceleration.

The average acceleration for any one particle will be its change in velocity at any one collision times the number of collisions it has each second with the walls of the container. That is, the product of the change in velocity per collision and the number of collisions per second is the change in velocity per second; and the change in velocity per second is the acceleration. Written in mathematical notation:

$$a_{av} = (\Delta v)(\text{number of collisions/sec})$$

But in discussing the temperature of the gas we have been referring to the velocity of the gas molecules, not to the change in the velocity Δv. Is it possible to relate the velocity to the change in velocity? Maybe we can, if we consider for a minute only those molecules that strike the wall at right angles, and if we assume that the temperature of the wall and the enclosed gas is the same. These assumptions will make it easier for us to make a statement about how the velocity is related to the change in velocity when a molecule rebounds from the wall of the container.

The molecule bouncing off the wall at right angles has its direction of travel reversed. If one direction is considered to be positive, then the other direction can be considered negative. Let us assume that the particle has a velocity $v_1 = -v$ before colliding with the wall (see Fig. 10-2). The same particle would then have a velocity $v_2 = +v$ after collision. Recall that the temperatures of the wall and

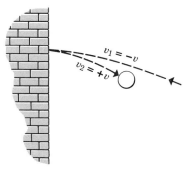

FIGURE 10-2
A ball bouncing off a wall.

gas are equal. The change of velocity Δv is given by

$$\Delta v = v_2 - v_1$$

Substituting our positive and negative values we find

$$\Delta v = +v - (-v)$$
$$\Delta v = +v + v$$
$$\Delta v = +2v$$

Consequently, the change in velocity is twice the final velocity of the molecule, and the final velocity differs from the initial velocity only in direction of travel (i.e., sign). So we are left to conclude that the change in velocity of a molecule during a collision with the wall of the container is indeed related to the average velocity of the molecules of the gas. And since we have already concluded that the temperature of a gas is related to the average velocity, it must also be related to that change in velocity we've been discussing.

A more complex and complete analysis would consider those many molecules that strike the wall obliquely; the results, however, are the same.

Now that we have a relationship between the change in velocity and the average velocity of the gas molecules, we must say something about how frequently this change in velocity occurs, for only then can we relate the change in velocity to the force exerted by the molecules on the wall.

What do we know about the relationship between the number of collisions a molecule makes against the wall of the container and the speed of that molecule? Surely if the speed is increased, the number of collisions per second will increase. In fact, whatever the dimensions of the container, if the speed of the molecules is doubled, the time between successive collisions will be halved. If the time between successive collisions is halved, the number of collisions per second will be doubled. By doubling the number of collisions per second, the average acceleration during collision is also doubled:

$$2a_{av} = (\Delta v)(2 \text{ times number of collisions/sec})$$

But we have also shown that (Δv) is proportional to the average velocity of the molecules (actually $\Delta v = +2v$), so if we double that average velocity, we would also double the change in velocity (Δv) during collisions with the wall. Therefore, the average acceleration would be doubled still again:

$$4 \, a_{av} = (2\Delta v)(2 \text{ times number of collisions/sec})$$

By doubling the average velocity we increase the average acceleration by a factor of four. If we were to triple the average velocity of the molecules, the average acceleration during a collision with the wall would be increased by a factor of 3 times 3, or 9. This can be expressed mathematically by writing

$$a_{av} \propto v^2$$

Knowing that the average acceleration is proportional to the average velocity squared, we can substitute v^2 for the average acceleration in Newton's second law $F = ma$, so long as we write that law as a proportionality and not an equality, and so long as we remember that it is the average force:

$$F_{av} \propto mv^2$$

So we have related the average force exerted by the air molecules against the walls of the container to mv^2. Now we must somehow relate this to the pressure of the gas.

The pressure of the gas is equal to the average force per unit area. Therefore, the pressure is proportional to the average force

$$P \propto F_{av}$$

Finally, by substituting mv^2 for F in this last proportionality, we see that

$$P \propto mv^2$$

Using Newton's laws of motion, we have been able to arrive at a proportionality that relates the pressure of a gas to the average velocity of the molecules composing that gas. Now we are left with the task of demonstrating that

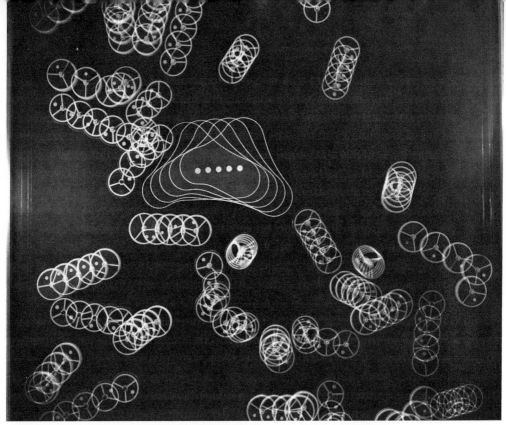

FIGURE 10-3 *Stroboscopic photograph of pucks moving on an air table and bombarding a large "Brownian motion particle." (Courtesy of The Ealing Corporation, Cambridge, Massachusetts)*

the temperature of a gas is also related somehow to the average velocity of the molecules of that gas in a manner which is more quantitative than our arguments about pucks on an air table.

We referred to Brownian motion on page 145. If particles that are much larger than molecules, but much smaller than grains of sand—say smoke particles—are suspended in air, they will be bombarded by the air molecules. Each particle will be bombarded millions of times each second. But since the motions of the gas molecules are random, the smoke particles are bombarded randomly. Sometimes they are bombarded more on one side than the other, and they are accelerated slightly by this uneven push. That uneven push is first in this direction, then in that. The net result is that the smoke particles themselves move about with random jumps first this way and then that way (see Fig. 10-3 and the photograph at the beginning of this chapter).

The motions of these smoke particles can be observed under a microscope. Detailed studies can reveal the average velocity of the smoke particles. By analogy with the pucks discussed in Section 10-2, we assume that the average kinetic energy of the smoke particles is equal to the average kinetic energy of the gas molecules. It is observed that if the temperature of the gas is increased, the average velocity of the smoke particles, and therefore the average kinetic energy of the smoke particles, increases. We thus conclude that the average kinetic energy of the gas molecules increases with the temperature. Detailed studies of this nature reveal that the temperature of the gas is proportional to the average kinetic energy of the gas molecules:

$$T \propto mv^2$$

where m is the mass of a molecule of the gas and T is the absolute temperature defined in Chapter 7.

But from Experiment 10-1 we discovered the relationship between the pressure and temperature of a gas. Writing it in terms of the absolute temperature, pressure is proportional to temperature:

$$P \propto T$$

Combining these two experimental results, we find that the pressure of a gas is proportional to the average kinetic energy of the gas molecules:

$$P \propto mv^2$$

This result agrees with the conclusion of our theoretical analysis.

The procedure that we followed in comparing our model of a gas with experimental results is of primary importance in all scientific work. We first observed something in an experiment—in Chapter 7, for example, we saw the increase in volume when a gas is heated and also the fact that gas spreads or diffuses throughout the air of a room. Then we wondered what a gas must be like in order to behave that way, and we made a mental model that seemed to be qualitatively consistent with those observations. In order to use the model to predict further experimental results, such as how pressure varies with temperature, we needed to know more about force. This knowledge we pursued by experimental investigation with rubber bands and carts in Chapter 8. Such quantitative concepts make possible the derivation of equations expressing relationships among measurable quantities, in this case pressure and temperature, by which the model can be tested. Our concepts of the physical world have been constructed in this manner. The particular concept we have just tested is called *the kinetic theory of gases*.

QUESTIONS

10-4 Suppose that the gas in a space vehicle is maintained at a pressure of 10 psi (pounds per square inch) and the temperature is 290°K (about 63°F). What will the pressure become if the temperature is increased to 310°K?

10-5 Suppose the pressure reading in the space vehicle of Question 10–4 was obtained by a gage that compared the pressure on the inside of the vehicle with that on the outside (essentially zero in outer space). What would that pressure gage read if the vehicle were brought back to the surface of the Earth, where the air pressure is approximately 15 psi?

10-6 A balloon expands as it rises in the air. This is also true of bubbles rising in beer or soda. Discuss the effects of pressure and temperature in both cases.

10-7 Some liquid sits in an open pan. As you know, the liquid will evaporate (at a rate depending on the surrounding humidity and temperature). Does the liquid become cooler, hotter, or remain at the same temperature? Try to write an explanation for each of the three alternatives. On the basis of these explanations, which would you choose to be correct?

10-8 Now is the time to turn back to Question 1–11 and answer that question again, but this time use the results of your further studies.

References   1. Baez, A. V., *The New College Physics*, W. H. Freeman & Co., 1967, Chapter 29, pages 365 to 377.

A discussion of the kinetic theory of gases.

2. Christiansen, G. S., and P. H. Garrett, *Structure and Change*, W. H. Freeman & Co., 1960, Chapter 9, Section 9–1, pages 118 to 119.

A spring is used as an analogy of a gas. Also, Chapter 11, Section 11–3, pages 153 to 154. A historical note on "quantity of motion."

CHAPTER 11

BONDING FORCES
WITHIN A CRYSTAL

Ice melts at 0° C; para-dichlorobenzene, used in Experiments 2–6, 2–7, and 7–2, melts at 53°C; naphthalene, used in those same experiments, melts at 80°C; tin melts at 231°C; iron at 1535°C; tungsten at 3370°C; and mercury at −39°C. Why does one substance melt at one temperature and another at a different one? In fact, why do solids melt at all?

But perhaps the question of why solids melt is not the most appropriate question yet. More fundamental problems are: Why do solids hold together? Why are solids rigid? Why do they hold their shape?

There must be forces within a solid holding it together; these forces holding all the particles together are called *bonding forces.* By extending our consideration of the kinetic theory of gases, you might conclude that the melting temperature of a solid depends on the bonding forces holding that solid together. Using this conclusion, would you expect ice to have stronger or weaker bonding forces than tungsten?

What is the nature of bonding forces? Are the bonding forces of sodium chloride the same as the bonding forces of iron? Do bonding forces differ in kind from one substance to another, or do they differ only in magnitude? In order to understand the structure of solids, we must answer these questions.

11-1   *Strength of a crystal*

In Experiment 5–3 we examined crystals of rock salt and concluded that they are composed of cubic units. One piece of evidence supporting this idea is the observation that the crystals cleave only in certain directions. Furthermore, considerable effort is required to break a crystal, even along the cleavage planes. Let us look at these crystals once more, this time thinking about the forces that hold the crystal together.

FIGURE 11-1 *(a) What forces prevent the string from breaking? (b) What forces prevent the pillar from collapsing? (Fundamental Photographs)*

Obtain a few grains of salt. Put them on a smooth, hard surface and examine them thoroughly. Push on one with a pencil point or the edge of your fingernail. It will take some effort to split or crush it.

The strength exhibited by this crystal suggests that the particles of which it is made are bound firmly in position by strong forces. Attempts to pull the particles apart (by trying to split the crystal) are met with considerable opposition, so the forces must be *attractive* (see Fig. 11–1a). Attempts to squeeze the particles together (by trying to compress the crystal) also meet with resistance, so the forces must also be *repulsive* (see Fig. 11–1b).

11-2   On understanding

Knowing more about bonding forces between particles of various materials can lead to some valuable and interesting results. In many situations knowledge leads to control.

Control of the chemical processes by which the composition of substances is rearranged had been developed through guesswork by artisans and alchemists since prehistoric times. For example, the Hittites in the 12th Century B.C. were able to extract iron from certain earthy substances. Iron gave them military superiority over their bronze-weaponed competitors, and thus shaped the course of history. But the spectacular growth of chemical science in the past hundred years far surpasses the total accomplishments of all the previous centuries. This growth has resulted from the ability to predict and control chemical changes— an ability that has been improved by an understanding of bonding forces.

These generalizations may seem far removed from our contemplation of the little salt crystals in front of us. However, just as the longest journey starts with a single step, our understanding of the structure of solids must start with the question: What is the nature of the forces that hold the particles of a crystal in place?

11-3   Requirements for the bonding force

Let us consider the nature of the forces holding a crystal together and make an assumption that will simplify our model of the crystal by assuming that the forces holding all of the particles of a particular crystal together are all the same kind of force. We will then want to set up some requirements for the nature of this kind of force. We may want to add other requirements later as the model becomes more complex, but for now we can set up three requirements. First, the forming of concepts in science is heavily influenced by the principle of universality. The application of this principle to the bonding forces that hold the particles of a solid together leads us to hope that all solid matter is held together by the same kind of force. We therefore assume that whether the solid matter is living or nonliving, whether it is a fish on the Earth or a rock on Mars, the bonding forces are all of a single kind. If different kinds of matter are held together by different kinds of forces, our model will become very complex, so we will assume (and hope) that the same kind of force holds all kinds of solid matter together. If this proves inadequate, then we will have to investigate the possibility of several different kinds of forces.

Second, it will further simplify our model if we assume that the one kind of bonding force accounts for both the forces of attraction that keep the crystal from being easily split apart and the forces of repulsion that keep it from being easily compressed.

Third, since science progresses from the known to the unknown, we will investigate forces that are already known before we postulate an entirely new kind of force. If none of the known forces satisfies the requirements demanded by these observations, we may have to consider a special force.

With your present information you are not able to make an exact calculation of the size of the bonding force between particles in a crystal, but a rough estimate should help you understand the nature of this force. This type of estimate is called an "order of magnitude" calculation. This means that if our result turns out to be, say, 7.8×10^{-18} newton, we cannot be at all certain of the 7.8 (it might be anywhere between 1.0 and 10), but we can be reasonably sure of the exponent -18 (it is either -17, -18, or -19, but certainly not -12). Such an order-of-magnitude calculation is fairly sure to put us on the right track and should help us in deciding which of the different kinds of forces we may consider the most reasonable ones to account for the bonding force.

Sodium chloride crystals are composed of layer upon layer of particles; within each layer the particles are in parallel rows. Calculating the bonding force on any one particle is an enormous task. We can simplify the process by using our imagination to strip away all the particles in the crystal except those in one vertical string, as in Figure 11–2. In order for the string to remain intact, the bonding

FIGURE 11-2
An imaginary string of sodium and chlorine particles.

force holding the bottom particle to the others must be at least as great as this particle's weight. Assume that the bottom particle has the mass of a sodium atom, which is about 4×10^{-26} kg. (Methods of determining the mass of an atom will be discussed in Chapter 13.) In Chapter 8 we saw that the weight of an object is equal to its mass in kilograms multiplied by the acceleration due to the gravitational force (9.8 meter/sec^2 at the surface of the Earth). Consequently, this particle has a weight of about 4×10^{-25} newton ($9.8 \times 4 \times 10^{-26}$). We need, therefore, to find a kind of force that meets the following requirements: when all of the individual forces operating between that bottom particle and those above it are added up, the net force must be greater than 4×10^{-25} newton and it must act in the upward direction to counteract the weight of the particle.

11-4 *Kinds of forces*

Ordinarily, people push and pull on things by touching, grabbing, shoving, and the like. We cannot accept "touching" or "contact force" as a possible bonding force between atoms. Careful consideration reveals that touching involves the force between the hand's atoms and the object's atoms, the very same force we are trying to explain. This force is merely another example of the one to be explained, rather than a possible explanation of the bonding force.

What kinds of force are you already familiar with? You are, no doubt, aware that objects are attracted to the Earth, iron filings cling to a magnet, and bits of paper are attracted by a comb. These examples suggest three kinds of forces: gravitational, magnetic, and electrical.

Perhaps you have heard of nuclear forces as well. The nuclear force is a very special kind. We will not consider it unless examination of the more familiar forces eliminates them as likely candidates for the bonding force between the particles in a crystal.

Question 11-1 Think of situations that involve force: grasping a book between one's fingers, the attraction of an iron nail to a magnet, the orbiting of a satellite, and the sticking of adhesive tape. Classify these and other examples that you think of under the following headings: gravitational, magnetic, electrical, uncertain. Forces in the last category might be examples of the kind of force we are trying to explain.

11-5 *Gravitational forces examined*

Hold a blackboard eraser at arm's length and let it go. It falls, and as it falls, its speed increases. The result is so familiar that we overlook its implications. What causes falling objects, including this eraser, to accelerate? If we accept Newton's first two laws of motion as valid descriptions of the way objects move, then we must conclude that a force acts on the eraser, causing it to pick up speed as it falls. Let us follow the logic more closely.

Newton's first law of motion says that an object will remain at rest or, if it is moving, it will continue in a straight line with a constant speed unless acted upon by something else. Our falling eraser may move in a straight line, but its speed surely is not constant.

Newton's second law of motion stipulates that any object, be it an eraser or a jet airplane, will change speed or direction of travel only if an unbalanced force acts on it.

The existence of a force that causes an object to gain speed as it falls was not recognized before Newton formulated his laws of motion. Newton's laws are concerned with the motion not only of a falling eraser, but also of a swinging pendulum, the Moon about the Earth, and the Earth about the Sun. None of these objects travels in a straight line with a constant speed. Therefore, we conclude that a force must act on each to cause a change in speed or direction of travel.

Newton concluded that the only object that could exert a force on the Moon to cause its motion to deviate from straight-line motion into a nearly circular path concentric with the Earth is the Earth itself. If the Earth pulls on the Moon, why not the Moon on the Earth? If the Earth exerts a force on an eraser, why shouldn't the eraser exert a force on the Earth? If the eraser exerts a gravitational force on the Earth, why not on another eraser or a book? Newton postulated that gravitation is not an exclusive characteristic of the Earth but is universally a property of all objects. Any two objects in the universe exert a mutual gravitational force on each other.

By comparing the motion of a falling apple with the motion of the Moon, Newton concluded that the magnitude of the gravitational force that two objects exert on one another depends upon the distance separating those objects. He amply demonstrated, and experiments have confirmed, that as the distance between two objects increases, their mutual gravitational force decreases, and in a way that is fairly easily described (see Fig. 11–3).

If the distance r between the centers of two objects is doubled, their mutual gravitational force becomes one-fourth

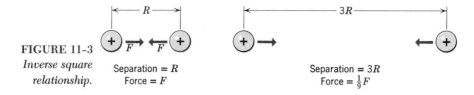

FIGURE 11-3
*Inverse square
relationship.*

Separation = R
Force = F

Separation = $3R$
Force = $\frac{1}{9}F$

as great; if the distance is tripled, the gravitational force becomes one-ninth as great; if the distance is quadrupled, the force is only one-sixteenth as great; etc. That is, the force is proportional to the *inverse square* of the distance separating the objects. The term inverse means that as the distance increases, the gravitational force decreases; the term square signifies that the distance r must be squared to establish the correct proportionality:

$$F \propto \frac{1}{r^2}$$

It appeared to Newton that gravitational forces are significant only when at least one of the objects involved is astronomically massive. From this he concluded that the gravitational force between two objects is related to the masses of the objects. How are the force and mass related? Newton was convinced that gravitational forces between objects are mutual. His third law of motion says that forces occur only in pairs, each pair consisting of equal and opposite forces. In an expression relating the magnitude of the force between two objects with the separation and the masses of the objects, the force should depend on each mass in the same manner. To involve the masses in the same mathematical way means that the masses, m_1 and m_2, could appear in an algebraic expression as a sum $(m_1 + m_2)$ or as a product $(m_1 \times m_2)$; our question is which.

The gravitational force exerted by the Earth on a 2-kg mass is twice as great as the force on a 1-kg mass. But the sum of the mass of the Earth and a 2-kg mass is not twice as large as the sum of the mass of the Earth and a 1-kg mass. On the other hand, replacing 1 kg by 2 kg will double the product. This establishes that in the mathematical expression relating gravitational force and mass, the product, not the sum, of the masses should appear. The expression is symmetrical with respect to the masses, as required by Newton's third law. Thus we have

$$F \propto m_1 m_2$$

This proportionality can be combined with the previous one and written

$$F \propto \frac{m_1 m_2}{r^2}$$

Proportionalities, however, are difficult to deal with so we can write it as an equality by including a *constant of proportionality*. For example, the circumference of a circle is proportional to the diameter:

$$C \propto D$$

By including the constant π, we can write it as an equation:

$$C = \pi D$$

The value of π tells us something about a circle. The constant of proportionality in the equation that expresses the law of gravity is G, and it tells us something about the strength of the gravitational force. The law of universal gravitation is written as:

$$F = G \frac{m_1 m_2}{r^2} \tag{11–1}$$

The actual numerical value of the constant depends on the units and must be determined experimentally. If the mass is given in kilograms, the distance in meters, and the force in newtons, $G = 6.7 \times 10^{-11}$ newton (meter)^2/kg^2.

We can calculate the force of gravitation between a pair of 1-kg masses placed 1 meter apart; $m_1 = 1$ kg, $m_2 = 1$ kg, $r = 1$ meter. The gravitational force is 6.7×10^{-11} newton. It is no wonder that it is difficult to measure.

We have been discussing Newton's law of gravitation in order to see whether this force plays a significant role in holding a crystal together. Now we will calculate the attractive force between the particles of a crystal. In Figure 11–2 we singled out a vertical string of particles for examination. We will determine the gravitational force acting on the bottom particle of this string and see if it is strong enough to prevent the particle from falling off. If that one particle falls off, then so should the bottom plane of the crystal of which that particle is a member.

X-ray diffraction techniques (see Chapter 6) show that the centers of adjacent particles in a sodium chloride crystal are separated by about 3×10^{-10} meter. This will be used as the distance r between adjacent particles in our imaginary string. We will assume that the mass of every other particle is equal to the mass of a sodium atom, $m_1 = 4 \times 10^{-26}$ kg, and that the mass of each of the remaining particles is equal to the mass of the chlorine atom, $m_2 = 6 \times 10^{-26}$ kg. Now we can calculate the gravitational force of attraction between the bottom particle and the one immediately above it. Using Newton's law of gravitation,

$$F = G \frac{m_1 m_2}{r^2}$$

and substituting the numerical values given above, we obtain

$$F = 7 \times 10^{-11} \frac{\text{newton(meter)}^2}{\text{kg}^2} \frac{(4 \times 10^{-26}\,\text{kg})(6 \times 10^{-26}\,\text{kg})}{(3 \times 10^{-10}\,\text{meter})^2}$$

$$F = \frac{7 \times 4 \times 6 \times 10^{-63}\,\text{newton}}{9 \times 10^{-20}}$$

$$F = 2 \times 10^{-42}\,\text{newton}$$

This force is not only smaller, but 10^{17} times smaller than the force of gravitation between the bottom particle of the chain and the Earth, about 10^{-25} newton ($10^{-42} \times 10^{17} = 10^{-25}$). We can conclude from this approximate calculation that the bonding force between the particles of a crystal is not a gravitational force. In fact, the gravitational force is unsatisfactory for a second reason; it is only attractive. Therefore, a gravitational force could not prevent a crystal from being crushed by your fingers. Nevertheless, this calculation of the force of gravity between these particles has been important, for now we know that the gravitational forces within a crystal are so small that we can neglect their influence entirely.

Question 11-2 When we calculated the gravitational force on the bottom particle of the vertical chain, we considered only the force between that particle and the one above it. Should we have considered the other particles in the string? For simplicity let us assume that each particle in the string has a mass $m = 6 \times 10^{-26}$ kg, and that the distance between the centers of adjacent particles is $r = 3 \times 10^{-10}$ meter. What is the gravitational force between the bottom particle and the second one above it? the third one?

Question 11-3 If there are six particles in the vertical string, what is the total gravitational force acting on the bottom particle? What fractional part of this force is exerted by the particle immediately adjacent to the bottom one?

11-6 *Magnetic forces examined*

You probably have enough experience with magnets to answer some leading questions. If you don't, you should acquire two magnets and experiment.

Question 11-4 Do magnets exert both repulsive and attractive forces? (A bar magnet and a compass can be used to provide sufficient observations for an answer.)

Question 11-5 Do nearly all objects respond to magnetic forces? Are particles of sugar, salt, and wood affected by a magnet?

Question 11-6 (a) Is the magnetic force between a small magnet and a nail great enough to support the nail?
(b) Is the magnetic force between a nail and a magnet weaker or stronger than the gravitational force between them?

Question 11-7 On the basis of the observations required to answer these questions, decide whether or not the magnetic force is a good candidate for the bonding force.

11-7 *Electrical forces examined*

The gravitational force does not satisfy the requirement of the bonding force in crystals on two counts: it is always attractive, never repulsive and, in addition, it is too weak.

The forces between two magnets are both attractive and repulsive, and they are much stronger than the force of gravity between those same two magnets. Magnetic forces, however, are effective on only a few substances. If magnetic forces act on atoms, then some respond sufficiently to magnetic forces (e.g., iron) and some do not (e.g., beryllium). Hence it appears unlikely that magnetic forces are the bonding forces in all crystals. We still have hope that all bonding forces are of one kind. What forces do we have left? How about the forces between electrical charges? Let us see how much we can find out about electrical forces experimentally.

EXPERIMENT 11-1   An introduction to electrical forces

Materials

Plastic strip (a plastic ruler will do), and clean, dry, paper towels. Miscellaneous light-weight objects, for example, salt, sugar, thread, powdered coffee, tea leaves, pepper, aluminum foil, bits of paper.

Procedure

Step One   Hold the plastic strip just above a small bit of paper lying on the table.

Step Two   Rub the plastic strip briskly with the paper towel. Again hold the plastic strip just above the small bit of paper. Is something happening which did not happen before you rubbed the plastic strip?

Step Three   Hold the freshly-rubbed plastic strip just above the fine hair on your arm. What happens?

Step Four   Hold the freshly-rubbed plastic strip closer and closer to your ear. What do you feel? What do you hear? You may try several times and still not detect anything.

Step Five   Put small bits of aluminum foil and other small objects, such as those enumerated in the list of materials, on separate sheets of clean dry paper. Hold a freshly-rubbed plastic strip above each material. Observation?

Step Six   Try rubbing the plastic strip with different

materials (cloth, hair) and repeat the experiment of step 5. Observation?

Question 11-8   What evidence do you have that rubbing the plastic strip gives it an ability to exert a force?

Question 11-9   Does the rubbed plastic exert a force on a bit of paper even if it doesn't touch the paper?

Question 11-10   Are any classes of materials unaffected by the electrical force?

Question 11-11   Is the force exerted by the plastic on the bit of paper as great as the gravitational force exerted by the Earth on the paper? What evidence do you have for your answer?

The forces observed in Experiment 11-1 have been given the name electrical forces. The word electric was taken centuries ago from the Greek word *elektron*, which means amber. When rubbed, amber behaves like the rubbed plastic strip, so objects which exhibit this behavior were classified with amber and called electrics. When they exhibit this property, they are said to have an *electric charge*. The act of rubbing something, be it amber or plastic strip, does something to the electric charges.

There is a close analogy between gravitational and electrical forces. Both are exerted at a distance without direct contact; and we will see later that, like gravitational forces, electrical forces are described by an inverse square relationship. Electrical forces are related to electric charges in the same way that gravitational forces are related to mass. But electrical forces between small objects can be much larger than gravitational forces between these same objects.

The characteristics of the electrical force thus far observed make it a good candidate for the bonding force. However, we have yet to discover whether it can be repulsive and whether it is strong enough.

EXPERIMENT 11-2   An investigation of types of electric charges

Materials
   A strip of acetate plastic, a plastic polyethylene bag, two graphite-covered plastic balls, and nylon thread.

Procedure
   Step One   With a drop of glue, attach to each plastic ball a nylon thread at least 24 inches long. Put the unattached ends of the threads together between the pages of a book and place the book on top of a table so that the balls hang side by side, about 18 in. below the edge of the table. The balls should touch. Avoid tangling the threads.
   Step Two   How do you know whether or not the balls are charged? If they are charged, discharge them by touching them.
   Step Three   Rub the plastic strip with the clean plastic bag. Test for a charge by bringing the plastic strip close to small bits of paper. In the same manner, test the plastic bag for a charge.
   Step Four   Rerub the plastic strip with the plastic bag, and retain the bag in your hand. Then bring the charged end of the strip into contact with the suspended balls. Gently rub the strip against the balls until they jump away. Describe the behavior of the balls with respect to the rubbed strip and each other.
   Step Five   Bring the plastic bag toward the balls. Observation? Compare the behavior of the balls with respect to the plastic strip and the plastic bag.

Question 11–12   What does this investigation show about the forces between the charged objects concerning (a) attraction and repulsion, and (b) the size of the force?

Question 11–13   What evidence do you have for the existence of two kinds of charge?

Question 11–14   In 1734, the French physicist Charles Dufay published a record of experimental work that led to the statement, "Like charges repel, unlike charges attract." Do your results confirm or refute this statement?

At this point it becomes useful to examine more closely what happens when a plastic strip is rubbed with a wiper. If we presume that neither the strip nor the wiper had an electric charge before being rubbed, then we can conclude that during the rubbing process one charge becomes lodged on the strip and the other on the wiper. One of these charges has been designated *positive* and the other *negative*. These terms were first applied by Benjamin Franklin. Although he used the terms to mean an excess or deficiency of a single electric fluid, they have come to identify the two different kinds of electric charges.

Franklin concluded that whenever a positive charge is produced by rubbing or some other process, an equal amount of negative charge is produced. This statement is called the *principle of conservation of electric charge* and, like all conservation principles, was derived from observations with a generous amount of intuition. It is now one of the best established of the conservation principles.

If neither the plastic strip nor the wiper was charged before being rubbed, we are faced with two possible explanations of the origin of the charges: (1) the charges were created during the rubbing process, or (2) initially the charges existed in equal amounts in both the strip and the wiper and were simply separated by the rubbing process. A choice between these two possibilities can be made only after a more detailed consideration of the solid state of matter.

One question remains: Is the electrical force strong enough to hold a crystal together? If we wish to calculate the magnitude of the force of attraction between the sodium particle and the chlorine particle in a salt crystal, we need an expression relating the magnitude of the electric force, the amount of each charge, and the distance between them. In 1785, Charles Coulomb, a French engineer, empirically derived the expression:

$$F = k\frac{Q_1Q_2}{r^2} \tag{11-2}$$

where F is the electrical force

Q_1 is the magnitude of one charge and Q_2 that of the other

r is the distance between the centers of the charged bodies

k is the constant of proportionality

Note the resemblance between this equation and the one used to calculate gravitational forces. In Equation 11–2, called Coulomb's law, the unit of charge Q is the coulomb. A *coulomb* is the amount of electric charge which, when placed 1 meter from an identical charge, will be repelled by an electrical force of 9×10^9 newtons. The size of the coulomb as a unit of charge was selected because of its convenience in other types of calculation. As a result of this selection, however, the constant k in Coulomb's equation must be 9×10^9 newton $(\text{meter})^2/(\text{coulomb})^2$.

Answering questions will help you to understand how this expression is used and give you some feeling for the magnitude of the charges with which you have been working.

Question 11–15 What is the magnitude of the force exerted by a 10-coulomb charge on a 2-coulomb charge when they are separated by 0.005 meter?

The magnitude of the force just calculated is very large. For comparison, recall that a 1-kilogram mass has a weight of about 10 newtons. The forces in our experiments were much less than this, hence the charges were also much smaller. How large were these charges? Let us consider the following example.

A small bit of paper is picked up by a charged strip. How large is the charge on the strip? The solution follows.

In order to estimate the charge on the strip, we must make simplifying approximations so that we can use Coulomb's law. Assume that (1) the magnitude of charge Q_1 on the strip is the same as that of charge Q_2 on the paper, and call each of these Q; (2) the separation r of the centers of the charges is 5×10^{-3} meter; (3) the magnitude of the electrical force F is the weight of the paper, about 10^{-3}

newton. Using Coulomb's law,

$$F = k\frac{Q_1 Q_2}{r^2}$$

$$10^{-3} \text{ newton} = 9 \times 10^9 \frac{\text{newton(meter)}^2}{\text{(coulomb)}^2} \frac{Q^2}{(5 \times 10^{-3} \text{ meter})^2}$$

$$Q^2 = \frac{10^{-3} \times 25 \times 10^{-6} \text{ (coulomb)}^2}{9 \times 10^9}$$

$$Q^2 = 2.66 \times 10^{-18} \text{ (coulomb)}^2$$

$$Q = 1.6 \times 10^{-9} \text{ coulomb}$$

This example indicates that the charge on a plastic strip rubbed with a wiper is only about 10^{-9} coulomb, a very small fraction of the charge considered in Question 11–15.

But we are interested in the forces holding individual particles together to make a crystal (or a plastic strip). We know that a crystal of sodium chloride is composed of particles of sodium and of chlorine. Let us now suppose that these particles are electrically charged. An atomic particle with a net electric charge is called an *ion*. An ion may be a single atom or a group of atoms with an electric charge. For example, a copper atom with an electric charge is more properly called a copper ion; the sulfate portion of copper sulfate is composed of 5 atoms, one atom of sulfur and 4 of oxygen, and it is called a sulfate ion.

Question 11–16   If the bottom particle of our sodium chloride chain is assumed to be a positively charged sodium ion, what kind of charge must the second particle have in order that the force be attractive?

In order to use Coulomb's law to calculate a force, we must know the separation of the ions and the amount of charge on each. In Section 11–5, we used 3×10^{-10} meter as the separation, but we still need to find a reasonable magnitude for the charge on these particles.

In the first decade of this century, the American physicist Robert Millikan measured the electric charge on tiny oil drops. He perfected a method equivalent to using an electrical force to balance the weight of the drop. (In our worked example the electrical force balanced the weight of the paper.) The smallest charge he found on oil drops was 1.6×10^{-19} coulomb. In thousands of attempts over a thirteen-year period, all charges measured on oil drops were either 1.6×10^{-19} coulomb, or a whole-number multiple of this minimum value. If we assume that the ions in sodium chloride crystals have this minimum charge, the calculated force will be a minimum.

We are now ready to make the calculation of the electrical force between two adjacent particles in the sodium chloride string of Figure 11–4. We will use the numerical values

$$k = \frac{9 \times 10^9 \text{ newton(meter)}^2}{(\text{coulomb})^2}$$

$$Q_1 = Q_2 = 1.6 \times 10^{-19} \text{ coulomb}$$

$$r = 3 \times 10^{-10} \text{ meter}$$

Substituting in the expression for Coulomb's law, we find

$$F = k \frac{Q_1 Q_2}{r^2}$$

$$F = \frac{9 \times 10^9 \text{ (newton)(meter)}^2}{(\text{coulomb})^2} \times \frac{(1.6 \times 10^{-19} \text{ coulomb})^2}{(3 \times 10^{-10} \text{ meter})^2}$$

$$F = 2.6 \times 10^{-9} \text{ newton}$$

The electrical force is 10^{16} times larger than the minimum value of the force required to support the bottom particle, which is 10^{-25} newton. If we consider the approximate nature of our calculations and remember that considerable

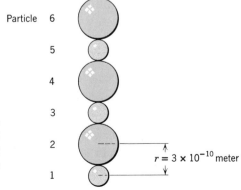

FIGURE 11-4

An imaginary string used to calculate gravitational force between particles of sodium and chlorine.

force is required to break a salt crystal, it appears that electrical forces are indeed capable of meeting all the requirements for the bonding force between atoms.

Question 11–17   How much electrical force do the rest of the ions in the salt string exert on the bottom ion? To approximate this, proceed by answering the following questions in order.

(a) What is the sign of the charge on the third particle from the bottom? the fourth?

(b) How does the separation of particles 1 and 3 (Fig. 11–4) compare with the separation of particles 1 and 2?

(c) Using the rule $F \propto 1/r^2$, find the ratio of the electrical force between particles 1 and 3 to that between 1 and 2.

(d) Is the force between particles 1 and 3 attractive or repulsive?

(e) Repeat Questions b through d for particles 1 and 4, 1 and 5, and 1 and 6.

(f) Is the net force exerted on particle 1 by particles 3, 4, 5, and 6 attractive or repulsive?

(g) Is the magnitude of the net force calculated in Question f more nearly 20%, 50%, or 100% of the electrical attraction between 1 and 2?

A more complete calculation would take into account the forces exerted by all the ions in the crystal acting on that one bottom particle. Surprisingly, the net force exerted by all the ions is not appreciably different from that exerted by the five ions considered in this question.

Question 11–18   The only valid test to determine whether an object has a charge is to see whether it is repelled by another suitably charged object. Justify this statement.

11–8 *Behavior of charges in materials*

In view of the fact that objects having unlike charges attract each other and those having like charges repel, how can we explain the observation that a charged plastic strip picks up an *uncharged* bit of paper, aluminum, or other materials? An answer to this question requires the construction of a suitable model.

Two possible models come to mind. One postulates the existence of electric charges within the material and supposes that these charges are free to move. If a charged strip is brought close to an uncharged object made of that material, the charges would move through the materials in response to the force exerted by the strip. Suppose, as in Figure 11–5a, that a positively charged strip is brought close to the object. Further, suppose that the object contains both positive and negative charges. The positive charges would migrate through the material and move as far as possible from the positively charged strip. The negative charges would migrate through the material and move as close as possible to the strip. Thus, one end of the object would become positively charged and the other end negatively charged, even though the object as a whole remains uncharged. This separation of charges would result in a net attractive force between the charged strip and the uncharged object. Why?

It is possible, of course, that if an object contains both positive and negative charges, only one of these kinds of charge is able to move through the material. For example, suppose that only the negative charges are able to move. They would be attracted by the positively charged rod and outnumber the positive charges in that portion of the object nearest the rod. The opposite end of the object would become positively charged, however, because of the local deficiency of negative charges.

If the situation were reversed by bringing a negatively charged strip near the uncharged object, the charges in the object would migrate in the opposite direction from that shown in Figure 11–5. But the net result would be the same. The portion of the object close to the negatively charged strip would become positively charged, the most distant portion would become negatively charged, and the object would be attracted to the negatively charged strip.

The second of our two models postulates that both positive and negative charges are present but neither is free to move. In this model the charges are tightly bound to each other within the particles constituting the material. Suppose,

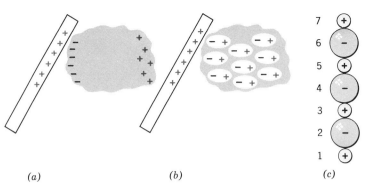

FIGURE 11-5

(a) Displacement of movable charges near a charged plastic strip. (b) Reorientation of fixed charges near a charged plastic strip. (c) Pairs of charged particles in an ion string.

(a)   (b)   (c)

however, that the positive and negative charges are paired to form neutral particles, but that the charges within a given particle can be slightly displaced when a charged strip is brought close to the material. If the charged strip is positive (see Fig. 11–5b), then within each particle the negative charge will move closer to the strip than the positive charge will.

The group of charges shown in Figure 11–5b is similar to the string of ions discussed in Question 11–17. Oppositely charged particles 2 and 3 (see Fig. 11–5c) can be considered as a pair of charges similar to those within each particle in Figure 11–5b. Just as particles 2 and 3 exert a net attractive force on particle 1, so each particle of the object exerts an attractive force on the strip. Because the force of each particle on the strip is attractive, the entire object must attract the strip. Consequently, the strip and the object will attract each other. As with the first model, changing the charge on the plastic strip will still result in an attractive force.

If the charges are free to move through the material, then that material is called a *conductor;* if the charges are bound to individual particles within the material, then it is called a *nonconductor.* The question is, which of these two models is more suitable for a given substance; that is, can we find substances that are conductors and nonconductors?

EXPERIMENT 11-3 An investigation of transfer of electric charges through various materials

Materials
1. One suspended ball (see Fig. 11-6a).
2. Charging strip and wipers.

3. A clean glass beaker or a styrofoam cup in which two notches have been placed in the rim at opposite ends of a diameter of the opening. Either the beaker or the cup will serve as an insulating stand.

4. Collection of various materials, including: glass stirring rod, piece of wood, plastic rods or strips, pencil lead (as from a mechanical pencil), various metals (such as coins, paper clips, aluminum foil, straight pin).

Procedure

A. *Discharging the ball*

Step One   Charge the ball by touching it with the rubbed plastic strip. Withdraw the strip. How do you know the ball is charged?

Step Two   Touch the ball with your finger.

Step Three   Bring the charged strip near, not touching, the ball. Has touching the charged ball changed its behavior with respect to the charged strip? What do you think happened when you touched the ball?

Step Four   Charge the ball again by touching it with the charged strip. Be sure it is charged. Now touch the ball with a very clean glass stirring rod. Remove the rod and bring the charged strip close to, but not touching, the ball. Observation? Did touching the charged ball with a clean glass rod change its behavior with respect to the charged strip?

Step Five   Roll the glass rod between the palms of your hands touching all the surface of the glass rod, and repeat Step 4. How do your observations differ from those in Step 4? Explain.

Step Six   Repeat Step 4 by replacing the glass rod with other materials (suggestions: metals, wood, plastics).

In Steps 3 through 6, you made observations that permit you to tell which materials transfer electric charge well. These materials are called conductors. Those that do not noticeably conduct charge are called nonconductors.

B. *Charging the ball*

We will now check these observations by seeing if the materials will conduct charge from a charged strip to an uncharged ball. The equipment is arranged as in Figure

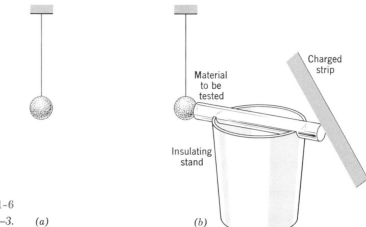

FIGURE 11-6
Experiment 11–3.   (a)   (b)

11–6*b*. The material is supported on the insulating stand (*clean* glass beaker or styrofoam cup with notches) so that it will not acquire charge from the table.

Step One   Start each test with the ball uncharged.

Step Two   Support the material to be tested on the insulating stand with one end touching the suspended ball. Gently rub the other end of the material with the charged strip. Remove the strip.

Step Three   Touch only the insulating stand, and carefully pull the material being tested away from the ball.

Step Four   Test the ball to determine if it is charged. Since the only valid test for a charge is a repulsive force, this might necessitate the use of both the charged strip and wiper (see Question 11–18).

Step Five   Classify each material as a good conductor, poor conductor, or nonconductor.

Question 11–19   In general, to what class of materials do the good conductors belong? Which model developed at the beginning of this section works better for conductors? for nonconductors?

Question 11–20   Why do you suppose that such materials as clean glass, hard rubber, and plastic are used to build up an electric charge by rubbing?

11-9 *A review and extension*

We can summarize our study of electric charges by the following statements.

1. There are two kinds of electric charge.

2. There is a force of attraction between oppositely charged objects, and a force of repulsion between similarly charged objects.

3. The size of the electric force between two charged objects is described by Coulomb's law:

$$F = k\frac{Q_1 Q_2}{r^2}$$

4. Electric charges move easily through one class of materials (good conductors), with difficulty through another class of materials (poor conductors), and negligibly through a third class of materials (nonconductors).

5. A charged body attracts uncharged objects.

6. Electric charge is conserved. Any change in positive charge is balanced by an equal change in negative charge.

7. An object may become charged by bringing it into close contact with a different kind of matter. Close contact at many points can most easily be accomplished by rubbing one substance with another.

Question 11–21 Which of these seven statements can you justify by Experiments 11–1, 11–2, 11–3?

You have observed on the *macroscopic* scale (i.e., with large objects) what happens during the charging process. However, to understand the process more thoroughly, we must consider what happens on the submicroscopic scale (i.e., with very small objects). When used in contrast with the term macroscopic, the term submicroscopic is generally abbreviated to simply *microscopic*. By using the term microscopic, however, we do not want to imply that the charge carriers can be seen with the aid of a microscope; they cannot be seen so easily as all that.

How can we describe on the microscopic scale what happens during the rubbing process to produce electrically charged objects? Do all objects contain both positive and negative charges or is there only one kind of charge with positive referring to an excess and negative to a deficiency, as Franklin suggested? During rubbing, is the charge transferred because it has a greater liking for one substance than the other? Are the charges produced by rubbing related somehow to the atoms of the two materials?

Answers to these questions are not possible without a more complete model that is closely tied to a model of atoms. Atomic models will be developed in Chapter 13.

QUESTIONS **11-22** If you hold two masses, $m_1 = 3.0$ kg and $m_2 = 2.0$ kg, at a distance of 2.0 meters from each other, with what force (in newtons) will they attract each other? What mass has a weight on Earth equal to the force you just calculated?

11-23 The Earth has a mass of 6×10^{24} kg; the Moon has a mass of 7×10^{22} kg, and its distance from the Earth is 4×10^{8} meter. What gravitational force do the Earth and Moon exert on each other?

11-24 In what ways have you observed the effects of electrical charges outside of the classroom?

11-25 If you hold a plastic rod with a negative charge of 1.0×10^{-6} coulomb (1 microcoulomb) in one hand and in the other hand you hold a glass rod with a positive charge of 1.0×10^{-6} coulomb, with what force will they attract each other? The distance between the charges is 2.0 meters.

11-26 It was stated in the text that in order to explain the strength of a crystal, the bonding force between the ions of the crystal must be both attractive and repulsive. Assuming that the bonding force is electrical in nature and that each ion is composed only of a single charge, can the force between two ions be both one of attraction and one of repulsion?

References   1. Baez, A. V., *The New College Physics*, W. H. Freeman & Co., 1967, page 457, Table 36.1.

   The triboelectric series (see page 456 or a good dictionary for a definition).

2. Feynman, R. P., R. B. Leighton, and M. Sands, *The Feynman Lectures on Physics*, Vol. 1, Addison-Wesley Publ. Co., 1963, Chapter 7, pages 7-1 to 7-11.

   A discussion of gravity, with some extensions from our treatment.

3. Lehrman, R. L., and C. Swartz, *Foundations of Physics*, Holt, Rinehart & Winston, New York, 1965, Chapter 13, pages 335 to 344.

   A simple discussion of part of the area of electric forces. Also note pages 359 and 360 for the answers to the text questions.

4. PSSC, *College Physics*, Raytheon Educ. Co. (D. C. Heath), 1968, Chapter 23, Sections 23–1 through 23–3, pages 420 to 423; Chapter 24, Sections 24–1, 24–2, pages 438 to 441.

   Electrostatics and Coulomb's law.

5. Shamos, M. H., *Great Experiments in Physics*, Henry Holt & Co., 1959, Chapter 5, pages 59 through 66.

   Charles Coulomb and the discovery of his law.

Boston Globe, Photograph by Philip Preston

CHAPTER 12

ELECTRIC CHARGES
IN MOTION

12-1 *The microscopic and the macroscopic views of electric charge*

In Chapter 11 we considered the possibility that the constituent particles in a crystal are held together by forces that are electrical in nature. We advanced the hypothesis that each particle has the minimum possible charge of 1.6×10^{-19} coulomb and found that the resultant force between particles is indeed large enough to hold the crystal together. If we are to verify this basic hypothesis we must look for other evidence that the constituent particles in a crystal have electric charges and find a means of estimating the magnitude of these charges.

We are now faced with two difficulties. First, we saw in Chapter 5 that the basic units which make up a crystal must be smaller than the wavelength of red light, about 6.5×10^{-7} meter. In Chapter 6 we became aware of the fact that x rays must be used to measure the distance between planes of particles within a crystal, so those particles must be very small indeed. Because the particles are so very small it is difficult to detect them individually. However, we can determine the properties of a macroscopic sample of material and, by assuming that these properties represent the average effect of a large number of particles, deduce the behavior of a single particle. This is the way the macroscopic charge on the plastic strip in the experiments in Chapter 11 is related to the microscopic charges on the constituent particles of the strip.

The second difficulty arises from the qualitative character and lack of reproducibility of the experiments with static charges performed in Chapter 11. If we wish to obtain quantitative results we must change the procedure. In Chapter 10 we discussed the relationship between the motion of the microscopic particles of a gas and the macroscopically observable properties of pressure, volume, and temperature that describe a body of gas. A similar relationship exists between the microscopic electric charges and the macroscopic observations of their effects; we can measure an electric current even though we cannot observe that the current is made up of individual charges. We will begin with these macroscopic observations and then apply our results to actual materials in the hope of finding a microscopic model consistent with the observations.

12-2 *An electric circuit*

We will begin our consideration of electric charges in motion by performing two experiments. The first is a thought experiment derived from the experiments of Chapter 11, and the second is a real experiment with an electric circuit. We wish to relate these experiments and to make the transition from the static electric charges described in the last chapter to the description of the moving charges in an electric circuit.

EXPERIMENT 12-1 Production of a flow of electric charges

This experiment is an extension of Experiment 11–3, Step 6, in which you put a metal rod or tube on an insulating stand. When you placed a charge on one end of the metal you saw that some charge was transferred to a graphite-coated ball at the other end. In this experiment we are concerned with the steps of the process and with the overall effect. Nothing would be gained by a repetition of the actual experiment so we reconstruct the experiment mentally. The four steps involved are illustrated in Figure 12–1*a–d*.

Step One   Charge the plastic strip and wiper. To start this experiment imagine rubbing the plastic strip with a wiper, and assume that the plastic strip takes on a negative charge and the wiper a positive charge (Fig. 12–1*a*).

Step Two   Seperate the charged plastic strip and the charged wiper, and move the strip so that it is near but not touching one end of the metal. In this step you have separated the two charges. Consider this separation process by answering the following questions (Fig. 12–1*b*).

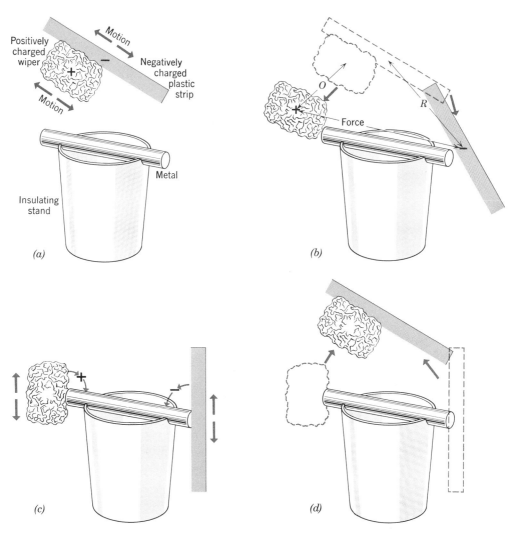

FIGURE 12-1 *(a) Charge the plastic strip and wiper. (b) Separate the charged strip and wiper. (c) Transfer the charges to the metal. (d) Return the strip and wiper for charging.*

Question 12-1 According to Coulomb's law, given in Chapter 11, how do the charged strip and wiper interact with each other? On what factors does this interaction depend? Did you exert a force to separate the strip and the wiper? Did you have to exert a force to hold them apart?

Question 12-2 Review the concepts of work done by an external agent, as discussed in Section 9–4. According to these concepts, what did you do in separating the strip and the wiper from each other? How would you calculate this quantity?

Question 12–3   Using the potential energy concept from Chapter 9, how can you describe the effect of separation on the wiper and strip? On what quantities would you expect potential energy to depend?

Notice that the answers to the first two questions describe what you had to do in order to separate the two charged objects. The third question, however, refers to a change in the energy state of the strip and wiper. Because the change in energy is due to a change in the relative positions of the two bodies rather than to a change in their motion, the separation can be described in terms of changes in potential energy. The potential energy discussed in Chapter 9 was due to work done by an external agent against mechanical forces (elastic and gravitational), but the potential energy involved in this experiment is due to work done by an external agent (you) against electrical forces.

Thus, we would answer Question 12–3 by saying that in separating the charged strip and wiper, the work done by you against electrical forces resulted in an increase in the electrical potential energy of the system.

Step Three   Transfer the electric charges to the metal by simultaneously rubbing the plastic strip over one end of the metal and the wiper over the other end. In this way the two ends of the metal are given opposite charges. Because the metal is a conductor of electricity the charges will not remain where you put them but will distribute themselves so that no part of the strip-metal-wiper system will have a net charge (Fig. 12–1c).

Question 12–4   What makes the charges move through the metal?

Question 12–5   By thinking about what happens when there is an electric current in an electric iron or in an electric light bulb, make a guess as to what happens to the electrical potential energy of the charges as they move through the metal.

Step Four   Return the wiper and plastic strip to their original positions. In this part of the experiment neither the strip nor the wiper has any net electric charge; consequently, there is no electrical force between them (Fig. 12–1*d*).

Question 12–6   What is the relationship between the amount of work you, the external agent, did on the wiper-strip system, and the change in energy of that system? What form of energy is involved?

Step Five   Repeat Steps 1 to 4 as rapidly as possible. If this repetition were quick enough, it would produce a continuous flow of charges through the metal.

Question 12–7   In this continuous process electrical potential energy, heat, and work done by an external agent are all involved. Where is each produced and what transformations occur between them?

Question 12–8   Try to design a system in which Steps 1 to 4 can be done continuously. Perhaps your instructor can show you such a system.

As we will see in this and succeeding chapters, there are many methods of giving energy to charges, other than the mechanical method described in Experiment 12–1. We need some general terms to describe the overall, macroscopic behavior of electric charges, even though we cannot see the microscopic (actually submicroscopic, of course) charges or analyze the forces acting on them. Let us perform another experiment that is almost an electric circuit analogue to the procedure of Experiment 12–1.

EXPERIMENT 12–2   A simple electric circuit

You will perform this experiment in seven steps with accompanying questions.

Step One   Use a magnifying glass to examine the filament of a flashlight bulb. Note the electrical connections supporting each end of the tiny, coiled filament wire. Look for electrical connections from the filament to the two metal parts on the outside of the base of the bulb.

Question 12–9 Is there a continuous conducting (metallic) path between the two metal parts on the outside of the base of the bulb? Sketch or describe this path.

Step Two Examine a flashlight battery.

Question 12–10 Describe the two separate conductors on the outside of the battery. One supplies charges like those on the plastic strip of Experiment 12–1, and the other supplies charges like those on the wiper. Does the battery indicate which conductor supplies which charge?

Question 12–11 Do you suppose that the battery actually makes the charges? What made the charges in Experiment 12–1? Could the charges actually have been present all along and been sorted out by some process?

Step Three Insert the bulb into a socket, and connect it to the battery so that the bulb becomes bright. Connections can be made by touching two conductors together.

Question 12–12 Is there a complete conducting path for the charges through the whole circuit? Sketch the form of the circuit you have assembled, using lines to represent wires, the symbol $\boxed{B^+}$ for the battery, and $-\!\bigcirc\!\!\!\!\!\!\wedge\!\!\!-$ for the light bulb.

Question 12–13 Does it matter which side of the bulb is connected to which end of the battery? Try reversing the connections and see.

Question 12–14 If the circuit is complete, what is happening in the filament to cause it to get hot? Compare your answer with that to Question 12–5.

Question 12–15 In the circuit you have drawn, which part serves the same function as the wiping and separation of charges in Experiment 12–1 (Steps 1 and 2)? What is happening to the charges in this part of the circuit?

Question 12–16 Which part of the circuit is analogous to the metallic conductor of Experiment 12–1 (Step 3)? What is happening to the charges in this part of the circuit?

Step Four Connect the flashlight bulb and two batteries so that the bulb glows more brightly than before. It may be helpful to place the batteries in the groove at the center of an open book.

Question 12–17   Answer Question 12–12 for this new arrangement.

Question 12–18   Why does the bulb glow more brightly when two batteries are used instead of one?

> *Step Five*   Arrange the two batteries in line again, but in such a way that the bulb *does not light,* even though the circuit is continuous.

Question 12–19   Why doesn't the bulb light in this circuit? Are any charges flowing? Is there any force on the charges? Do the charges on one side of the battery combination differ in any way from those on the other side of the combination?

> *Step Six*   Using the arrangement of Step 4, replace the light bulb with the bare wire. Feel the wire.

Question 12–20   Where does the heat come from?

> *Step Seven*   Connect the light bulb in the circuit of Step 6 at the following points.
>
> (a) Between the bare wire and one end of the batteries.
> (b) Between the bare wire and the other end of the batteries.
> (c) Between the two batteries.
>
> Compare the brightness of the bulb at the three places.

Question 12–21   If the brightness of the bulb is a measure of the charge per second passing through it, what observation can you make about the charge per second passing through each of the three places?

Question 12–22   Now what can you deduce about the way in which charges accumulate at any place in a circuit?

Question 12–23   Suggest an experiment to demonstrate that whatever flows in the circuit of Experiment 12–2 also flows through the conductor in Step 3 of Experiment 12–1.

12–3   *Electric current and potential difference*

The observations of Experiment 12–2 were based on the brightness of the light bulb. The brightness is a measure of the energy released as light per unit of time. Your answers to the questions (your conclusions from the experiment) describe the circuit in terms of: (a) the charge flowing per unit of time and (b) the energy gained or lost by the charge. Consider a section *ab* of a circuit. It may contain a battery, a light bulb, a motor, or some other piece of equipment. It is represented in Figure 12–2 by a piece of wire with a charge Q coming in at *a*, passing *c*, and going out at *b*. (It could be the metal in Experiment 12–1.) Suppose that this charge Q flows through the wire in an interval of time t and that it gains an amount of potential energy $W_{a \to b}$. If $W_{a \to b}$ turns out to be negative, it follows that there must be a loss in potential energy.

FIGURE 12-2
A section of an electric circuit.

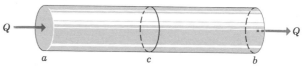

Question 12–24 If a positive charge flows from *a* to *b* in Figure 12–2, what sort of electrical equipment must there be between *a* and *b* if $W_{a \to b}$ is positive? if $W_{a \to b}$ is negative?

In order to simplify further discussion we will use the terms *electric current* and *difference of potential*, which are closely related to your answers to the questions in the last experiment and to a macroscopic description of the situation just outlined. The definitions of these terms depend on the definitions of charge (see p. 242) and the definition of the coulomb (see p. 245).

The electric current I at a point (*a*, *b*, or *c*) in a circuit is the charge Q that flows through the whole cross section of the circuit at that point, divided by the time interval t during which it flows:

$$I = \frac{Q}{t} \tag{12–1}$$

If Q is measured in coulombs and t in seconds, then I will be measured in coulombs per second. This combination of units is given a name of its own, the *ampere* (abbreviated amp):

$$1 \text{ ampere} = 1 \text{ coulomb/second}$$

The difference of potential $V_{a \to b}$ between two points (a and b) in an electric circuit is equal to the gain in potential energy $W_{a \to b}$ of a charge Q as it moves from a to b, divided by the charge:

$$V_{a \to b} = \frac{W_{a \to b}}{Q} \tag{12-2}$$

If $W_{a \to b}$ is measured in joules (defined in Chapter 9) and Q in coulombs, then $V_{a \to b}$ will be measured in joule per coulomb. Another name, the *volt*, is given to this combination of units:

$$1 \text{ volt} = 1 \text{ joule/coulomb}$$

By making use of the five quantities represented by the symbols Q, I, $V_{a \to b}$, $W_{a \to b}$, and t, we can give a concise statement of our conclusions from Experiment 12–2. Although these statements apply specifically to the circuit used, they apply equally to other electric circuits.

A few general statements about electric circuits will help to clarify many of the ideas that have been presented.

(a) An electric circuit must provide a continuous path through which the charges can move (Experiment 12–2, Steps 3 and 4).

(b) In a continuous electric circuit with no branches, the current in all parts of the circuit is the same. Charges neither pile up nor disappear at any point in the circuit (Experiment 12–2, Step 7).

(c) In any part of an electric circuit, the charge will either gain or lose potential energy, depending upon the nature of that part. It will gain potential energy in those parts in which work is done by chemical reactions in batteries and by mechanical forces in generators. The charge will lose potential energy in those parts of the circuit in which it does work. When a moving charge in a circuit loses energy, that energy is converted into heat, light, mechanical energy, or chemical energy (Experiment 12–2, Steps 3 and 6).

(d) When a charge traverses a complete electric circuit,

it must gain as much energy as it loses. This follows from the application of the law of conservation of energy. In Experiment 12–2, Step 4, each coulomb of charge must gain as much potential energy in the battery as it loses in the wire.

Question 12–25   Justify the circuit statements (a, b, c, and d) in terms of your observations in Experiment 12–2.

Question 12–26   Comment on the statement "the batteries in a flashlight are the source of the charges that flow through the bulb."

Question 12–27   What is the function of a 110-volt electric outlet in your home? How does it differ from a 220-volt outlet?

12–4   *Energy and power in an electric circuit*

Of the electrical quantities used to describe a circuit, the electric current and the difference of potential are the simplest to measure. They can be used to determine other useful information about the circuit. For example, we can find the amount of energy supplied to a section of a circuit between points a and b by finding the loss in potential energy of the charge Q as it flows from a to b. If we multiply both sides of Equation 12–2 by Q we obtain:

$$W_{a \to b} = QV_{a \to b}$$

We can now substitute It for Q (multiply both sides of Equation 12–1 by t to obtain $Q = It$):

$$W_{a \to b} = ItV_{a \to b} \tag{12–3}$$

In Equation 12–2, and consequently in Equation 12–3, the energy $W_{a \to b}$ and the difference of potential $V_{a \to b}$ were both defined as positive when work is done so that the charges gain electrical potential energy. On the other hand, if the charges deliver energy to a system as heat or mechanical energy, there must be a decrease in the electrical potential energy of the charges, so $V_{a \to b}$ must be negative and, by Equation 12–3, $W_{a \to b}$ must also be negative. We can avoid any difficulty about signs for the situations we will consider if, in the equation given, we let $W_{a \to b}$ represent the work done *by* the charges, and consider $V_{a \to b}$ to be a decrease in potential. This will be a useful procedure whenever electrical energy is transformed into heat, light, mechanical energy, etc.

The following example illustrates the use of these concepts. When a particular light bulb is connected to a 110-volt outlet it is found that the current is 0.5 amp. How much energy is transformed into heat and light if this bulb is left on for 5 min? We are asked to find $W_{a \to b}$, the decrease in the electrical potential energy, so the 110-volt difference of potential must be a decrease. We can find $W_{a \to b}$ by substituting the values

$$V_{a \to b} = 110 \text{ volt} = 110 \text{ joule/coulomb}$$
$$I = 0.5 \text{ amp} = 0.5 \text{ coulomb/sec}$$
$$t = 5 \text{ min} = (5 \text{ min})(60 \text{ sec/min}) = 300 \text{ sec}$$

into Equation 12–3. We obtain

$$W_{a \to b} = (110 \text{ joule/coulomb})(0.5 \text{ coulomb/sec})(300 \text{ sec})$$
$$W_{a \to b} = 16,500 \text{ joule}$$

Notice that all the units except joules cancel. As we saw in Chapter 9, the joule is the mks unit in which energy is measured.

Let us carry our consideration of the energy in an electric circuit one step further. If we divide both sides of Equation 12–3 by t we obtain

$$W_{a \to b}/t = IV_{a \to b}$$

The quantity on the left in this equation is the rate at which the potential energy of the charges is converted into heat and other forms of energy. The rate at which energy is converted is defined as the *power* $P_{a \to b}$. This definition of power may be written as an equation:

$$P_{a \to b} = W_{a \to b}/t \tag{12–4}$$

If $W_{a \to b}$ is measured in joules and t in seconds, then $P_{a \to b}$ will be measured in joules per second. The *watt* is the name given to this combination of units:

$$1 \text{ watt} = 1 \text{ joule/sec}$$

By combining Equation 12–4 with the equation just before

it, we obtain

$$P_{a \to b} = IV_{a \to b} \qquad (12\text{--}5)$$

This equation tells us that the power supplied to the section of the circuit between a and b (expressed in watts) is equal to the product of the current in the section (expressed in amperes), and the decrease in potential between the two points (expressed in volts). These relationships will be more meaningful if you answer Question 12–28.

Question 12–28   A light bulb has printed on it the statement "60 watts, 120 volts." Assuming that the bulb is used as it was designed to be used, what do these two numbers tell you about the following?

(a) The difference of potential across the bulb.
(b) The potential energy lost by each coulomb of charge as it passes through the bulb.
(c) The current in the bulb.
(d) The number of coulombs passing through the bulb each second.

12-5   *What determines the current in a conductor?*

Although our discussions of electric current may seem to have no relation to the structure of solids, we are now able to describe the energy losses as a charge flows through a solid conductor. But what kind of charge—positive, negative, or both? Without some model for the means by which charge is transferred through solids, we cannot answer these questions. In fact, this model is an important part of an understanding of the structure of solids.

Let us assume that electric current is a flow of positive charge that gains potential energy as it moves through a battery from the negative to the positive terminal and then loses potential energy as it moves through the circuit back to the negative terminal of the battery. If the best model for charge conduction shows that the moving charge is negative, then the direction of charge flow must be reversed, but the description of energy changes will be the same. If both types of charge move, then our assumption is at least half right.

The following experiment is designed to help us see how the nature of a metal conductor determines the rate of flow of charge through it.

EXPERIMENT 12-3   Using an electric current to study materials

Part A   Metallic conductors

In this experiment we will observe the current in two different wires connected to the same set of batteries. One wire is made of copper; the other one, which has the same dimensions, is made of nichrome, a substance used in the construction of electrical heating elements.

The electric circuit is shown in Figure 12–3. Potential

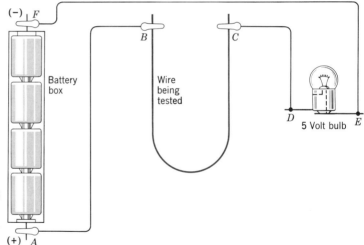

FIGURE 12-3
Circuit for observing the effect of a wire on the current.

energy is supplied at the rate of about 6 joules per coulomb of charge by the four flashlight batteries, that is, the potential difference across the batteries is about 6 volts. The wire being tested is connected to the circuit by the clips at *B* and *C*. If the moving charge is positive, it gains potential energy as it moves from *F* to *A*. Part of this energy is lost in the wire between *B* and *C*, part in the light bulb between

D and E, and the rest in the connecting wires (A to B, C to D, and E to F). Our experiment involves noting the effect of the wire between B and C by observing the brightness of the flashlight bulb.

Step One Connect the circuit as shown in Figure 12–3 but without the wire between B and C. Momentarily touch clip B to clip C and observe the brightness of the bulb.

Step Two Connect clip B to one end of the copper wire and momentarily touch clip C to the other end. Observe the bulb.

Question 12–29 Compare the brightness of the bulb in Steps 1 and 2. Does inserting the copper wire into the circuit noticeably influence the current through the bulb?

Step Three Replace the copper wire between B and C with nichrome wire. Observe the brightness of the bulb.

Question 12–30 Is as much energy per unit time transformed in the light bulb in Step 3 as in the previous two steps? How could you measure the amount of heat given off by either wire? Can you use the same technique to measure the total amount of energy given off by the bulb?

Question 12–31 Compare the current in the circuit in Steps 1, 2, and 3.

Step Four Slide clip C from one end of the nichrome wire to the middle of the wire, observing the brightness of the bulb as you do so.

Question 12–32 How does changing the length of wire affect the current in the bulb?

Question 12–33 Suppose that the light bulb and connectors were removed and that first the copper wire, then the whole nichrome wire, and finally half the nichrome wire were connected directly from A to F. Estimate the relative current in each step.

Let us now consider another situation in which we have two different conductors—two incandescent light bulbs, each designed to operate at 120 volts.

Question 12–34   What physical properties would cause the two bulbs to carry different currents when connected to 120 volts? (Experiment 12–2, Step 1, and Experiment 12–3, Part A, might suggest an answer.)

Suppose that one bulb is marked 60 watt and the other 120 watt, and that they are both connected to a difference of potential of 120 volts. By using Equation 12–5 (and dividing both sides by $V_{a \to b}$) we can calculate the current in the 60-watt bulb.

$$I_{60w} = \frac{P_{a \to b}}{V_{a \to b}} = \frac{60 \text{ watt}}{120 \text{ volt}} = \frac{60 \text{ joule/sec}}{120 \text{ joule/coulomb}}$$

$$I_{60w} = 0.5 \frac{\text{coulomb}}{\text{sec}} = 0.5 \text{ amp}$$

In the same manner we find that the current in the 120-watt bulb is

$$I_{120w} = \frac{120 \text{ watt}}{120 \text{ volt}} = 1 \text{ amp}$$

There must be some characteristic of each of the filaments such that when the difference of potential across each is 120 volts, the current in one is 0.5 amp and in the other is 1.0 amp. This characteristic of a conductor is called its *conductance*, which is a measure of the ease with which charge can flow through a conductor.

Question 12–35   How does the conductance of a piece of copper wire compare with the conductance of a similar piece of nichrome wire? (See Question 12–33.)

Quantitatively, the conductance $C_{a \to b}$ of a conductor is equal to the current I divided by the difference of potential $V_{a \to b}$ between its ends (a and b).

$$C_{a \to b} = I/V_{a \to b} \tag{12–6}$$

If I is measured in amperes and $V_{a \to b}$ is given in volts, then $C_{a \to b}$ is in *mhos* (as in "mows the lawn"):

$$1 \text{ mho} = 1 \text{ amp/volt}$$

We can now describe a characteristic of the filament of the 60-watt bulb if we find its conductance.

$$C_{a \to b} = \frac{I}{V_{a \to b}} = \frac{0.5 \text{ amp}}{120 \text{ volt}} = \frac{0.5 \text{ amp}}{0.12 \times 10^3 \text{ volt}}$$
$$= 4.2 \times 10^{-3} \text{ mho} = 4 \times 10^{-3} \text{ mho}$$

Question 12–36   Find the conductance of the 120-watt bulb. How does it compare with that of the 60-watt bulb?

We can summarize the concepts leading up to and including the concept of conductance by referring to the circuit described in Question 12–33, and by assuming that a battery with a fixed difference of potential is placed in that circuit. The copper wire, having a high conductance, carries a large current and, by Equation 12–5, expends a lot of power. The nichrome wire, having a low conductance, carries a small current and expends little power. When the amount of electric current directed through the wire is

Table 12–1   Summary and Definitions of Electrical Terms

What Is Measured	Unit of Measurement	Definition of Unit	Quantitative Example
Force (mass × acceleration)	1 newton	The force required to accelerate 1 kg 1 meter/sec/sec	Approximately the gravitational force between the Earth and an apple near the surface of the Earth
Work or energy (force × distance)	1 joule	The work done by a force of 1 newton acting through 1 meter	Approximately the work (or energy) required to lift an apple from the ground to your pocket
Power, the rate of doing work (work/time)	1 watt	1 joule/sec	Approximately the power you expend in lifting an apple to your pocket in 1 sec
Charge	1 coulomb	Two 1-coulomb charges 1 meter apart exert a force on each other of 9×10^9 newton	In the experiments in Chapter 11, the charged objects probably carried charges of about 10^{-8} coulomb
Current (charge/time)	1 ampere	1 coulomb/sec passing through surface (such as the cross section of a wire)	Desk lamp: about $\frac{3}{4}$ amp Stove: about 10 amp
Potential difference (change of potential energy/charge)	1 volt	1 joule/coulomb	Flashlight battery: $1\frac{1}{2}$ volt Household outlet: 120 volt
Conductance (current/potential difference)	1 mho	1 amp/volt	Electric heater: about 0.1 mho Typical 5-ft lamp cord: 20 mho

actually measured, it is found that half the length of wire conducts twice the current that the entire length of wire conducts. Therefore half the length of wire expends twice the power.

There are many different electrical quantities, and each is related in some way to the units used in the studies of force and energy. Table 12–1 summarizes these various units for you.

Question 12–37   Two wires having the same dimensions but made of different materials have conductances of 0.10 mho and 1.0 mho. A difference of potential of 10 volts is applied to their ends. Compare the current and the power developed in the two wires.

12-6   *The conductivity of materials*

We have seen that conductance is the characteristic of a conductor that permits us to determine the current in the conductor when a known difference of potential is applied to its ends. We also saw in Experiment 12–3, Part A, that a copper wire and a nichrome wire with the same dimensions have quite different conductances. The difference must be in the different properties of the materials. We have analyzed this characteristic by comparing the conductances of two wires that have the same dimensions but that are made of different materials. This property is called *conductivity,* and it depends only upon the material of which the conductor is made; the conductance depends on both the material and its dimensions. By common agreement the conductance and conductivity are equal for a wire of unit length and unit cross-sectional area.

In the following experiment we will compare the relative conductivities by using the same circuit we used in Experiment 12–3, Part A, and compare the currents by observing the brightness of the light bulb. The brightness increases with the current, and the current increases with the conductivity of the material being tested. For a good conductor (high conductivity, as in the copper wire) the current will be large and the lamp bright; for a poor conductor (lower conductivity, as in the nichrome wire) the current will be smaller and the lamp not as bright; for a nonconductor (very small conductivity) the current will be extremely small and the lamp will not light.

EXPERIMENT 12-3   Using an electric current to study materials

Part B   Relative Conductivities of Materials

We will use the same circuit as in Part A (Fig. 12–3), but we will replace the wire from *B* to *C* with a conductivity cell. This cell serves the purpose of applying a difference of potential to two points of the sample that are a fixed distance apart. By keeping the electrodes (see Fig. 12–4) the same distance apart, we always make use of samples of the same length.

Step One   Assemble the conductivity cell as shown in Figure 12–4.

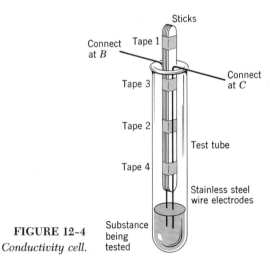

FIGURE 12-4
Conductivity cell.

(a) Fasten four sticks together with transparent tape at points 1 and 2.
(b) You will be given two pieces of stainless steel wire. Bend both at a point 9 cm from one end to form a right angle.

(c) Use tape to fasten the wires to the opposite sides of the bundle of sticks at points 3 and 4 so that 1.5 cm of the 9-cm length of wire extends beyond the end of the sticks.

(d) Make sure that this assembly, when placed in a 13 × 100-mm test tube, will rest on the connecting wires with the lower ends of the wires within a centimeter of the bottom of the tube.

Step Two Using the clips, make the connections shown in Figure 12–3 to complete the circuit. Make sure that the electrodes do not touch each other at any point.

Step Three Make a *continuity check*. Bend a paper clip so that it can simultaneously touch both electrodes near their upper ends. If there are no faulty connections in the circuit, the lamp will burn brightly. If the lamp does not light, check all connections; make sure the bulb is screwed in tightly, and check the batteries and the bulb.

The continuity check is used to make sure that the circuit is operational when the lower ends of the electrodes are in contact with the substance being tested. The paper clip provides a high conductance path through which the majority of the charge will move even if only a small charge flows through the low conductance path in the cell. Thus, when this check is made the lamp will glow brightly if the circuit is continuous.

Step Four Test the conductivity of the solids provided.

To test the conductivity of large pieces of solids it is necessary only to press the extended ends of the two electrodes against the surface of the solid and observe the light bulb.

For granular and powdery solids, and for liquids, you should place a sample of the substance to be tested to a depth of 2–2½ cm in a clean, dry test tube. Fasten the tube to the pegboard, and leave room for an alcohol burner underneath. Insert the electrodes until they are well into the sample contained in the test tube.

The length of the sample tested depends on the separation of the electrodes, but the cross-sectional area depends on how the electrodes make contact with the sample. However, in this experiment the variations in conductivities far exceed the variations in cross-sectional area.

Connect the battery and observe the light bulb. If it shines brightly, the sample has *high conductivity*. If it glows dimly, the sample is a poor conductor or there is a loose connection or a weak battery. Check the last two possibilities by using the paper clip to make a continuity check while the cell is still in place. If the lamp shines brightly in the continuity test, but dimly with the cell between *B* and *C*, we can conclude that the material has *low conductivity*. Finally, if the bulb does not glow at all with the cell in the circuit but burns brightly in the continuity test, we can conclude that the material has *negligible conductivity*.

Using the procedure described, classify the following materials according to their conductivity (high, low, or negligible). *The test tubes must be clean and dry for each test.*

1. Sodium chloride
2. Sodium acetate
3. Zinc
4. Zinc chloride
5. Tin
6. Sugar
7. Iodine (do not heat)
8. *Para*-dichlorobenzene
9. Paraffin
10. Water

   (a) Test the conductivity of each substance at room temperature.

   (b) Heat each solid, except iodine, until it melts or reaches the maximum temperature obtainable with the alcohol burner; retest its conductivity.

During this heating period it is a good idea to keep the circuit connected so that any change in the conductivity can be immediately detected. If a substance begins to melt very quickly, the application of heat should be moderated by moving the flame back and forth under the sample until melting is complete. Some of the substances that melt easily have a tendency, on strong heating, either to discolor or char (indicating decomposition), or to boil and even catch fire. So be careful!

   (c) After each substance has cooled, try to dissolve some

of it in half a test tube of distilled water (do this with the iodine as well, even though you did not heat it). How can you tell whether the substance is dissolving? Support the test tube on the peg-board, connect the circuit, make the continuity test to see that the circuit is operational, and then slowly dip the electrodes into the solution. Does the solution conduct? Wash the electrodes carefully and test a sample of distilled water. Does pure water conduct?

You will find it convenient to record all of your results in a single table in your notebook. The table should have four columns: the first for the name of the substance, the second for its conductivity in the solid state, the third for the con-ductivity after heating, and the fourth (for those substances which are soluble in water) for the conductivity of the water solution.

The results of this experiment form the basis for some significant ideas leading to a model that describes how charge is conducted.

Question 12–38   Which substances are conductors in the solid state? What other properties do they have in common?

Question 12–39   Are any of the substances that are nonconductors in the solid state, conductors when they are melted?

Question 12–40   Do all liquids conduct electricity? Do gases conduct elec-tricity? Does a charge move through them under any special circumstances?

Question 12–41   If neither pure water nor pure solid salt conducts electricity, what model can you suggest that is consistent with the fact that a solution of salt in water does conduct electricity?

Some substances conduct well, others poorly, some not at all; sometimes a solid does not conduct an electric current, whereas its melt does; some substances are nonconductors even when melted; metals conduct well in the solid state. These are some of the phenomena you have observed in your conductivity tests. How can we utilize this information to learn about the structure of solids and the forces that held them together?

12-7 *A preliminary microscopic model of conduction in solids*

In Chapters 6 and 11 we hypothesized that a crystal is composed of particles held in a regular array by forces that appear to be electrical. We have just seen that the electrical conductivity of a substance depends on the material and on its state (i.e., solid, liquid, or solution). Let us now combine these two ideas in order to develop a microscopic model that will account for electrical conduction in solids.

The kinetic theory model we developed in Chapter 10 suggests that a solid consists of microscopic particles; each particle is in motion but cannot move far from a fixed point. The model developed in Chapter 11 suggests that the force which pulls the particle toward its fixed point is probably electrical in origin. For any one particle the kinetic energy depends on its distance from its own particular fixed point (in a manner similar to that in which the kinetic energy of a swinging pendulum depends on the distance of the bob from its lowest point). For a single particle the kinetic energy continually changes, but for a large number of particles the average kinetic energy depends on the temperature and is constant if the temperature is constant.

We have a number of possible models to explain the nature of an electric current: the charge may be a single continuous fluid (as suggested by Benjamin Franklin), or it may be two continuous fluids, or it may be something that is carried only by particles. There are several reasons for choosing the particle model. First, in order to account for the bonding forces, we have already assumed that the particles of a solid carry an electric charge. Second, the history of physics reveals that every time a continuous fluid theory has been advanced to explain a macroscopic phenomenon, an examination on the microscopic level has shown that the fluid must consist of discrete particles. So let us assume that electric charges are always carried by particles. What particles? What kind of charge—positive, negative, or both? How large a charge? We are not yet able to answer these questions.

Lacking any knowledge of the charge-carrying particles, what can we say about them? For a gas or a liquid they may be the same particles whose existence we postulated in developing the kinetic theory. In gases and liquids these particles move about through the whole volume, so perhaps we can describe the conduction of electricity in gases and liquids if we can formulate a mechanism by which some of the particles have an electric charge. Then we would be able to account for the differences in the conductivities of the liquids examined in Experiment 12–3 by saying that some of the particles of the conducting liquids may have a net charge, but that none of the particles of the nonconducting liquids has.

So far, our kinetic theory model will not account for the conduction of electricity in solids, since we have already postulated that the motion of the particles of a solid is restricted by electrical forces to small regions around a fixed point. Thus, if a charge moves through a solid, and we observed in Experiment 12–3 that in some solids it may, it must be carried by particles different from those described so far. If our model is to be fruitful, we must be able to account for the differences in the electrical conductivity of solids in terms of the presence or absence of these additional particles and by variations in the freedom with which they move through a solid.

It should be obvious that we have gone about as far as we can go in formulating a model to account for the conductivity of various materials. We cannot pursue this any further without a knowledge of the structure of the particles that comprise matter. In Chapter 13 we will turn our attention to the structure of these basic particles. But before leaving the subject of the conduction of electric charges in solids, we can review the concepts we have developed and the model we have formulated by performing an experiment with a mechanical analogue to our model.

EXPERIMENT 12–4   A mechanical analogue to conduction in solids
(a demonstration)

In this analogue the volume occupied by the solid conductor is represented by the space immediately above a board. The side boundaries of the conductor are represented by vertical sheets attached to the edges of the board, and the particles of the solid are represented by nails driven into the board in a regular pattern. The analogue of the charged particles, whose nature we have not yet established, is a bunch of marbles.

Step One   With the board in a horizontal position, place a number of marbles among the nails on the board.

Question 12–42   The marbles do not move when the board is horizontal. What can you say about the resultant force acting on each marble? What is the analogous situation for the charged particles we have postulated to exist in a solid conductor?

Step Two   Raise one end of the board slightly and observe the motion of the marbles.

Question 12–43   Describe the motion of the marbles. Does the speed remain constant? Does it increase uniformly? Is the speed ever zero? Is there a maximum speed that the marbles never exceed? On what factors does the speed of the marbles depend?

Question 12–44   Suppose that this experiment had been performed by using a board without nails. On the basis of your experience, describe the speeds of the marbles in this case. How would the final speed of a marble depend on the height at which it started? What changes take place in the energy of a marble? (If you can't answer these questions, borrow a marble and use a book instead of the board.)

Question 12–45   What happens to the kinetic energy of a marble when it strikes a nail? Describe the energy transformations that take place as the marble rolls down the board.

Question 12–46   In a conductor the fixed particles are analogous to the nails. When a charged particle moves through the conductor and strikes the bound particles, what energy transfer takes place? How would you expect to observe macroscopically the effect of this energy transfer? Does this agree with any macroscopic observations you have made?

Step Three   Repeat Step 2, but this time remove the marbles as they reach the bottom of the board and place them, a few at a time, at the top of the board. Maintain a steady flow of marbles down the board.

Question 12–47   In terms of energy, what was done to the marbles to get them to the top of the board? Who did it? How does the amount of energy per marble depend on the height? On the mass of a marble?

Question 12–48   In terms of the circuit of Figure 12–3, what was done to the charged particles that was analogous to lifting the marbles? What part of the circuit accomplished this? On what factors does the energy given to each charged particle depend?

Question 12–49   According to Equation 12–5, the rate at which energy in a conductor is dissipated in the form of heat is equal to the product of the difference of potential across the conductor and the current through the conductor. What are the corresponding factors in the mechanical analogue for the rate at which energy is absorbed by the nails?

QUESTIONS   **12–50**   Each month the electric company sends us a bill reporting that we have used a certain number of *kilowatt-hours* (kwh). Take these words apart (for example, kilo means 1000, etc.) and use Table 12–1 to find the physical quantity that a kwh measures.

This physical quantity is usually expressed in a different unit. How many of these units does 1 kwh equal? It will help you to know that 1 hour is equal to 3600 seconds.

12–51   When an electric stove is in operation, the current in the heating element is 10 amp and the difference of potential across it is 220 volts.

(a)   How many coulombs pass through the stove in one minute?

(b) How much potential energy is lost by each coulomb in the heating element?

(c) How much potential energy is lost each minute by the charges flowing through the heating element?

(d) What should the label on the stove state as the power of the heating element?

12-52   Suppose that further experiments on the conductivity of metals demonstrated that the moving particles in a metallic circuit carry a negative charge rather than a positive charge as we assumed in describing the energy changes in the circuit of Figure 12–3. If the moving charge is negative, in which direction does it move through the battery box? Will it have more potential energy at *A* or at *F*? Trace the path of the charge through the rest of the circuit and describe the energy changes.

12-53   If you were asked to measure the conductance of a metal rod, what measurements would you make? What additional measurements would be required if you wished to determine the conductivity of the metal?

12-54   It was observed in a conductivity experiment similar to that of Experiment 12–3, Part B, that the current was 0.2 amp when the difference of potential across the cell was 40 volts. What was the conductance of the cell? At what rate was electrical energy used in the cell?

References   1. PSSC, *College Physics,* Raytheon Educ. Co. (D. C. Heath), 1968, Chapter 25, Sections 25–2 through 25–5, pages 467 to 473.
Electric current, electric forces, and electric energy.

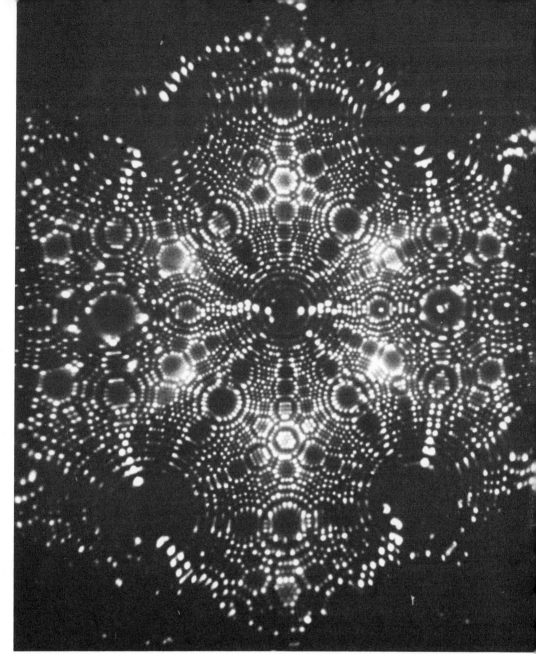

CHAPTER 13

MODELS OF ATOMS

In Chapter 12 you investigated the electrical conductivity of some solids, melts, and solutions. You found that all the metals you tested conduct electricity and that some materials, such as zinc chloride, do not conduct electricity in the solid state but are good conductors when melted. Many other substances, such as nylon, glass, and rubber, do not conduct electricity but, if rubbed, can be made to attract or repel other objects; this behavior is attributed to the acquisition of electric charge.

Chapter 12 concluded with a preliminary microscopic model of conduction in solids which postulated minute charge carriers that are free to move through the fixed array of particles in the solid. What are these charge carriers? At present we cannot say, but we suspect they may be part of the atoms composing a metal.

In this chapter we seek a model of the atom that will be consistent with all relevant observations and that will help us understand how particles hang together to form substances. Scientists have long been thinking about atoms and have constructed atomic models to guide them. Models are aids to thinking and are used to help tie together data from many experiments; they are not meant to be literal pictures. If a model becomes inconsistent with new relevant experimental data, the model, to be useful, must be modified or it becomes of historical interest only. It is important that you recognize the tentative nature of models.

We will present the highlights of the history of a model that was developed to account for the behavior of atoms both individually and collectively. This model should be able to answer such questions as: What is the difference between a metal and a nonmetal atom? Why do solids have different electrical conductivities?

We will, in fact, find it necessary to discuss several variations of our basic atomic model, but we will not present all the details of any one. Instead, we hope that as you study this chapter you will participate vicariously in the unfinished search for a useful atomic model and that you will understand the qualitative features of this model. We hope you will see the way those features help us understand the qualitative nature of this model and the way those features help us understand the structure and characteristics of solids.

13-1 *The beginnings*

The belief that complex substances are composed of simple elemental components existed in ancient Greece. During several centuries a number of elemental substances were suggested, for example, water, fire, earth, and air. In the fifth century B.C., the Greek Democritus taught that all substances are composed of tiny indivisible atoms (in Greek *atomos* means uncut or indivisible). He believed that the infinite variety of observable things could be explained by combinations of atoms of different shapes and sizes. The atoms themselves, he supposed, remain unchanged even when combined and later separated.

Although these earlier speculations were unsupported by experiments, similar ideas occurred to people throughout the centuries after the time of Democritus.

In the first decade of the nineteenth century, John Dalton, an English chemist, initiated the development of atomic theory as we know it today. Dalton described the atom as follows.

There must be some point beyond which we cannot go in the division of matter. The existence of ultimate particles of matter can scarcely be doubted, though they are probably much too small even to be exhibited by microscopic improvement.

I have chosen the word "atom" to signify those ultimate particles, in preference to particle, molecule, or any other diminutive term, because I conceive it is much more expressive; it includes in itself the notion of indivisible, which the other terms do not. . . .°

During the first half of the nineteenth century, chemists, applying Dalton's postulates, generally pictured atoms as tiny indestructible spheres. They believed that atoms of one

° John Dalton, *A New System of Chemical Philosophy*, printed by S. Russell for R. Bickerstaff, London (1808–1810).

kind, such as those of hydrogen, are alike in all respects and can combine with atoms of another kind to form more complex substances that have new properties.

However, many questions were left unanswered. What holds atoms together to form these more complex substances? Why do iron atoms combine with oxygen atoms to form rust? For that matter, why do iron atoms combine with iron atoms to form solid iron? What determines how many atoms of one substance will combine with one atom of another substance? Is this amount always the same for a particular substance?

The work of two Englishmen, Sir Humphry Davy and Michael Faraday, was important in the search for answers to these and related questions. Early in the nineteenth century they decomposed substances by passing an electric current through them. This gave strong support to the idea that atoms are somehow electrical in nature. Faraday wrote:

> Although we know nothing of what an atom is . . . there is an immensity of facts which justifies us in believing that the atoms of matter are in some way endowed or associated with electrical powers, to which they owe their most striking qualities, and amongst them, their mutual chemical affinity.

Many of the early discoveries concerning the nature of both electricity and atoms were made by experimental work on conduction of electricity through gases. One type of apparatus used is shown schematically in Figure 13–1. Electrodes (E in Fig. 13–1) are sealed into the ends of the tube so that a difference of potential can be established between them. This difference of potential can be measured

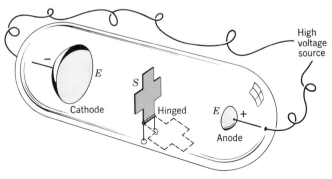

FIGURE 13–1
Gas-discharge apparatus.

by a voltmeter. The tube is connected to a vacuum pump to reduce the pressure inside.

When the pressure in the tube is atmospheric, a very large potential difference is required to produce an electrical discharge—a spark—between the electrodes (about 3×10^5 volts are needed when the electrodes are 10 centimeters apart and in air). As the air is pumped out of the tube the spark changes to a "ray," because the air begins to glow along a narrow path. At very low pressures only the glass of the tube glows. For most glasses the glow is a greenish color.

If an object such as a metal cross is placed inside the tube (S in Fig. 13–1), a distinct shadow of the object is cast on the end of the tube farthest from the cathode (the negative electrode). This shadow partially obscures the greenish glow of the glass. It is easily deduced from the arrangement of the cathode, the metal cross, and the shadow that the glow is caused by something emitted as rays from the cathode; hence the name *cathode rays*.

The formation of the shadow indicates the direction of the cathode rays; furthermore, the fact that the glass on the far end of the tube glows indicates that energy is being transported from the cathode to the glass. How is that energy transported? As seen in Chapter 3, waves transport energy, but so does a stream of particles. Is the cathode ray a stream of particles or the passage of a wave? Or can the cathode ray be something other than a wave or a stream of particles? It doesn't seem likely; cathode rays do carry energy, and no one has yet thought of any method other than waves or particles by which energy can be transported from one place to another.

During the many experiments that were performed to decide whether cathode rays are composed of waves or particles, many other pertinent properties of these rays were discovered. It was soon learned, for example, that cathode rays carry an electric charge. A simplified version of the experiment that revealed this is shown schematically in Figure 13–2. An opening in the positive electrode (the one on the right) permits cathode rays to go right through. After passing through the opening, the rays enter a collect-

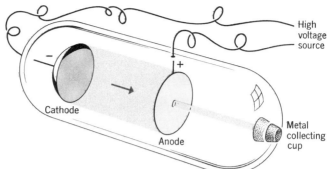

FIGURE 13-2
Cathode rays charge a
collecting cup.

ing cup. When the tube is in operation the collecting cup becomes charged. Tests like the ones you performed in Chapter 11 show that the charge on the cup is negative.

How can a negative charge be produced on an object such as the collecting cup? You produced charges on objects during the experiments of Chapter 11 by rubbing two materials together. At that time we did not have enough information to decide whether a charge can be created during a rubbing process, or whether the two charges already exist in equal amounts and can be simply separated by the rubbing process. Correspondingly, it is possible to explain the negative charge on the collecting cup by either of two hypotheses: (1) negative charge is transported to the initially neutral collecting cup by cathode rays, or (2) positive charge is caused to leave the collecting cup when the cathode rays strike it. This experiment does not yield enough information to enable you to decide which of these two hypotheses is more acceptable.

In order to determine a closer relationship between the cathode ray and the electric charge, other experiments were performed. These experiments show that a beam of cathode rays is deflected either by an electric field or a magnetic field (see Fig. 13–3). When, for example, the beam is passed between a pair of oppositely charged plates, it is deflected from the negative toward the positive plate. This observation indicates that the beam itself carries a negative charge and supports the hypothesis that the negative charge is transported to the cup by the beam.

Since light-like waves are not altered by either an electric

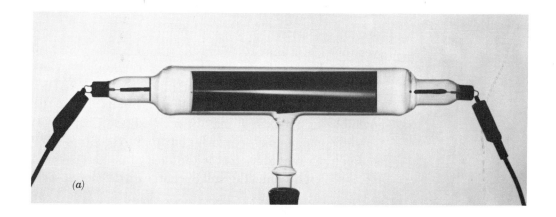

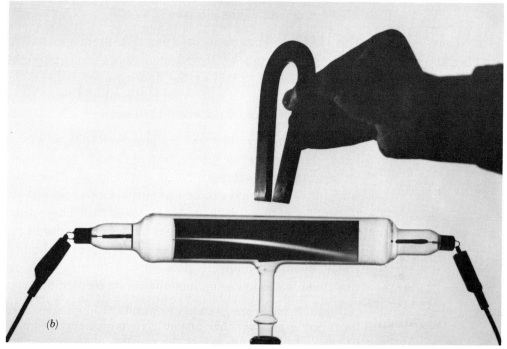

FIGURE 13-3 *(a) Cathode ray beam. (b) Cathode ray beam deflected by a magnet. (Fundamental Photographs)*

or a magnetic field, it can be concluded that cathode rays are not light-like. In addition, J. J. Thomson, an English physicist, noted in 1897 that the beam is deflected by an electric and a magnetic field as if it were composed of high velocity particles, each with a very small negative charge.

Thomson was able to measure the ratio of the charge of each particle to its mass. His results indicate clearly that every particle in any cathode ray has the same charge-to-mass ratio; these particles came to be called *electrons*.

Measurements by Thomson and others have revealed that the charge-to-mass ratio of the electron is 1.8×10^{11} coulomb/kg. Millikan, it may be recalled (see Chapter 11), found the minimum charge on his oil drops to be 1.6×10^{-19} coulomb. Presumably, this is the charge on the electron, and as such it becomes the unit of charge on the atomic scale.

Question 13–1   If the charge of the electron is $e = 1.6 \times 10^{-19}$ coulomb, and the charge-to-mass ratio is $e/m = 1.8 \times 10^{11}$ coulomb/kg, what is the mass of the electron?

Question 13–2   The charge-to-mass ratio of a beam of hydrogen ions is 9.5×10^7 coulomb/kg. If we assume that the charge on the electron is equal to that on the hydrogen ion, what is the mass of the hydrogen ion? What is the ratio of the mass of the hydrogen ion to the mass of the electron?

At this point you may well ask how the experiments that led to the discovery of the electron are related to our study of the historical development of an atomic model. Thomson's study of cathode rays resulted in his inquiry into the possible sources of those particles. He reasoned that the electrons were coming from either the gas or the electrodes in the discharge tube. His belief that electrons are part of atoms rested upon his ability to detect them in other ways, as described in the following quotation from his writings.

They (electrons) are given out by metals when raised to a red heat. Any substance when heated gives out electrons to some extent; indeed, we can detect the emission of them from any substances, such as rubidium and the alloy of sodium and potassium, even when they are cold; they are also given out by metals and other bodies, but especially the alkali metals when these are exposed to light; they are produced in large quantities when salts are put into flames; and it is perhaps allowable to suppose that there is some emission by all substances.

Because electrons are the lightest charged particles known, they are good candidates for the charge carriers responsible for electrical conductivity in metals. Thomson, convinced that electrons can be removed from all substances, postulated that they are part of all atoms. But if common materials are electrically neutral, then presumably individual atoms are also neutral; and if atoms are composed, in part, of electrons, they must contain enough positive charge to neutralize the electrons' negative charge.

Experiments indicate that hydrogen ions have a positive charge. In Question 13-2 you calculated the mass of the hydrogen ion to be about 1.7×10^{-27} kg; in Question 13-1 you calculated the mass of the electron to be about 9×10^{-31} kg. So we conclude that the bulk of the mass of the hydrogen atom has a positive charge. In fact, that bulk is nearly 2000 times the mass of the electron. If most of the mass of the hydrogen atom has a positive charge, then presumably so does most of the mass of every other kind of atom.

Question 13-3   A hydrogen atom is composed of a hydrogen ion and an electron. What is the mass of a hydrogen atom?

How are electrons related to atoms? How is the negative charge distributed within the atom? How is the positive charge distributed? To answer these questions, we must start to build a model of the atom.

Many experiments, including Thomson's measurement of the charge-to-mass ratio of the electron, led to the conclusion that the negatively charged electron is a tiny particle, much smaller than an atom. Consequently, we can only surmise that the negative charge within the atom is somehow confined to individual electrons. But what about the more massive portion of the atom, the positive charge?

It is conceivable that the positive charge within the atom is (1) distributed uniformly throughout the atom, (2) concentrated into one small bundle, or (3) distributed among a number of bundles. A decade after Thomson's work, Ernest Rutherford, an English physicist, obtained experimental evidence that clearly indicates how the positive charge is distributed within the atom.

13-2   *Rutherford's scattering experiments*

Radioactivity—the spontaneous emission of radiation—was discovered during the last decade of the nineteenth century. Radiation can be detected by its ability to expose photographic film (recall Experiment 3–1). A particularly good source of it is radium, which was used by Rutherford to identify the distribution of positive charge within the atom. In order to form a more complete concept of this radiation, Rutherford prepared an experiment which clearly indicated that it has more than one component.

Apparatus that would yield essentially the same results as Rutherford's experiments is shown schematically in Figure 13–4. Radium is placed in a lead box fitted with a narrow

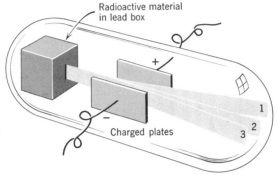

FIGURE 13-4
Schematic drawing of apparatus for studying the radiation from radium.

slit at one side so that the radiation is emitted in only one direction. The radiation passes between the metal plates and is detected when it exposes a photographic film. In the absence of a charge on the plates, the film is exposed at point 2 only. When the plates are charged, as in Figure 13–4, the film is exposed in three different regions, labeled 1, 2, and 3. Apparently, some of the radiation is attracted by the negative plate, some is attracted by the positive plate, and some is unaffected by the electric charges. These rays are called α, β, and γ, respectively (the first three letters of the Greek alphabet). Experiments in the early years of the 20th century revealed that alpha (α) radiation is a stream of helium ions, each with a charge twice that of a hydrogen ion; beta (β) radiation is a stream of electrons, like cathode rays; gamma (γ) radiation is like x rays but has shorter wavelengths.

The recognition that both alpha rays and beta rays are composed of particles has given rise to the terms alpha particles and beta particles. An *alpha particle* is a charged helium atom; a *beta particle* is an electron.

Question 13–4 What kind of a charge, positive or negative, does the alpha particle have?

Question 13–5 Gamma rays are not deflected by charged plates. Does it necessarily follow that they are like x rays?

You have learned something about solid matter by heating a number of different materials; you also observed a reaction by directing light on one particular substance (Experiment 3–1). Rutherford reasoned that he could learn something of the nature of solid matter by bombarding material with a beam of alpha particles. It is possible that alpha particles might cause a change in the material, or the material might somehow alter the beam of alpha particles. Little was known about the structure of matter before he performed this series of experiments, and he did not really know what results to expect.

Studies of the material can be made to see if it has been changed by the bombardment of alpha particles, but how can an alteration of the beam of alpha particles be detected?

Whenever a single alpha particle strikes a screen painted with a fluorescent* material such as zinc oxide, a tiny flash of light, called a *scintillation,* can be observed from the point where the particle hits the screen.

A beam of alpha particles can be formed by placing a source of alpha particles inside a lead box and then placing that lead box inside a glass container. If the inside of that glass container is painted with a fluorescent material, and if the air inside the container is evacuated by a vacuum pump, then the alpha particles leaving the slit-shaped opening will travel the length of the evacuated glass container and strike the far end. The beam of alpha particles will cause many scintillations, which will form a sharp narrow rectangular pattern the same size and shape as the slit-shaped opening in the box (see Fig. 13–5).

* Fluorescent material gives off light when suitable rays or particles hit it.

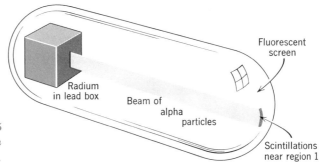

FIGURE 13-5
*Scintillations from a beam
of alpha particles.*

If air is then permitted to enter the glass container, the sharp pattern becomes a bit fuzzy at the edges. This change indicates that the air somehow deflects a small portion of the alpha particles from their original direction of travel. The deflection, it should be noted, is very slight; the original sharp slit-shaped pattern does not smear out into a diffuse blob, it only becomes slightly fuzzy at the edges. Rutherford interpreted this observation by supposing that whenever an individual alpha particle strikes a molecule in the air, the direction of travel of that alpha particle is slightly altered. This alteration in the direction of travel is called *scattering*. When a beam of alpha particles is projected through air, only a slight scattering occurs, so only a very small fraction of the alpha particles must collide with molecules in the air.

Now what would happen, Rutherford and his German student Hans Geiger asked, if they were to place a thin foil of gold in the path of the beam of alpha particles? If the atoms of gold are hard and solid like the matter they make up, and if the individual atoms are packed close together, as crystal studies suggest, then one would expect that the beam of alpha particles would be completely deflected. Alpha particles should not be able to penetrate right through the foil, although perhaps a few might make it through, since gold foil is thin.

In 1908 Geiger did place a foil of gold into the beam of alpha particles all inside of a glass container from which the air had been evacuated. Much to his surprise, the pattern of scintillations was the same as it was with only air in the tube (i.e., without the gold foil)—the slit-shaped pattern was a bit fuzzy at the edges. It was obvious that the thin gold foil permitted most of the alpha particles to pass right on through; only a few of the particles were scattered from their initial direction of travel.

How did the alpha particles get through the gold foil? Were the atoms of gold not hard little spheres? Or were they not packed closely together? Since Rutherford was convinced that the gold atoms were closely packed, he chose to suggest that they were not hard little spheres. But then what were they? How could an alpha particle pass right through a gold atom? Are gold atoms soft and spongy?

In 1909 Rutherford suggested to Geiger that a student, Ernest Marsden, repeat the alpha-particle experiment but extend it by looking for alpha particles that might be scattered backwards—that is, that have their direction of travel altered by more than 90 degrees (Fig. 13-6). Because the gold foil in previous experiments had not noticeably weakened the beam of alpha particles and had apparently caused only small-angle scattering, Rutherford did not really expect to find any scattering over angles larger than 90 degrees. But Marsden observed the unexpected.

Over 99% of the scintillations on the screen were near region *1* (Fig. 13-6). Some alpha particles were scattered slightly, striking the screen between regions *1* and *2*. But, Marsden also saw a few scintillations near region *3*, and even an occasional one near region *4*. This indicated that a few particles were scattered over large angles. This unexpected observation required an explanation.

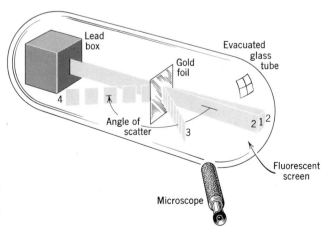

FIGURE 13-6
*Scattering of alpha
particles by gold foil.*

Gold atoms have electrons associated with them some-how or other, and it is reasonable to ask if it is these electrons that cause the large angle scattering of the alpha particles. A negatively charged electron would attract the positive alpha particle and, for some incoming paths, the attraction might be just enough to cause the alpha particle to swing around the electron and proceed in the general direction from which it came. However, such an event would require that the electron be much more massive than the alpha particle.

What about the mass of the electron relative to the mass of the alpha particle? By 1909 the masses of both the electron and the alpha particle were known: the mass of the electron is 9×10^{-31} kg, and that of the alpha particle is 7×10^{-27} kg. The ratio of these masses is $10^{-27}/10^{-31} = 10^4$ or 10,000! It would be just as likely for a 5-ton truck to bounce backwards after colliding with a pound of butter as for an alpha-particle to bounce backwards after colliding with an electron.

What about the positively charged portion of the atom? Could *it* be responsible for the large-angle scattering? Yes, the positively charged portion of the gold atom was known to have a mass 50 times that of the alpha particle. But while a large mass is necessary, it is not sufficient to account for backscattering. It is also necessary that a very large force be exerted on the alpha particle by that positively charged portion of the gold atom. The only explanation Rutherford could offer to account for the large force was to propose that the positively charged portion of the gold atom must have a volume much smaller than that of the total atom. He assumed that all of the positive charge is concentrated in a very small volume at the center of the atom, and he called that portion the *nucleus*. His results indicate that the atomic nuclei have a diameter of about 10^{-14} meter. The diameter of the entire atom is about 10^{-10} meter.

The following experiment is designed to give you confidence that sizes of objects can indeed be determined by means of a scattering type experiment such as the one Marsden performed.

Determining the size of an object from collision probabilities

You are going to try to determine the size of marbles by rolling one, the *bombarding marble,* repeatedly at a group of *target marbles.* However, since the bombarding marbles are vastly different from alpha particles, and the target marbles are equally different from the gold atoms, this experiment is not completely analogous to the alpha particle scattering experiment. Yet in both experiments, the size of objects can be determined by a study of collisions. To be sure, the size of a marble can be measured directly with a meter stick, but this is not so for gold nuclei.

a. Reasoning Behind the Experiment

The experimental arrangement is shown in Figure 13–7.

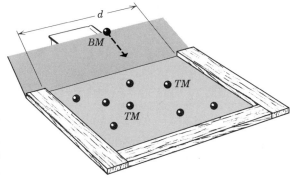

FIGURE 13-7

Arrangement of marbles, walls, and inclined plane for scattering experiment.

The bombarding marble will be rolled many times from the left, parallel to the sides of the target area, to simulate a beam of alpha particles. You can aim the bombarding marble at a particular target marble, but alpha particles cannot be aimed at a particular gold nucleus. Therefore, to simulate the alpha-particle experiment more closely, start the marble rolling from points selected randomly across the opening of the target area.

Suppose you place eight marbles randomly within the given area and roll the bombarding marble toward that area. What does the probability of a collision depend on? Surely it would depend on the size of the target marbles. If they are big and occupy most of the given area, you are apt to score a hit. However, if each of those eight marbles is small, you might very well miss them all.

The probability of scoring a hit also depends on the size of the bombarding marble. If it is as big as the given area you are certain of scoring a hit; but if it is very small you might miss all the target marbles.

It is convenient to classify the results of each roll of the bombarding marble as either a *hit* or a *miss*. If the bombarding marble reaches the back of the area without a collision, we call it a miss. If it collides with any number of target marbles, for the sake of simplicity we will call it only one hit.

In order to score a hit, the center of the bombarding marble must come within a distance of the center of the target marble equal to the sum of the two radii $(R + r)$, where R is the radius of the target marble and r is the radius of the bombarding marble (see Fig. 13–8). Because the

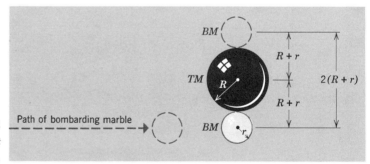

FIGURE 13-8
Target size of two different size marbles.

bombarding marble may hit either side of the target marble, there will be a hit if the center of the bombarding marble passes through a distance $2(R + r)$, which we will call the *target size*. The target size of any target marble is centered on that marble.

What else will affect the probability of scoring a hit? Look again at Figure 13-7. If the target marbles are placed randomly throughout the area between the walls, the probability of scoring a hit will depend on the size of the opening d. The larger the opening, the larger the distance between the marbles, and the less frequently you will score a hit.

You might argue, however, that we could put more marbles in the area and increase the chances of scoring a hit. That is true and, consequently, the probability of scoring a hit also depends on the number of target marbles N. The probability of scoring a hit depends on both d and N and, furthermore, the probability increases as N increases but decreases as d increases. For want of a better relationship, we will assume that the probability is proportional to the ratio N/d. This assumption will permit us to derive an expression relating the probability to the variables in the experiment. Once an expression is derived, we can test it with actual examples to see if it is a good one.

To illustrate how overlapping target sizes would affect the probability of scoring a hit, consider what would happen if we doubled the number of target marbles. If the target sizes did not overlap, the probability of scoring a hit would also be doubled. However, if we placed each one of these additional marbles directly behind one already there, the probability of scoring a hit would not be changed even though the number of marbles would be doubled. To account not only for complete overlapping, but also for partial overlapping would complicate our analysis considerably, and we would like to avoid such complications. Therefore, during the performance of the experiment make sure that no target sizes overlap. A sure way to do this is to see to it that the bombarding marble can roll between any two

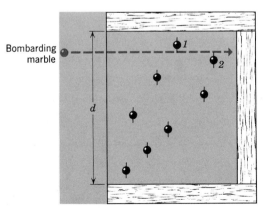

FIGURE 13-9
*Overlapping of target sizes,
as for marbles 1 and 2,
will yield poor results.*

marbles without hitting either (Fig. 13–9). One way to do this is to use marbles that are much smaller than the opening. If these very small marbles were distributed randomly throughout an area with a large opening, the chances of overlapping would be greatly reduced.

We define the probability V of scoring a hit as the ratio of the number of hits H to the total number of trials T:

$$P = \frac{H}{T}$$

But we have seen that the probability P of scoring a hit increases as $2(R + r)$ increases; it also increases as the ratio N/d increases. In fact, P is directly proportional to each of those quantities. Therefore, we have

$$P \propto \frac{2(R + r)N}{d}$$

In support of this proportion it should be noted that the ratio $2(R + r)/d$ is the fractional part of the total opening occupied by the target size of one marble. For example, if the target size is 1 centimeter and the width of the opening is 60 centimeters,

$$\frac{2(R + r)}{d} = \frac{2}{60} = \frac{1}{30}$$

This ratio multiplied by N, the number of marbles, is the fractional part of the total opening occupied by all of the target sizes (assuming no overlapping) of all the marbles. To continue with our same example, suppose there are 10 marbles placed in the 60-centimeter target area with no overlapping. Then we have

$$\frac{2(R + r)N}{d} = \frac{10}{30} = \frac{1}{3}$$

The target sizes of all 10 marbles would occupy $\frac{1}{3}$ the width of the total opening. If one marble were rolled 30 times, without aiming, at the target area, we would expect 10 hits and 20 misses.

If the ratio $2(R + r)N/d$ were one, we would be absolutely sure of a hit, because the entire opening would be covered with the target sizes of all the marbles. If that ratio were one-half, we would expect one hit for every two rolls of the bombarding marble, because one-half of the opening would be covered with the target sizes of all the marbles. Consequently, the proportion does describe the probability of scoring a hit and, in fact, it can be turned into an equation. (The constant of proportionality is taken to be 1.)

If our reasoning is correct, the two quantities that are each equal to P:

$$P = \frac{H}{T}$$

$$P = \frac{2(R + r)N}{d}$$

should be equal to each other:

$$2(R + r)\frac{N}{d} = \frac{H}{T}$$

We can now calculate $(R + r)$ by determining the values of T, H, N, and d, and solving the equation for $(R + r)$. Dividing both sides of this equation by 2 and by N, and then multiplying both sides by d, we obtain

$$(R + r) = \frac{dH}{2NT} \tag{13-1}$$

Remember that the purpose here is to simulate Rutherford's scattering experiment. Therefore we must pretend that, like Rutherford, we cannot measure the radius of either the bombarding or the target marbles. However, after our experiment we can check our results by directly measuring the diameters and thus the radii of the marbles. Rutherford was not permitted such a luxury, nor has any scientist since his time been able to directly measure the diameter of a nucleus. The radii of atomic nuclei have been measured indirectly by other experiments, and the results are consistent with Rutherford's.

b. Performance of the Experiment

You will be given nine target marbles and one additional marble of the same size with which to bombard the target marbles. You will also be given a rectangular masonite board about 15 cm × 60 cm. This board is smooth on one side and can be used as an inclined plane to roll the bombarding marble down. If its length (about 60 cm) is placed across the width of the target opening, you can arrange eight of the target marbles inside of a three-sided enclosure (Fig. 13–7). The two long edges of the board will both be horizontal; the two short edges should make an angle of about 30° with the horizontal. A book can be used to prop up the high side.

Repeatedly roll the bombarding marble down the incline toward the target marbles, but make sure that you do not aim it at either a marble or a space between two marbles. In fact, you might cover the target area so that you can't see the marbles, and then place the bombarding marble at various places along the top of the inclined masonite board. After a large number of rolls, it will have rolled down and entered the target area at many different places across its 60-centimeter opening.

Remember that if the bombarding marble collides with any number of target marbles before hitting the back wall, you have one hit. Don't count any collision after the marble bounces off the back wall. The target marbles may be placed in any random position, but no target sizes of any two marbles should overlap. To achieve this, some of the target marbles probably will have to be rearranged after each hit. Attempt to roll the bombarding marble parallel to the sides of the target area, and also roll it randomly.

The results improve with the number of rolls and may be poor for fewer than 200. However, experience indicates that accuracy does not improve significantly beyond 300 rolls.

Keep an accurate count of the number of rolls T and the number of hits H. Record them in tabular form in your laboratory notebook, along with the number of target marbles N and the measured distance d.

You can now calculate the diameter D of the target marbles by applying Equation 13–1. Since the target and bombarding marbles all have the same diameter, $(R + r)$ simply equals the diameter of any one of them.

As a check on your determination of the diameter of a target marble, place 10 marbles in the center groove of an open book and, using a meter stick, measure the length of 10 diameters. Now calculate the length of one diameter.

Question 13–6   What is the numerical difference between the radius of a target marble as determined by Equation 13–1 and as measured by using the meter stick? What percent of the radius measured by the meter stick is this numerical difference?

Question 13–7   Which method do you think yields the better result?

Question 13–8   List as many reasons (sources of error) as you can to explain why the radii found by the two different methods are not exactly the same.

Question 13–9   In what ways does the situation in this experiment differ from that in which alpha particles were used to bombard the gold foil target?

Question 13–10   Would the probability that a collision will take place increase or decrease if some of the target marbles were too close together to let the bombarding marble go through? Explain the reason for your answer.

Question 13–11   What is the total range of possible results for the probability of scoring a hit?

Now that you have seen that the sizes of objects may be measured indirectly, let us return to some questions about the atoms used in Rutherford's experiment, and also ask some questions that occur to us in considering the Rutherford model of this atom.

Question 13–12   If a gold atom has a diameter of 2.8 Å, how many layers of gold atoms are needed to make a foil 1.1×10^{-5} cm thick?

Question 13–13   The mathematical theory Rutherford developed did not permit him to state for sure whether the nucleus of an atom is positive or negative. What evidence supports the idea that the nucleus is positively charged?

Question 13–14   Compute the density of the gold nucleus, assuming its diameter is 2×10^{-12} cm and its mass is 3.3×10^{-22} grams. Assume the nucleus is a sphere with a volume of

$(4/3)\pi r^3$. Compare the density of the nucleus and the density of water.

The atomic model presented thus far is incomplete. It postulates that all the positive charge and nearly all the mass of the atom are concentrated in a nucleus having a diameter of the order of 10^{-14} meter. The negative charge, in the form of electrons, is somewhere within the atom. But there are a number of unanswered questions for which a satisfactory model should provide answers. Do all atoms contain the same number of electrons? If many electrons are present in an atom, what is their arrangement? Because electrons are attracted by the positive nucleus, do they fall into the nucleus unless they are moving? Do electrons move about the nucleus as planets move about the Sun? What keeps the nucleus—and, for that matter, the electron—from flying apart?

13-3   *Atomic mass*

The experimental work of Rutherford and his colleagues resulted in the proposal and ready acceptance of the nuclear atom as a useful though incomplete atomic model, and the scattering experiment appeared to be the deciding factor in its acceptance. However, since this experiment gives no information concerning the actual number, arrangement, and motion of electrons, evidence was sought from other sources. Studies of chemical reactions and of spectra provided such evidence, and so we now look in these directions.

Recall that Dalton, at the beginning of the nineteenth century, extended the concept that matter is composed of atoms. He and others were aware that when two substances combine to form a third, the mass of this new substance equals the sum of the masses of the original substances. For example, 1 gram of hydrogen gas combines with 8 grams of oxygen gas to form 9 grams of water.

Dalton reasoned that since the masses of hydrogen and oxygen are equal to that of water, mass is conserved. Furthermore, since the atoms were supposedly indestructible, the masses of individual atoms must also remain unchanged. Such arguments about atoms resulted in a great deal of interest in the problem of determining their masses.

Dalton and others believed that individual atoms could not be weighed because they were so tiny that a balance sensitive enough to detect the weight of a single atom could not be made. Nevertheless, they were able to determine the *relative masses* of atoms.

At this point we need to differentiate between the language of chemical reactions on the macroscopic and on the microscopic level. On the macroscopic level, *chemical elements* are the simplest substances; these elements can combine to form more complex substances called *chemical compounds*. On the microscopic level, *atoms* can combine to form *molecules*. For example, macroscopically we know that 1 gram of the element hydrogen combines with 8 grams of the element oxygen to form 9 grams of the compound water. Microscopically, we know that 2 atoms of hydrogen combine with 1 atom of oxygen to form 1 molecule of water. On the macroscopic scale, the *formula* H_2O represents a definite quantity of water (that quantity, a mole, is defined in Section 14–3); on the microscopic scale, H_2O represents 1 molecule of water.

We can say that for the number N of water molecules that have a combined mass of 9 grams, that same number N of oxygen atoms have a total mass of 8 grams, and $2N$ hydrogen atoms have a mass of 1 gram. The mass of one oxygen atom, therefore, is given by:

$$\text{mass of 1 oxygen atom} = \frac{8 \text{ grams}}{N \text{ atoms}}$$

The mass of one hydrogen atom, correspondingly, is given by:

$$\text{mass of 1 hydrogen atom} = \frac{1 \text{ gram}}{2N \text{ atoms}}$$

The mass of the oxygen atom relative to the mass of one hydrogen atom is called the *relative mass* of oxygen, and is given by the ratio of the mass of one oxygen atom to the mass of one hydrogen atom, or

$$\text{relative mass of oxygen} = \frac{\dfrac{8 \text{ grams}}{N \text{ atoms}}}{\dfrac{1 \text{ gram}}{2N \text{ atoms}}} = \frac{16}{1}$$

The relative mass gives us the mass of 1 oxygen atom relative to the mass of 1 hydrogen atom. It has no units such as grams or kilograms because the units cancel. Hydrogen is the simplest and least massive of all the atoms, so we can arbitrarily assign it an atomic mass of 1 unit. Therefore, the *atomic mass* of any atom will be equal to its mass relative to that of the hydrogen atom, that is, its relative mass. For example, in the foregoing calculation we see that the atomic mass of oxygen is 16. For historical reasons, atomic mass is often called atomic weight.

Knowing the atomic masses of atoms of all the elements, and knowing the number of atoms of each element that combine to form 1 molecule of a compound, we ought to be able to predict how much of one element will combine with a given mass of another to form that compound. It was this hope that prompted much of the early interest in determining atomic masses.

EXPERIMENT 13-2   Determination of the atomic mass of magnesium

Historically, the determination of atomic masses was accomplished in much the same way that you will go about finding the atomic mass of magnesium in this experiment. When a sample of magnesium is heated, it combines with the oxygen in the air to form magnesium oxide. One atom of magnesium combines with 1 atom of oxygen to form 1 molecule of magnesium oxide. The experimenters who first determined the atomic mass of magnesium did not, of course, know the formula of magnesium oxide (MgO). We need not concern ourselves now with the details of arriving at that formula, but we will make use of the information it contains to determine the atomic mass of magnesium.

If you know that the atomic mass of oxygen is 16, how can you determine the atomic mass of magnesium? Clearly, the mass of magnesium in grams that combines with 16 grams of oxygen is equivalent in number to the atomic mass of magnesium. Proceed as follows.

Step One   Dry Crucible and lid. Obtain a clean crucible and lid, and arrange the apparatus as shown in Fig. 13–10.

FIGURE 13-10
*Apparatus for atomic
weight determination.*

A triangle is used to support the crucible. Strongly heat the crucible and lid with a Bunsen burner for 5 to 10 minutes to drive off traces of moisture, and then allow them to cool to room temperature.

Step Two   Determine mass of crucible and lid. Before the mass of the crucible and lid can be determined you will have to balance the beam of your equal arm balance. To do this, place one of the paper cups supplied with the balance in the center of each pan. The mass of the paper cups varies, so they cannot be used interchangeably. Make sure you mark them so that one always stays on the left and the other on the right. Remove the rider from the right arm of the beam, and move the rider on the left arm so that the pointer on the beam will swing an equal number of divisions on each side of the center of the pointer scale. The beam is now balanced. Crimp the rider on the left arm so that it does not move during the experiment and interfere with your measurements.

You can make surprisingly accurate measurements of mass with this equal-arm balance if you calibrate the arm that has been so carefully balanced. Use the other rider (or a piece of paper clip 4.5 centimeters long) on the right side of the arm for calibration.

Place a bead in the left pan and move the second rider on the right side of the arm until the equal-arm balance is again balanced. The distance from the center of the arm to where the movable rider now rests corresponds to the mass of one bead. If that distance is divided up into 10 equal divisions, each division when marked will correspond to a mass one-tenth that of a bead. The rider placed at any one of these marks will indicate a mass only a fraction of a bead. For example, suppose a pebble is placed in the left pan, and 11 beads in the right pan nearly balance the pebble, but not quite. A more accurate mass of the pebble can be obtained by moving the right rider until balance is achieved. If it rests on the 6th mark from the center of the arm, that pebble has a mass 11.6 times the mass of one bead.

Place the crucible and lid on the left pan. Add strips of beads and single beads to the cup on the right pan until it almost balances the crucible and lid. The beam should be close enough to the balancing point so that one additional bead would make the right side of the beam too heavy. Now move the rider on the right arm until the beam is balanced. Record in your notebook the number of beads used and the position of the right hand rider needed to obtain the balanced beam. Lable this entry "mass of crucible and lid"; it will be needed later.

Step Three   Determine mass of magnesium. Next obtain a length of magnesium ribbon (about 20 cm) and coil it tightly around a pencil. Then spread the coil into the shape of a cone, but keep it small enough to fit completely into the crucible. The coil is spread to permit good contact of the air with the magnesium. Determine the mass of the crucible, lid, and contents using the procedure outlined in Step Two. (Make sure you use the same cups on the same pans throughout the experiment.) Record this mass as you did the other. By subtraction, determine the mass of the magnesium and record it.

Step Four   Heat magnesium to a constant mass.

(a) Place the covered crucible on the triangle and heat it strongly for 20 min.

(b) Use forceps to momentarily lift one side of the lid *very slightly* to allow more air to enter the crucible. *Do not remove the lid under any circumstances.* Hot magnesium burns violently in air. If the cover is removed, white smoke consisting of magnesium oxide will escape. Why would this be undesirable? Continue heating for about 5 min more, lifting the lid slightly from time to time (at least 5 times) as described. Remove the burner. Now, using forceps, place the lid slightly ajar so that you can see a small crack between the top edge of the crucible and the lip of the lid; leave it in this position. Replace the burner and strongly heat the crucible for another 20 min.

(c) Remove the burner, allow the crucible to cool to room temperature, and then determine the mass of the crucible, lid, and contents.

(d) Replace the crucible over the burner, set the lid ajar, and heat it again for 5 min.

(e) Cool it, and determine the mass as before.

(f) Repeat steps *d* and *e* until two successive heatings give the same mass within one-tenth of a bead. You may then assume that all of the magnesium has been burned to magnesium oxide. Record the final mass of the magnesium oxide.

Arrange all your data in a neat table. Determine the mass of oxygen that combined with the magnesium, and calculate the ratio of the mass of magnesium to the mass of oxygen. Then, use this ratio to calculate the mass of magnesium required to combine with 16 mass units of oxygen. In this experiment each mass will be expressed in units of beads. The ratio of masses has no dimensions; a number of beads divided by another number of beads is a pure number. Your answer is your determination of the atomic mass of magnesium.

Question 13–15   If you spilled some magnesium oxide before determining the mass of the crucible and its contents after heating, would the atomic mass of magnesium that you determined be too high or too low? Justify your answer.

Question 13–16   List possible sources of error in this experiment.

Question 13–17   Suppose the original strip of magnesium contained some impurities that did not react with oxygen and were not driven off during the heating process. Would the atomic mass of magnesium be too high or too low? Justify your answer.

As atomic masses became known, it became possible to arrange the elements in order of increasing atomic mass. When they were so arranged, each element was given a number that indicated its position in the list. This is called the *atomic number*. The first twenty elements in this list are given in Table 13–1.

Table 13–1   Elements with Atomic Numbers 1 through 20

Element	Symbol	Atomic Mass	Atomic Number
Hydrogen	H	1.0	1
Helium	He	4.0	2
Lithium	Li	6.9	3
Beryllium	Be	9.0	4
Boron	B	10.8	5
Carbon	C	12.0	6
Nitrogen	N	14.0	7
Oxygen	O	16.0	8
Fluorine	F	19.0	9
Neon	Ne	20.2	10
Sodium	Na	23.0	11
Magnesium	Mg	24.3	12
Aluminum	Al	27.0	13
Silicon	Si	28.1	14
Phosphorus	P	31.0	15
Sulfur	S	32.1	16
Chlorine	Cl	35.5	17
Argon	Ar	40.0	18
Potassium	K	39.1	19
Calcium	Ca	40.1	20

You may wonder how the symbols were chosen. In general, the symbol for an element is the capitalized first letter of its name (e.g., H for hydrogen, N for nitrogen). If the letter has already been taken, two letters may be used, in which case the second is not capitalized (e.g., He stands for helium, Ne for neon). Some of the symbols are derived from the Latin names of the elements. The symbol Na for sodium comes from its Latin name, natrium; the symbol K for potassium comes from its Latin name, kalium.

Chemists of the 19th century were not only busy determining atomic masses, but were also investigating other properties of the elements. In the process of classifying them by properties (as you did with some substances in Experiment 1–3), they discovered that some had properties similar to those of others. For example, the elements lithium, sodium, and potassium make up one group having similar properties. One property they all share is their ability to react vigorously with water.

Count the number of elements in the list between lithium and sodium, and again between sodium and potassium. What have you discovered? The early chemists also discovered that elements with similar properties occur periodically when all the elements are considered in order of increasing atomic number. Their arrangement in a table to exhibit this periodicity is called the Periodic Table of Elements. A version of the modern Periodic Table is included in Appendix C.

Let us look briefly at the way the elements are arranged in the Periodic Table. It is designed to show both the periodic repetition of properties and the gradual change in properties as atomic number increases. For example, similarities in behavior and properties of lithium, sodium, and potassium have already been mentioned. Fluorine (atomic number 9), chlorine (17), bromine (35), iodine (53), and astatine (85) are shown in the same vertical column in the Periodic Table. These elements form compounds with similar properties when combined with the same element, for example, sodium. Because these elements have similar properties, they are said to be in the same family. Another example of periodic recurrence of properties is illustrated by a group of gases: helium (2), neon (10), argon (18), krypton (36), xenon (54), and radon (86). One property shared by these elements is that they do not readily react with other elements to form compounds.

Question 13–18 What is the difference between the atomic numbers of: (a) helium and neon, (b) neon and argon, (c) fluorine and chlorine?

Question 13–19 What is the difference between the atomic numbers of: (a) chlorine and bromine, (b) bromine and iodine, (c) argon and krypton?

Question 13–20 Use the Periodic Table to find at least two places where the sequence of atomic numbers is different from a sequence based on increasing atomic mass.

This periodic arrangement of elements is strictly on the basis of atomic number and similarities of properties. We would like to understand why it is that elements with similar properties occur at intervals in the Periodic Table. First we have to determine how, according to our atomic model, one element differs from another. Assume that the hydrogen nucleus has one positive charge and, therefore, the hydrogen atom has one electron (see Section 13–1); and that the helium nucleus (the alpha particle) has twice the charge of a hydrogen nucleus and, therefore, the helium atom has two electrons. Could it be that the charge on the nucleus, and therefore the number of electrons in the atom, is the property that differentiates an atom of one element from an atom of another? In fact, could it be that the number of electrons in an atom is equal to the atomic number of the element? More evidence is necessary to answer these questions.

13–4   *Spectra and energy levels*

The key that unlocked a good many of the secrets of the atom was the light those atoms emitted.

Although the first observations were made in the eighteenth century, a number of observers in the first part of the nineteenth century described what has come to be called a *bright-line spectrum*. One of these descriptions was made by the German physicist, Joseph Fraunhofer, who reported seeing two bright yellow lines when a flame was viewed through a prism. By performing Experiment 13–3 you can become better acquainted with the bright-line spectrum and the way in which it can be produced.

EXPERIMENT 13–3   Observation of various spectra

You may recall from Chapter 4 that a diffraction grating separates white light into its component colors, that is, into beams that are spread out at various angles according to their wavelengths. If the light source is broad, the spectral colors produced by a diffraction grating may overlap. It is for this reason that we usually restrict the light beam by passing it through a slit before it reaches the grating. Some light sources emit light of only a few colors (wavelengths). The spectrum observed from these sources, a bright-line spectrum, consists of several separate lines, each of a different color. However, some light sources emit white light comprising all wavelengths, which the diffraction grating spreads out into a spectrum with no gaps. This spectrum is called a *continuous spectrum.*

Your cardboard-tube spectroscope is provided with a diffraction grating at one end and a slit at the other. In order for the spectrum to be clearly visible, the fine lines of the diffraction grating must be aligned with the slit. You cannot see the lines of the diffraction grating, so the best way to achieve this alignment is as follows. With the grating end near your eye, and the slit at the far end in a vertical position, look through the spectroscope at a sunlit object or a bright light bulb. Do not look directly at the sun because it might injure your eyes. While looking at the light, slowly turn the black cylindrical piece that holds the diffraction grating until the colors of the continuous spectrum take a rectangular shape along the inner wall of the tube to the left and right of the slit. With the slit vertical, the colors should appear spread out horizontally. Once a satisfactory spectrum has been attained, the black rim that holds the grating should be left in position and not rotated again. You will observe a number of different spectra with the spectroscope. It will be easier to compare them with each other if you always have the slit in the same orientation.

With your spectroscope you can analyze not only the light generated by a source such as a lamp, but also the fainter light reflected by various other objects. You will find some surprises as you examine the spectra of variously colored objects in bright sunlight. In Experiment 3–2 you observed through colored filters the light coming from colored marks on white paper. What does the spectrum of white light look like after the light has passed through such a filter?

Some possible sources of light to examine are:

1. An incandescent source such as a light bulb.
2. A neon sign.
3. Other light sources of various colors.
4. Fluorescent lights.
5. A candle flame.

Answer the following questions for some of the more interesting sources that you observed.

Question 13–21   Describe the source.

Question 13–22   Where does the light source get its energy?

Question 13–23   Describe the spectrum observed and give the sequence of colors and their relative intensities.

Question 13–24   What is the predominant color (or colors) in the spectrum, and how does this compare with the color of the source when viewed without the diffraction grating?

You have observed various kinds of spectra. Some of them were bright-line spectra; some were like the simple rainbow, the continuous spectrum. You also found that different substances produce different bright-line spectra. The spectrum of sodium chloride has a bright yellow line (actually two lines very close together), neon has many red lines, etc. Were you to proceed systematically through all of the elements, observing the bright-line spectrum of each, you would find that each has its own individual and unique bright-line spectrum. The uniqueness of each bright-line spectrum provides a means of identifying elements and is used by chemists for just this purpose. And it was the uniqueness of the bright-line spectrum of each element that permitted chemists and physicists to learn details of the electron structure within the atom and how that structure is related to the chemical properties of the element. Because the hydrogen atom has only one electron, its spectrum is the easiest to interpret. The portion of the spectrum of hydrogen first analyzed successfully is shown in Figure

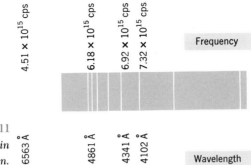

FIGURE 13-11

A series of spectral lines in the spectrum of hydrogen.

13–11. This analysis revealed a simple mathematical relationship among the wavelengths of the lines.

The explanation of bright-line spectra required new ideas about the way matter emits and absorbs energy, and these ideas in turn helped to connect a number of observations. The initial contribution to these new ideas was provided by Max Planck in 1900. Planck guessed that electromagnetic radiation (including light) is emitted and absorbed in energy packets, which can be thought of as particles.

Then in 1905, Einstein extended the energy-packet idea to explain the way metals emit electrons when light shines on them. This extension included the concept that light is composed of particles called *photons*. The idea that light is composed of particles is an apparent contradiction to the wave model of light discussed in Chapters 3 and 4. The wave model and the particle model describe different aspects of the behavior of light. These two models were found to be related by the expression

$$E = hf \tag{13–2}$$

The quantity E specifies the energy carried by the particle; the quantity f specifies the frequency of the wave. The quantity h is the constant of proportionality and is called Planck's constant. The value of this constant has been determined experimentally. In the mks system of units the value of h is 6.6×10^{-34} joule-sec.

In 1913 Niels Bohr used the energy-packet idea to explain the bright-line spectrum of hydrogen. He postulated that the hydrogen atom can exist with only certain discrete amounts of energy. From this model Bohr was able to calculate the amount of energy for each discrete state of the hydrogen atom. These states are called *energy levels* and can be represented by a diagram such as the one shown in Figure 13–12.

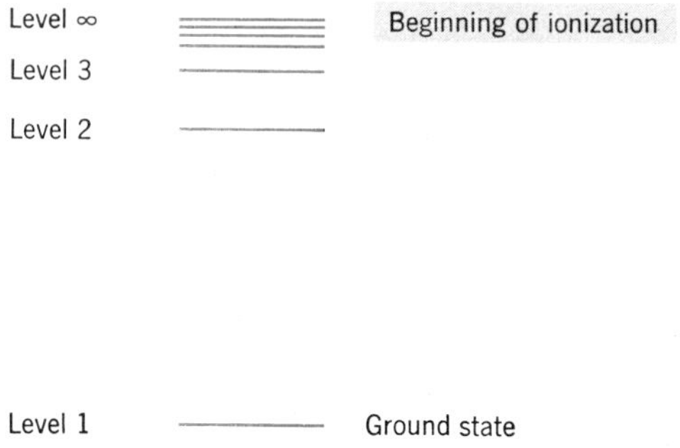

Level ∞   ═══════   Beginning of ionization

Level 3   ───────

Level 2   ───────

FIGURE 13-12

Energy-level diagram for
the hydrogen atom.   Level 1   ───────   Ground state

It may help you to think of the energy levels of an atom as analogous to the possible values of the gravitational potential energy of a cup on the various shelves of a cupboard. The energy of the cup depends on the height of the shelf on which the cup rests. It cannot rest between shelves. If it is lifted from one shelf to a higher one, its energy will increase by a definite amount that must be supplied by an external agent—you, for example.

Bohr postulated that the hydrogen atom can change to any higher energy level by absorbing just the right amount of energy. However, it will not remain at this higher energy level, but will spontaneously emit just the right amount of energy in the form of a photon of light by changing from one energy level to any lower level. These photons have certain discrete energies and account for the bright lines of the hydrogen spectrum. The connection between the energy of the atom before and after emission, the energy of the photon, and the frequencies of the emitted light is given by the equation

$$(E_{before} - E_{after})_{atom} = E_{photon} = hf_{wave}$$

This equation tells us that an atom loses energy when it emits light. In Experiment 13–3 you raised the atoms to higher energy levels by heating the substance with a flame or by supplying electrical energy. The photons emitted as the atoms returned to lower energy levels produced the spectra you observed. Each element has its own characteristic energy levels and consequently its own characteristic spectrum.

13-5   *Removing electrons from atoms*

It is possible to give an atom enough energy to remove an electron entirely; then the atom becomes an ion. In fact, with the exception of hydrogen, if you give an atom enough energy you can remove more than one electron. The amount of energy that is required to remove an electron from an atom is not the same for all the different kinds of atoms or for all the electrons in the same atom. The amount of energy involved, however, is not of the same order of magnitude as the energies we've discussed so far. The unit we use for energy is the joule, but the energy emitted by an atom when one of its electrons changes from one energy level to another of lower energy is so much less than 1 joule that it will be convenient to use another, smaller unit.

In Chapter 12 you learned that a particle of charge Q can have work $W_{a \to b}$ done on it by a difference of electrical potential $V_{a \to b}$:

$$W_{a \to b} = Q V_{a \to b}$$

If the particle moves freely, that is, without bumping into other particles, that work increases the particle's kinetic energy by an amount,

$$\Delta KE = W_{a \to b}$$

Therefore the change in kinetic energy can be related to the charge and difference of potential:

$$\Delta KE = Q V_{a \to b} \tag{13-3}$$

Suppose an electron, initially at rest and with a charge $Q = 1.6 \times 10^{-19}$ coulomb, is caused to move in a vacuum through a difference in potential of 1.0 volt. We can readily calculate its kinetic energy by Equation 13–3:

$$KE = QV_{a \to b}$$

$$KE = (1.6 \times 10^{-19} \text{ coulomb})\left(1.0 \ \frac{\text{joule}}{\text{coulomb}}\right)$$

$$KE = 1.6 \times 10^{-19} \text{ joule}$$

In a discussion of atomic reactions such as ionization, energies of the order of 10^{-19} joule frequently occur. Therefore, it is convenient to define a new unit of energy, the *electron volt* (abbreviated eV), which is the change in kinetic energy of an electron when it moves unimpeded through an electrical potential difference of 1.0 volt. From the foregoing calculation we see that

$$1 \text{ eV} = 1.6 \times 10^{-19} \text{ joule}$$

Question 13–25 The energy required to remove the electron from a hydrogen atom is 13.6 eV. How many joules is this?

Table 13–2 lists the energy, in electron volts, necessary to remove the least tightly bound electron from an atom of each of the elements of atomic numbers 1 through 20. This is called the *ionization energy.*

Question 13–26 Plot the data of Table 13–2 with atomic number along the abscissa and ionization energy along the ordinate. What is the difference between the atomic numbers of successive peaks? Compare this answer with your answer to Question 13–18.

Question 13–27 Chemical families have been mentioned. In which family (list the members) do the members have high ionization energies? In which family do the members have low ionization energies?

From Table 13–2 we see that it is easier to remove an electron from some atoms than from others. The force holding the electrons to the nucleus is electrical, and this force decreases with increasing separation of the electrons from the nucleus. Therefore, the electrons that are easy to remove are outermost. Consequently we can offer a reasonable hypothesis about the location of the least tightly bound electron in an atom if we know how difficult it is to remove

Table 13–2

Atomic Number	Element	Symbol	Ionization Energy (eV)
1	Hydrogen	H	13.6
2	Helium	He	24.6
3	Lithium	Li	5.4
4	Beryllium	Be	9.3
5	Boron	B	8.3
6	Carbon	C	11.3
7	Nitrogen	N	14.5
8	Oxygen	O	13.6
9	Fluorine	F	17.4
10	Neon	Ne	21.6
11	Sodium	Na	5.1
12	Magnesium	Mg	7.6
13	Aluminum	Al	6.0
14	Silicon	Si	8.1
15	Phosphorus	P	10.5
16	Sulfur	S	10.4
17	Chlorine	Cl	13.0
18	Argon	Ar	15.8
19	Potassium	K	4.3
20	Calcium	Ca	6.1

that electron from an atom. This suggests that we might learn more about the arrangements of all the electrons in an atom by finding out how hard it is to remove additional electrons, one after another. The experimental values for the energies needed to remove successive electrons from atoms of elements with atomic numbers 1 through 20 are listed in Table 13–3.

The photograph that opens this chapter is the result of a new technique and shows atoms at the tip of a metal needle with a very sharp point. The needle is enclosed in one end of a tube with a fluorescent screen at the other end, and a vacuum in between. A high positive charge is placed on the needle and helium atoms are fired at the needle's tip. As each helium atom strikes the tip it loses one of its electrons to the positively charged needle. The helium ion is then repelled by the needle and follows a straight-line path to the fluorescent screen where it produces a scintillation. A helium atom loses an electron to a metal atom; many helium ions leaving the tip—each from a particular metal atom—will produce many scintillations on the screen.

Table 13–3   Energy (*eV*) Required to Remove Successive Electrons from Atoms of Elements 1 through 20

Atomic Number	Element	Symbol	First Electron	Second Electron	Third Electron	Fourth Electron	Fifth Electron	Sixth Electron
1	Hydrogen	H	13.6					
2	Helium	He	24.6	54.4				
3	Lithium	Li	5.4	75.6	122.5			
4	Beryllium	Be	9.3	18.2	153.9	217.7		
5	Boron	B	8.3	25.2	37.9	259.4	340.2	
6	Carbon	C	11.3	24.4	47.9	64.5	392.1	490.0
7	Nitrogen	N	14.5	29.6	47.4	77.5	97.9	552.1
8	Oxygen	O	13.6	35.1	54.9	77.4	113.9	138.1
9	Fluorine	F	17.4	35.0	62.7	87.1	114.2	157.2
10	Neon	Ne	21.6	41.1	63.4	97.1	126.2	157.9
11	Sodium	Na	5.1	47.3	71.7	98.9	138.4	172.1
12	Magnesium	Mg	7.6	15.0	80.1	109.3	141.3	186.5
13	Aluminum	Al	6.0	18.8	28.4	120.0	153.8	190.5
14	Silicon	Si	8.2	16.3	33.5	45.1	166.8	205.2
15	Phosphorous	P	10.5	19.7	30.2	51.5	65.0	220.5
16	Sulfur	S	10.4	23.4	35.0	47.3	72.7	88.0
17	Chlorine	Cl	13.0	23.8	39.9	53.5	67.6	97.0
18	Argon	Ar	15.8	27.6	40.9	59.7	75.2	91.2
19	Potassium	K	4.3	31.7	45.8	61.1	82.7	100.0
20	Calcium	Ca	6.1	11.9	51.2	67.3	84.5	108.8

But each helium ion that leaves a given metal atom will strike the screen in the same place. Consequently, each bright spot on the screen represents a particular atom in the tip of the metal needle. That the metal is composed of atoms spaced in a regular crystalline lattice is clear.

13-6   *The shell model and cloud model of the atom*

Can we discover any pattern in the data in Table 13–3? Clearly, lithium holds one electron much less strongly than it holds the other two. Examination of elements numbered 4, 5, and 6 shows the beginnings of a pattern. From beryllium, two of the four electrons are easily removed; from boron, three of the five; from carbon, four of the six. Two are always held very tightly.

Now examine the data for atoms numbered 11, 12, 13, 19, and 20. From sodium, one of the eleven electrons is easily removed; from magnesium, two of the twelve; from aluminum, three of the thirteen. In these elements, ten electrons are held very tightly. From potassium, one of the nineteen electrons is easy to remove; from calcium, two of the twenty. In these two elements, eighteen electrons are held very tightly.

Observations summarized in these two paragraphs lead to the following conclusions. In atoms with more than two electrons, two are held more tightly to the nucleus. In atoms with more than ten electrons, two are held very tightly, eight somewhat less tightly, and the remaining ones still less tightly. In atoms with more than eighteen electrons, two are held extremely tightly, eight somewhat less tightly, another eight still less tightly, and the remaining electrons are relatively easy to remove. These groups of electrons about a nucleus have been likened to *shells;* the first shell can hold two electrons; the second, eight, etc. These conclusions are summarized in Table 13–4, which also illustrates the hypothesis that two electrons form a tightly bound arrangement within atoms, and eight electrons form another tightly bound arrangement. This hypothesis forms the basis of the *shell model* of the atom. The word shell is

Table 13–4   Arrangement of Electrons Within Atoms

Symbol	Number of Electrons	First Shell	Second Shell	Third Shell	Fourth Shell
H	1	1			
He	2	2			
Li	3	2	1		
Be	4	2	2		
B	5	2	3		
C	6	2	4		
N	7	2	5		
O	8	2	6		
F	9	2	7		
Ne	10	2	8		
Na	11	2	8	1	
Mg	12	2	8	2	
Al	13	2	8	3	
Si	14	2	8	4	
P	15	2	8	5	
S	16	2	8	6	
Cl	17	2	8	7	
Ar	18	2	8	8	
K	19	2	8	8	1
Ca	20	2	8	8	2

used to designate a group of electrons within an atom which are all comparably bound to the nucleus of that atom. Apparently there is a limit to the number of electrons in a given shell. The first shell is filled when it contains two electrons; the second can hold no more than eight electrons; and higher shells also have limits.

We have arrived at the shell model from a consideration of ionization energies. Does this model help us account for other properties of elements, such as their chemical behavior? The shell hypothesis has been useful in the prediction of chemical properties of many elements and helps to account for some differences and similarities among elements. In Chapter 14 we will use this model to help us understand the chemical reactions by which elements combine to form compounds.

Let us review our atomic model as we have developed it so far. The Rutherford scattering experiment established that the positive charge is localized in a very small region, called the nucleus, which contains most of the mass of the atom. The negative charge of an atom is associated with particles called electrons, which can be removed from the atom one at a time, leaving behind a positive ion. The energies required to remove successive electrons exhibit a periodicity that has led us to propose that the electrons within an atom are grouped in shells.

At this point, it is natural to ask where the shells within the atom are, and where the electrons within a shell are. These questions cannot be answered precisely. Even though we don't know exactly where the electron is at any particular instant, we can say where it spends most of its time and therefore where it is most apt to be at any given time.

To help visualize the uncertainty of locating an electron, consider the following analogue. Imagine that we use a rubber rope to tie a very active white dog to a post in a field of dark green grass. High above the post we suspend a camera with a very insensitive film so that we can leave the shutter open and take a long time exposure. When we develop the film, what will we see? A blurred image that might be called a "dog cloud."

What can we learn from this dog cloud? We can assume that the dog spent most of his time where the cloud is densest. Notice that the cloud will not tell us about the path along which the dog moves, nor where we could be sure to find him at any particular instant; but it will tell us where he spent most of his time, and therefore where he was most apt to be at any given time.

The probability of locating the dog is analogous to the probability of locating the electron in a hydrogen atom. Figure 13–13 shows the probability of finding the electron

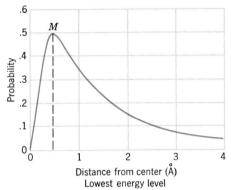

FIGURE 13-13

Probability of finding the electron at a given distance from the center of a hydrogen atom.

at a given distance from the center of the hydrogen atom when the atom is in its lowest energy level (see Fig. 13–12). This curve was obtained from a mathematical theory that correctly predicts the behavior of atoms. The maximum on the curve, labeled M, predicts the most probable distance of the electron from the nucleus. For the hydrogen atom this distance is about 0.529 Å. This figure refers to the atom at its lowest energy, which is known as the ground state. The probability curves for hydrogen atoms in higher levels, and for other atoms, are different. Evidence that the shape of the electron cloud within atoms has a great deal to do with the way atoms bond together to form molecules and crystals is presented in Chapters 16 and 17.

QUESTIONS   13-28   What do scientists mean when they speak of a model? What function does a model serve in science?

13-29   What evidence did Thomson have for the existence of electrons as constituent parts of all atoms?

13-30   How many electrons does the sodium atom have? If one electron is removed from a sodium atom, is it still sodium?

13-31   How was it decided that hydrogen should be element number 1?

13-32   An electron "falls through" 50 volts. What is its kinetic energy? Express your answer in joules and in electron volts.

13-33   Consider the hypothesis "The charge carriers responsible for conduction in metals are electrons." What evidence has been presented in Chapter 13 in support of this hypothesis?

References   1. Baez, A. V., *The New College Physics*, W. H. Freeman & Co., 1967, Chapter 48, Sections 48–1 through 48–3, pages 603 to 609.
   Rutherford's experiment.
2. Chem Study, *Chemistry, an Experimental Science*, W. H. Freeman & Co., 1963, Chapter 15, Sections 15–1 through 15–1.4, pages 252 to 260.
   Energy in the hydrogen atom. These sections contain a simple analogy to illustrate energy levels. A new unit system is used for energy, but if it is remembered that only the *ratios* of the energy values are significant, there should be no trouble.
3. Greenstone, A. W., F. X. Sutman, and L. B. Hollingworth, *Concepts in Chemistry*, Harcourt, Brace & World, New York, 1966.
   Insert facing page 84. Spectra of some elements. Also, 3rd page of insert between pages 548 and 549. Flame color of some elements.
4. Lehrman, R. L., and C. Swartz, *Foundations of Physics*, Holt, Rinehart & Winston, New York, 1965, Chapter 21, pages 565 to 569.
   Models of atoms.

   ALSO OF INTEREST

5. Anderson, D. L., *The Discovery of the Electron*, D. Van Nostrand, Princeton, New Jersey, 1964.

A history of the development of the atomic concept of electricity, the cathode ray, and electrons.

6. Andrade, E. N. da C., "The Birth of the Nuclear Atom," *Scientific American*, November 1956, page 93.

An account of the development of the nuclear atom in the days of Rutherford. It deals not with technical details but with scientific reactions to Rutherford's work.

7. Andrade, E. N. da C., *Rutherford and the Nature of the Atom*, a Doubleday Anchor paperback, Garden City, New York, 1964.

A more extensive treatment of Rutherford's work than that in the article in *Scientific American* (Reference 6). This book is very lucid and describes much of the atmosphere of Rutherford's laboratory from first-hand experience.

Courtesy of Bell Telephone Labs

CHAPTER 14

IONS

*In Experiment 12-3 you investigated the elec-*trical conductivity of a number of substances, both metals and nonmetals. Some of them were tested in the melted state as well as in the solid form. We were interested in finding out which of the substances conducted an electric current, and how well they did so. Since then, you have learned more about electric charges and about matter, and we will now take a closer look at what happens when an electric current passes through a substance. First let us review some of the ideas that may be useful to you in interpreting your observations.

14-1   *A look back and a look forward*

In Chapter 11 evidence was presented that led to the hypothesis that the bonding forces in sodium chloride are electrical. Then you studied various aspects of the behavior of electric charges. An electric current, such as you will use in the next experiment, was defined in Chapter 12 as a flow of electric charges.

In Chapter 13 you learned of a model for atoms that includes a nucleus with a positive electric charge, surrounded in some way by negative electric charges. If the atoms and ions of which matter is composed are endowed with positive and negative electric charge, then the hypothesis that bonding forces between these particles are electrical is strengthened. Indeed, since there is so much evidence in favor of it and none against it, this hypothesis is widely accepted, although the exact way in which electrical forces operate is not clearly understood in all cases.

When this bonding force between the particles that make up a particular substance is strong, the particles are held together rigidly and the substance is solid. If the particles that compose the substance are bonded in a regular and orderly pattern, then the macroscopic shape of the substance reveals a regularity such as the crystalline "whiskers" which appear in the photograph that opens this chapter. With enough thermal agitation, the substance may melt. The particles then tumble about in disorderly array, each one being bonded now to one of its fellows, now to another, so that they readily move past each other. The substance, now a liquid, can be poured from one container to another. In a gas the particles fly about in all directions and are not bonded to their neighbors.

As you do the next experiment then, keep the following things in mind.

(a) An electric current is defined as a flow of electric charges.

(b) The outer part of an atom, the part that would seem most likely to be affected (or changed) by outside forces, is composed of negative charges, although the total charge of the atom is zero.

(c) Since the state of a substance is related to the mobility of the particles of which it is composed, it seems likely that if a given substance conducts an electric current in the molten state but not in the solid state, conductivity is related to the mobility of charged particles.

14-2 *Ions in melts*

For the experiment that follows we want to choose a simple substance that melts at a convenient temperature and conducts electricity only in the molten state. It is not easy to find a suitable one. Those with low melting points, such as zinc chloride, take up water from the air and hold on to it very tightly. A substance that does this is said to be *hygroscopic.* Any water that is present will make the solid difficult to melt without violent spattering, and would also make uncertain the influence of the unremoved water on the results of the experiment. Some other substances combine with oxygen or decompose. Those that have none of these inconvenient characteristics have melting points too high for simple equipment.

The substance we shall use is lead chloride. It decomposes very little, and it is not hygroscopic. A disadvantage is that its melting point is inconveniently high. We can partly overcome this by adding a small amount of sodium chloride. A mixture of one part by weight of sodium chloride to ten parts of lead chloride will melt at 411°C, while the melting point of pure lead chloride is 501°C. You probably have observed a similar lowering of the melting point when salt is added to ice. Salt added to ice on the streets will melt the ice even though the temperature is somewhat below the freezing point of pure water. See also Experiment 7–2, Part B.

EXPERIMENT 14-1   The effect of an electric current on lead chloride

Set up the apparatus as shown in the photograph in Figure 14–1 and in the schematic diagram in Figure 14–2. The clamp with the asbestos covering should be mounted near the edge of the board, with a sheet of asbestos behind the clamp. Fill the U-tube with the chloride mixture, and tamp it down tightly by using the carbon electrode. After the

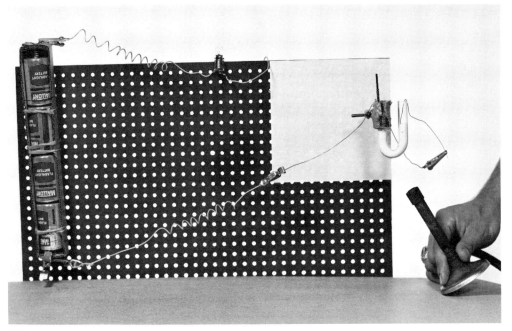

FIGURE 14-1   *Photograph of equipment arrangement for Experiment 14–1.*

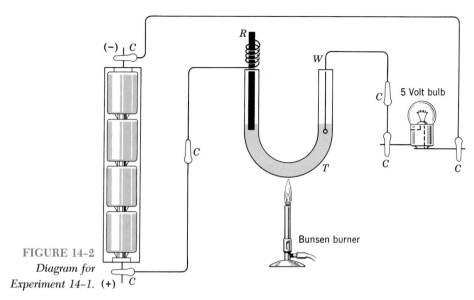

FIGURE 14-2
*Diagram for
Experiment 14–1.*

tamping, the height of the material should be about 2 in. in each arm of the tube. The U-tube is now mounted in the asbestos-covered clamp. With the clamp mounted at the edge of the board, the tube can be heated from both front and rear without burning the board. Heat the tube with a Bunsen burner until all of the chloride mixture is melted. Start heating at the top of the tube, and gradually work down. Move the flame alternately from arm to arm and also from back to front.

After the material has melted, the top level of the melt in the tube will be considerably lower than the original level of the dry mixture. Turn off the burner, and let the U-tube and its contents cool for a few minutes. The melt will quickly solidify. While the U-tube is cooling prepare the electrodes.

The negative electrode is made from a 20-cm piece of No. 22 gage, bare copper wire. A small loop is made at one end of the wire. The diameter of this loop should be no more than 1 mm. The wire (W in Fig. 14–2) is to be bent as shown. The loop end is placed in one arm of the U-tube so that it is about $\frac{1}{4}$ in. above the surface of the solidified mixture. Bend the wire at the top of the tube so that the loop will remain at this depth. Lay the wire electrode aside until later.

The positive electrode R is a carbon rod. Wrap one end of the electrode with 4 or 5 turns of a 15-cm piece of copper wire. The spiral wrap must be tight enough to support the carbon rod. The carbon electrode should extend down into the arm of the U-tube the same distance as the copper electrode. This distance can be obtained by sliding the spiral coil up or down the carbon rod. Connect the clip lead from the positive electrode of the battery to the wire that is wrapped around the carbon rod (see Fig. 14–2). Connect the clip lead from the light bulb to the copper wire electrode. Both clip leads should be as far from the U-tube as possible to avoid damage from heating. Place both electrodes on the table, so that they won't touch each other until you are ready to pass a current through the melt.

Back to filling the U-tube. It should now be cool enough to handle. To each arm of the U-tube add about 2 inches more of the chloride mixture. Tamp it down compactly. Replace the tube in the clamp on the board.

Now heat the U-tube and its contents again, using the same technique as before. The tube should be heated until all of the dry material is completely melted. The resulting liquid will turn light yellow in color because of slight decomposition of the lead chloride, which is not harmful to the experiment. When all the chloride mixture is in the molten state, remove the burner for a few seconds, then quickly place the carbon electrode in one arm of the U-tube and the copper electrode in the other arm. Both electrodes should dip a little below the surface of the liquid. An eighth of an inch is about right. If they do not extend far enough they will have to be removed and their length readjusted. Return the burner, and continue to heat the tube while this adjustment is being made. When the electrodes are the proper length, replace them in the U-tube as before. The burner should be removed only while you are replacing the electrodes.

The current should be permitted to pass through the melt for a minute or two. Continually heat the U-tube during this time. Observe each electrode for any reaction that may be taking place. A light brown liquid may diffuse downward from one of the electrodes. When this liquid reaches the bottom of the arm of the U-tube, remove the burner and stop the current by removing the carbon electrode: grasp the clip on the end of the wire attached to the carbon rod, and lift the rod from the U-tube. Similarly, remove the copper electrode but hold it so that the loop at the end is about $\frac{1}{2}$ in. above the melt for 15 or 20 sec before you remove it completely. Observe each electrode for any changes which may have occurred. If a deposit is found, scrape it with a knife or some other sharp edge. Place the electrode in water to see if the deposit reacts.

Question 14-1 What did the glowing light bulb indicate?

Question 14-2 On which electrode did bubbles of gas appear?

Question 14-3 What could the gas be that formed the bubbles?

Question 14-4 Why do you suppose the melt in one arm of the U-tube turned light brown? Which electrode was in this arm?

Question 14-5 Did the appearance of either electrode change? Describe what you saw.

Now let's review the results, one at a time. The fact that the flashlight bulb did light showed that electric charges were moving through the wires. But there was no wire in the U tube connecting the copper electrode to the carbon electrode. Therefore, we must conclude that charges moved through the liquid. When the charges moved through the wires and the flashlight bulb, there was no evidence of any new product being formed. However, there were new products formed at the electrodes. What, then, was going on in the liquid?

You found a drop of gray metal deposited on the loop of the copper electrode. The metal was soft enough so that some of it could be easily scraped off with a knife. It does not react with either air or water; it was deposited on the negative electrode. There was no metal in the liquid, so the metal must have been formed at the surface of the electrode.

The negative electrode attracts positive charges; therefore the metal there must have resulted from positively charged particles in the liquid. You learned in Chapter 13 that metals form positively charged particles by easily losing one or more electrons, and that these particles are called ions. When they come into contact with the negative electrode, they regain their lost electrons and deposit on the electrode as metal atoms. The melt was composed of only lead chloride and sodium chloride, so the metal we saw on the copper is lead, sodium, or a mixture of both.

To determine the composition of the deposited metal, we must look at the properties of lead and sodium. You may be familiar with some of the properties of metallic lead. It is soft enough to scrape easily, does not react with water, does not corrode in air, and conducts an electric current. This description agrees with the results you obtained when you tested the metal that was deposited on the copper electrode. On the other hand, sodium metal reacts vigorously with water and is so active with the oxygen in the air that it must be kept submerged in a light oil such as kerosene.

Gas bubbles were observed collecting on the carbon electrode; the carbon rod was connected to the positive terminal of the battery; and the gas was not in the liquid to start with. It was only when something came into contact with the positively charged carbon rod that the gas was formed. What can it be? Unfortunately, the volume of gas evolved is not large enough for you to make identifying tests. A faint odor may have been detected near the arm of the U-tube containing the carbon rod immediately after the burner was removed. However, even without testing we can identify the gas by deduction. In the mixture of lead chloride and sodium chloride only three constituents are present: lead, sodium, and chlorine. We have accounted for the metals; therefore the gas must be chlorine.

If we conclude that the gas is chlorine we must consider how it existed in the melt. The gas appeared on the positively charged electrode, which indicates that it was formed from negatively charged particles. To assume a negative charge, the chlorine atom would have to add an electron.

Is this consistent with our shell model of the atom? Referring to Table 13–4, you can see how many electrons are in the outer shell of chlorine and conclude for yourself that it is apt to add one electron. If these chlorine ions were in the melted chloride mixture they would have been attracted to the positively charged carbon rod, and there lose their extra electrons to become chlorine atoms.

A picture of molten lead chloride as made up of lead ions bearing a positive charge and chlorine ions bearing a negative charge, each free to move, seems consistent with your observations. It is a reasonable extension of this model to describe the structure of solid lead chloride as made up of these same ions but with the freedom of motion gone. Certainly the solid does not contain lead metal or chlorine gas. But we must say that we haven't yet given unequivocal evidence for the existence of ions in the crystalline lead chloride.

14-3   *Introduction to the use of chemical equations*

In the discussion of what happened during the electrolysis of molten lead chloride, the reaction which took place at each electrode would be written by a chemist as follows:

At the positive electrode: $\quad\quad\quad\quad Cl^- \rightarrow Cl + e^-$
At the negative electrode: $Pb^{+2} + 2e^- \rightarrow Pb$

This is a kind of "shorthand" method to say that a chlorine ion yields a chlorine atom plus an electron, and a lead ion plus 2 electrons yields a lead atom. Because this shorthand is used in this and succeeding chapters (and all chemical literature), you will need to understand it. Therefore, we present other examples of chemical shorthand in the following paragraphs.

In Experiment 13–2 you found that a number of mass units of oxygen combined with a number of mass units of magnesium. You were given the information that the ratio of the number of atoms of magnesium to the number of atoms of oxygen is $1:1$. By using the symbols for magnesium (Mg) and oxygen (O) we can write the chemical for-

mula for magnesium oxide as MgO. A chemical formula
expresses an experimentally determined ratio of the numbers
of atoms of elements that are combined in a compound.
Magnesium oxide is an example of a chemical compound. A
chemical compound is a substance that consists of two or
more kinds of atoms combined in a definite ratio. This ratio
is not always 1:1 as in MgO. For reasons we will discuss
later, magnesium chloride, for example, is found with a
combining ratio of one atom of magnesium to two atoms of
chlorine. We write this formula $MgCl_2$, the subscript 2 indicating that two atoms of chlorine are involved. A subscript 1
for magnesium is not written because it is understood.

The changes that take place when elements or compounds react are shown by means of equations. These equations show the initial and final substances, as well as the
composition and relative proportions of the substances involved in the reaction. Here, for example, is the equation
for the combination of magnesium with oxygen to yield
magnesium oxide:

$$2Mg + O_2 \rightarrow 2MgO$$

How did we know what to write? What does it all mean?
The 2 in front of the Mg, for example, refers to the amount
of magnesium entering the reaction. It means 2 moles of
magnesium are used. A *mole* is 6.02×10^{23} chemical units
of whatever is being considered, for example, atoms, ions,
molecules, electrons, etc. This number 6.02×10^{23} is called
Avogadro's number and is discussed in Chapter 15.

The subscript 2, as in O_2, represents the fact that the element oxygen (like many gaseous elements) ordinarily does
not exist in the form of individual atoms, but rather as pairs
of atoms. Any group of atoms combined into a unit, such
as O_2 or H_2O, is called a *molecule.*

When translated into words, then, the above equation is
usually read as "2 moles of magnesium react with 1 mole of
oxygen to yield 2 moles of magnesium oxide." This says that
a given number (a mole) of oxygen molecules, when combined with twice that number (2 moles) of magnesium atoms,
will yield twice that number (2 moles) of combined chemical units of magnesium oxide MgO. The reason that we

avoid speaking of molecules of magnesium oxide is that in the crystalline magnesium oxide that results from the reaction we discussed, the magnesium and oxygen ions are arranged alternatively, like the red and black squares of a checkerboard, and no one oxygen ion is paired with any particular magnesium ion in the close and exclusive association we call a molecule.

How did we decide to place the 2 in front of the Mg and the MgO? Atoms can neither be created nor destroyed in a reaction, and therefore the total number of atoms of each element on one side of the equation must equal the total number of atoms of the same element on the other side. We can represent this by placing the correct number in front of each formula in the equation; this is known as *balancing the equation*. (This phrase, though commonly used, is a little misleading. An expression is not an equation until it is balanced.) Keep in mind that the number in front of the formula applies to the whole formula, but the subscript belongs only to the symbol immediately preceding it. For example, in NH_3, the 3 refers only to the H.

As another example, consider the reaction of nitrogen with hydrogen, which forms ammonia gas. Chemical analysis shows that nitrogen and hydrogen atoms combine in the ratio of $1:3$ to form ammonia. We can also determine, by methods which will be shown later, that nitrogen and hydrogen, like oxygen, are diatomic molecules (molecules composed of two atoms), and also that the molecule of ammonia has one atom of nitrogen and three of hydrogen. Using formulas, we can write

$$N_2 + H_2 \rightarrow NH_3$$

How can we balance this equation? Recognizing that the subscripts must remain unchanged, we need to insert numbers in front of appropriate formulas to indicate that the number of atoms of each substance does not change.

Consider N first: there are two atoms on the left, so we need two atoms on the right. How do we manage this? Place a 2 in front of the formula on the right side.

$$N_2 + H_2 \rightarrow 2NH_3$$

But now we have a total of six atoms of H on the right, so we need six atoms of H on the left. Place a three in front of the H_2 on the left.

$$N_2 + 3H_2 \rightarrow 2NH_3$$

Question 14-6   The equation for the combination of hydrogen and oxygen to form water is: $2H_2 + O_2 \rightarrow 2H_2O$. List all the information you obtain by inspecting this equation.

Question 14-7   Magnesium reacts at high temperatures with the nitrogen of the air, forming magnesium nitride Mg_3N_2. Write an equation for the reaction of magnesium with nitrogen.

14-4   *Ions in solution*

In order to write balanced equations for chemical reactions we must identify both the reactants and the final products. One means of identification is observation of colors. To illustrate this, your instructor will dissolve a little of each of the following substances in different containers of water:

> nitric acid (HNO_3)
> copper nitrate ($Cu(NO_3)_2$)
> potassium bromide (KBr)
> copper bromide ($CuBr_2$)

All these substances dissociate into ions in solution in the following ways:

$$HNO_3 \rightarrow H^+ + NO_3^-$$
$$Cu(NO_3)_2 \rightarrow Cu^{+2} + 2NO_3^-$$
$$KBr \rightarrow K^+ + Br^-$$
$$CuBr_2 \rightarrow Cu^{+2} + 2Br^-$$

Some ions give a color to the solution and some do not. Which of the ions just mentioned are colored in solution and which are colorless?

EXPERIMENT 14-2   Migration of ions

An electrolysis apparatus similar to that used in Experiment 14–1 is used here, but we replace the U-tube and

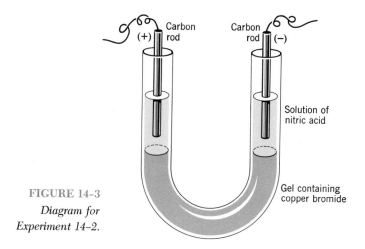

FIGURE 14-3
*Diagram for
Experiment 14-2.*

electrodes with the arrangement shown in Figure 14-3. The electric current is passed through a solution of copper bromide held in a gel (pronounced "jell"). The reason the copper bromide is held in the gel is that the gelatin forms a kind of network that partly traps water molecules and ions in solution and hinders their motion. The gel is made by dissolving gelatine in a warm-water solution of the copper bromide. Like any gelatin preparation, it will stiffen on cooling.

Place some of the gel in the U-tube, and allow it to set. Then add a solution of 2% nitric acid to each arm of the tube so that it rests on top of the gel (see Fig. 14-3). The purpose of the nitric acid is to provide a colorless medium for conduction of ions. A short time after you complete the circuit, observe what happens and record your observations, paying particular attention to any color changes that take place.

Question 14-8 Explain what you have observed.

14-5 *Electrolysis of solutions*

We have established that lead ions, sodium ions, and chlorine ions exist in a melt, and that copper ions and bromine ions exist in a solution. We can extend our knowledge of ions in solutions by studying two more crystalline. substances. For example, do we find positive lead and sodium ions, and negative chlorine ions in an aqueous solution of these elements, as we did in a melt? If these solutions do conduct an electric current, then we can conclude that they contain ions.

As you know, sodium chloride is quite soluble in water, and therefore we can easily make a strong solution with it. However, lead chloride is only slightly soluble in water at room temperature; a solution of it is too weak to carry an appreciable amount of electric current. Consequently, we will substitute another lead compound, lead nitrate, which is soluble, and try passing a current through it. This will tell us whether or not lead ions exist in solution and behave like lead ions in a melt. We can also try passing a current through an aqueous solution of sodium chloride to obtain information about the existence and behavior of sodium and chlorine ions in solution.

The following experiment will be done in two parts. Part A will be the electrolysis of an aqueous solution of lead nitrate. Part B will be the electrolysis of an aqueous solution of sodium chloride. An important phenomenon is observed in the reaction that takes place at the negative electrode in each part of the experiment. Watch both electrodes carefully.

EXPERIMENT 14-3 Electrolysis

Part A Electrolysis of Lead Nitrate

We will use the same arrangement as in Experiment 14–1 (Figs. 14–1 and 14–2) except that we will fill the U-tube about half full of a water solution of lead nitrate. The solution is made by dissolving 30 grams of lead nitrate in each 100-ml portion of water. For better results, shape the copper electrode differently; rather than a simple loop, form it into a spiral coil. Three or four turns about $\frac{1}{4}$ in. in diameter should be made on the end of the wire that dips below the surface of the liquid. The coil is easily made by wrapping the wire around a pencil.

When all is ready, complete the circuit as directed in Experiment 14–1, and leave it connected for about 20 min. Watch carefully. Observe the flashlight bulb, the electrodes, and the solution.

Part B Electrolysis of Sodium Chloride

The procedure for this part of the experiment is the same as for Part A, but a different solution is used. The solution used here is made by dissolving 10 grams of sodium chloride in each 100-ml portion of water.

You should make two tests, however, that were not made in Part A. After the current has been on for about 10 min, carefully bring a lighted match to the mouth of the arm of the U-tube that contains the copper wire electrode. Keep your face as far as possible from the U-tube. Repeat the test. Hold a piece of moist potassium iodide-starch paper in each arm of the U-tube. Make notes of your observations. The following questions may help you organize your thinking.

Question 14–9 Did you notice any difference between the solution in the negative (copper wire) electrode side and that in the positive (carbon rod) electrode side in Part A? in Part B?

Question 14–10 Was the product that formed at the copper electrode the same for the lead nitrate solution as for the lead chloride melt?

Question 14–11 In Parts A and B, at which electrodes did you observe an evolution of gas?

Question 14–12 Look at the diagram in Figure 14–2 and trace the path of the current from the battery through the bulb and back to the battery.

Question 14–13 The symbol for a lead ion is Pb^{+2}. The addition of two electrons converts the lead ion to a lead atom. Write an equation to show this reaction.

Question 14–14 The expression $H_2O + e^- \rightarrow H_2 + (OH)^-$ represents the reaction at one of the electrodes. Write it as a balanced equation.

Question 14–15   Identify the carriers by means of which the charges are conducted through the various parts of the circuit, excluding the battery, in the electrolysis of the sodium chloride solution.

When we compare the results of the electrolyses of molten lead chloride, aqueous lead nitrate, and aqueous sodium chloride, the most obvious difference is the product formed at the negative (copper) electrode. Metallic lead was deposited on the copper wire with both molten lead chloride and aqueous lead nitrate. However, no metal was deposited on the copper wire with aqueous sodium chloride; instead a gas was formed. Why wasn't sodium metal deposited on the wire as lead metal was? In all three experiments a gas evolved around the positive (carbon) electrode. Is this gas the same in all three cases? Is the gas formed at the negative (copper) electrode in the electrolysis of an aqueous solution of sodium chloride the same as any or all of the gases formed at the positive electrode?

Let us first consider the questions about the gases. The identity of a gas can be established only by testing it. Testing a substance involves determining a sufficient number of its properties to permit identification. Must one determine all the properties of a gas in order to identify it? Generally, no. Before testing to determine which gas is evolved, it is best to examine the experimental situation in order to have a basis for predicting which substance might be produced in the reaction. Then suitable tests can be performed to distinguish among the possible products.

We deduced that chlorine gas was the only gas that could possibly be evolved at the positive electrode during the electrolysis of molten lead chloride. When the aqueous solution of sodium chloride was electrolyzed, however, a yellowish-green gas appeared in the arm of the U-tube containing the positively charged carbon rod, and the solution around the carbon rod turned the same color as the gas. This suggests that some of the gas dissolved in the solution. Chlorine gas will cause moist potassium iodide-starch paper to turn dark. Did your paper turn dark? It appears that chlorine gas was evolved, but we should consider other possibilities before drawing a final conclusion.

Water is made up of atoms of hydrogen and oxygen. Could the gas be either hydrogen or oxygen? No, both hydrogen and oxygen are colorless and insoluble, and neither affects moist potassium iodide-starch paper. The yellowish-green gas which was observed must be chlorine.

During the electrolysis of lead nitrate, $Pb(NO_3)_2$, a colorless gas was evolved from the positive electrode. We have already established that lead was deposited on the negative electrode. The colorless gas then must come from the nitrate ions $(NO_3)^-$ or from the water molecules H_2O. One of the experiments in Chapter 16 will help you identify the colorless gas and determine its source. We will postpone any further discussion until that time, but for now, try guessing what the gas is.

We have yet to answer the question about the lack of sodium metal deposited on the negative (copper) electrode during the electrolysis of aqueous sodium chloride. In both of your experiments involving lead compounds, lead chloride and lead nitrate, lead was deposited on the negative electrode. By analogy one would expect that sodium metal would also be deposited on the negative electrode. However, during the electrolysis of sodium chloride we observed a colorless gas being evolved, with no indication of any metal. This raises several questions. What is the gas? What is its source? Why was sodium metal not deposited?

The gas must be chlorine, hydrogen, oxygen, or compounds containing these elements, since only the chlorine ions and water molecules in the solution can form gases. We have just shown that chlorine is colored and is evolved at the positive electrode; therefore we can eliminate it as a possibility. Of the remaining possibilities only hydrogen and oxygen are not soluble in water. In addition hydrogen and oxygen are both colorless. When you held a lighted match at the mouth of the arm of the U-tube containing the negative electrode, a small pop was heard. This is a property of hydrogen. We conclude, therefore, that hydrogen gas is given off at the negative electrode. Further evidence for this is given in Experiment 16–1.

What happened to the sodium and oxygen? To answer this question we must consider what is required to liberate a substance during electrolysis. For example, to liberate chlorine gas, two chlorine ions must each lose an electron and then combine together to form a chlorine molecule Cl_2. This process requires energy. During electrolysis, that energy is supplied by the battery.

However, suppose that two reactions are possible. They must then compete with each other for the available energy. Furthermore, suppose that one of these two reactions requires more energy than the other. Which will proceed more rapidly? The one that requires the lesser amount of energy will proceed more rapidly than the other one. In fact, it is possible that the one requiring the lesser amount of energy may proceed while the other one does not.

The amount of energy required to deposit sodium metal at the negative electrode is much greater than that needed to liberate hydrogen. The reaction that requires the least energy will take place first, therefore hydrogen, not sodium, was produced at the copper wire electrode. The equations for the two possible negative electrode reactions are:

$$Na^+ + e^- \rightarrow Na \text{ (metal) (large amount of energy needed)}$$
$$2H_2O + 2e^- \rightarrow H_2 \text{ (gas)} + 2(OH)^- \text{ (small amount of energy needed)}$$

The same principle applies for the two possible reactions at the positive electrode, where we observed chlorine given off:

$$2Cl^- \rightarrow Cl_2 \text{ (gas)} + 2e^- \text{ (small amount of energy needed)}$$
$$2H_2O \rightarrow O_2 + 4H^+ + 4e^- \text{ (large amount of energy needed)}$$

Look at the first pair of equations. Sodium ions Na^+ present in the solution and the $(OH)^-$ ions formed in the reaction shown in the second equation remain in solution. If the electrolysis is permitted to continue long enough, the solution will contain a strong concentration of Na^+ ions and $(OH)^-$ ions. The water can be evaporated and solid sodium hydroxide NaOH will remain. This is the commercial method used for the production of sodium hydroxide, one of our most abundantly used chemical compounds.

Table 14–1

Substances Electrolyzed	Product at the (−) Electrode	Product at the (+) Electrode
Molten lead chloride and sodium chloride	Lead metal	Chlorine
Solution of lead nitrate	Lead metal	Colorless gas (unidentified)
Solution of sodium chloride	Hydrogen	Chlorine

Table 14–1 summarizes the reactions at the electrodes for the three substances that were electrolyzed.

In Section 14–8 a further study is made of the idea that some ions are deposited in preference to others. The result of this study will enable us to draw some conclusions as to the relative chemical activity of some common chemical elements.

14-6 *Chemical properties of ions and atoms*

Both positive and negative ions have characteristic properties that are quite different from those of the atoms from which they are formed. Chlorine ions, for example, do not escape from the solution as a gas, even on boiling, as chlorine molecules do. These ions do not react with hydrogen or metals as the molecules do; they are inert or practically so.

Why should the ion be inert and the atom active? In Chapter 13 it was suggested that chemical activity may be dependent on the electron structure of an atom. How many electrons are in the outer shell of the chlorine atom? the chlorine ions? (See Table 13–4.) Could it be that when the outer shell has 8 electrons, the atom—or the ion—is chemically inactive?

Which atoms have outer shells with 8 electrons? We see in Table 13–4 that both neon and argon do. We know that these are both chemically inert. Therefore, we can conclude that the chlorine ion is inactive because its outer shell is filled with 8 electrons.

The properties of sodium ions also differ widely from those of sodium atoms. A piece of sodium metal reacts vigorously with water, oxygen, and chlorine, but sodium ions do not. The sodium atom becomes an inactive sodium ion by losing its outermost electron. Because this electron is the only one in the outermost shell of the sodium *atom*, the sodium *ion* is left with an 8-electron shell. This is still another example of chemical inactivity associated with 8 electrons in an outer shell.

You eat sodium and chlorine ions every day but not the atoms. If you put a piece of sodium metal in your mouth, it would react violently with your saliva; chlorine is a poison gas, one of the first used in warfare; but table salt is added to food to improve the taste.

14-7   *Behavior of ions*

You have now found evidence suggesting that indeed there are charged particles in lead chloride, sodium chloride, and lead nitrate. By directing an electric current through the melted mixture of sodium chloride and lead chloride, you were able to separate metallic lead and gaseous chlorine (although the latter may have been hard to identify). The fact that you were able to separate these materials by means of an electric current suggested that the lead, sodium, and chlorine atoms were charged, that is, that they were ions.

Were these particles already ions when they were in the solid, or did they become ions when the solid was melted, or did they become ions when an electric current was passed through the melt?

When you caused an electric current to pass through a water solution of lead nitrate, you once more obtained metallic lead; and from a solution of sodium chloride you obtained gaseous chlorine as well as evidence that sodium ions were present. Again, lead, sodium, and chlorine seemed to have been present as ions. Were they ions in the original solids or were they somehow converted to ions during the experiment?

The best evidence you have on which to base an answer to these questions is that the solids did stick together, and rather tightly at that. So there certainly are strong bonding forces in solid sodium chloride, and we have been working on the hypothesis that these forces are electrical. Thus, we can now assume that the particles in solid sodium chloride are *ions* of sodium and chlorine, and in Chapter 15 we will investigate their arrangement.

But before we consider this solid structure, let us turn to an investigation of the general behavior of ions. So far we have seen that, as we would have expected, ions are attracted to oppositely charged objects (the electrodes in your experimental apparatus). We have assumed that oppositely charged ions are attracted to each other, and we have explained how an ion can become an atom. Using the atomic model developed in Chapter 13, we have partially answered the question, "How does an atom become an ion?" The data given to support the model consisted primarily of ionization potentials. We may now ask the question a little differently: "How can atoms be changed to ions in an ordinary laboratory situation?" We may also ask, "What happens when atoms are changed to ions?" These questions will be considered in the remaining sections of this chapter.

14-8 *Replacement of ions*

One way to change atoms to ions and ions to atoms is to place a piece of metal in contact with a solution containing ions of a different metal.

EXPERIMENT 14-4 Replacement of ions

You will be given three small envelopes containing crystals of lead nitrate, copper nitrate, and silver nitrate.

> *Caution.* All of these compounds are poisonous. Do not leave the solids or their solutions in a place where a child, or anyone, may find them and take them internally. When you have finished, flush the solutions down a drain but certainly *not* the kitchen sink. Silver nitrate will turn your skin dark brown or black on contact. The color comes from a deposit of finely divided silver. It is not harmful, and although it will not wash off, it will wear off.

Dissolve each solid in about 60 ml (2 oz or $\frac{1}{4}$ cup) of water in a container such as a paper cup, a small glass jar, or a plastic container. Now let us test for possible reactions of these solutions with metals. Procure three pennies and three silver dimes. (New dimes are copper clad with nickel alloy, so you may want to use copper and silver foil instead of the coins.) The coins or the foil should be cleaned with a little tooth paste and elbow grease. You also will be given three short pieces of wire made of lead.

Devise a table that shows all possible combinations of the three metals with the three solutions. Divide each of your three solutions into three approximately equal parts. Label the nine containers, and place one piece of metal in each container so that each kind of metal is in contact with each kind of solution. Record your observations in the table. Some pieces of metal will have to remain in contact with the solution for as long as two hours.

When you discard the solutions, it is possible to restore the coins or foils by washing them in water, and drying them with a paper towel.

Note that two experiments of this type were performed early in this course (Experiment 1–3, Part B). Some students added water to a vial containing copper sulfate and an iron nail, and others added water to a vial containing silver nitrate and a copper wire. Review the observations that were made during that experiment.

Question 14–16   Which of the metals (lead, silver, and copper) was coated with another metal when it was taken from the lead nitrate solution? the silver nitrate solution? the copper nitrate solution?

Question 14–17   Which of the solutions changed color?

Question 14–18   Give a reason for any color change in the solutions.

Question 14–19   If a piece of iron were placed in a silver nitrate solution, do you think it would become coated with silver?

In this experiment, there was a competition for electrons in each solution containing a metal. The positive metal ions in solution needed electrons if they were to come out of solution as atoms of metal. The atoms of the piece of metal you put in contact with the solution needed to hold on to their electrons if they were to remain atoms and not become positive ions in solution. In each solution one of the metals won the competition for electrons. From experiments such as these it is possible to set up an electron-acquisition competition list that shows whether a particular element will win or lose the competition for electrons against another particular element. Elements that give up electrons more readily are called more *active* and are placed higher in the list than those that give them up less readily.

You now have observations in your notes that will enable you to construct such a list for copper, silver, and lead, with respect to each other. Construct your list, placing the most active element at the top and continuing so that in the list each element will replace the ones below it from solutions. To extend your understanding of the atomic model, we will now examine more reactions in which atoms form ions and ions are converted to atoms.

14-9 *Combination*

The experiments performed so far in this chapter have resulted in the breaking up of chemical compounds. Is it possible to reverse the process and cause elements to combine spontaneously to form compounds?

You may remember that energy, in the form of electric energy, was required to change lead chloride to lead and chlorine. If lead and chlorine combine spontaneously to form lead chloride, would you expect energy to be absorbed from or released to the surroundings during the reaction? We will examine a similar reaction in the following experiment.

EXPERIMENT 14-5   Combining zinc and iodine (a demonstration)

Grind 3.2 g of iodine crystals in a mortar until the particle size is very small. (Be careful not to inhale the iodine vapor while grinding.) Pour the powdered iodine onto a square of paper, and add to it 0.8 g of powdered zinc. Mix the iodine and zinc thoroughly with a spatula, and pour the mixture into a 6-inch test tube. Place the test tube in a clamp on a pegboard, and have ready a stopper that fits the mouth of the test tube.

Before proceeding with the experiment, however, try to predict what will happen when you add 2 or 3 drops of water to the mixture in the test tube. Write a chemical equation for what you expect to happen.

Now cautiously add 2 or 3 drops of water to the mixture in the test tube, and quickly place the stopper in the tube. If no change is observed immediately, add a few more drops of water and replace the stopper.

Question 14-20   Was there a chemical reaction? What visible evidence supports your conclusion?

Question 14-21   Was energy liberated? What evidence do you have for your answer?

Question 14-22   What caused the purple color?

Question 14-23   Can you guess what the product of the reaction might be?

Question 14-24   How might one decompose the product formed during this demonstration? Would an energy change be a part of that reaction?

The product of the reaction in this demonstration is a compound containing zinc ions and iodine ions, which has been formed by a process of transfer of electrons from zinc to iodine. Why should an element such as zinc give up electrons to form a positive ion? Table 13–3 shows that energy must be put into a metal to ionize it. Zinc is not included in that table, but like all other elements, it requires energy to form a positive ion. Let us explore where this energy comes from.

One other part of the process is that of forming a negative ion when an atom of an element adds an electron. Why do some elements add electrons? We have seen in

Chapter 13 that there is a special stability associated with a shell of two electrons for the simplest elements, and with a shell of 8 electrons for elements of higher atomic number. How many total electrons has fluorine, atomic number 9? How would these be distributed in shells? What *change* in the total number of electrons would give fluorine an outer shell of 8 electrons? This change, the process of adding one electron to a fluorine atom to form a fluorine ion, $F + e^- \rightarrow F^-$, is one which liberates energy. Chlorine, bromine, and iodine atoms are all more complex (higher atomic number) than fluorine, but all are alike in having seven electrons in their outermost electron shells. Therefore, each can form an ion by adding an electron and for each, energy is given up in the process. This energy helps supply the energy required to form the positive ion, but that is not the whole story.

In an ordered collection of particles, some with positive and some with negative charges, there is energy associated with the attraction of unlike charges. This energy can be calculated in the same general way as the force between charged particles was calculated in Chapter 11.

Thus we see that the overall energy change in such a reaction can be considered in three parts: (1) energy added to make positive ions from atoms, such as $Na \rightarrow Na^+ + e^-$; (2) energy released when negative ions are formed from atoms, such as $Cl + e^- \rightarrow Cl^-$; and (3) energy change because of changes in position of intermixed positive and negative ions. The energy of position, which we earlier called *potential energy*, of intermixed positive and negative ions is the largest of the three, especially in a crystalline solid. Patterns are much less regular in a liquid, and the energy of position is somewhat less, but still is so high that it is the dominant energy of the three that contribute to the total energy of the process. This discussion of the energy changes involved in the formation of ionic crystals is taken up again in Section 15–7.

Do all reactions involving the combination of atoms to form ionic crystals occur with the exchange of only one electron per atom? Refer to Table 13–3. How many electrons can be easily removed from a magnesium atom (atomic number 12)? The process of ionization of magnesium may then be written

$$Mg \rightarrow Mg^{+2} + 2e^-$$

If magnesium metal is combined with chlorine, each magnesium ion formed in the process will have a charge of $+2$. Each chlorine ion will have a charge of -1. Each magnesium ion, then, will attract and hold *two* chlorine ions, and the compounds formed is represented by the formula $MgCl_2$.

Like magnesium, zinc and calcium atoms have two electrons in their outer shells; and when these electrons are removed, the resulting ions have a $+2$ charge.

Question 14–25   Write an equation for the process of forming a zinc ion from a zinc atom.

Question 14–26   Write an equation for the process of forming an iodine ion from an iodine atom.

Question 14–27   Modify the equation you wrote as an answer to Question 14–26 to take account of the fact that the iodine molecule has two atoms.

Question 14–28   Combine the two equations (from Questions 14–25 and 14–27) by addition to obtain an equation for the overall reaction.

Question 14–29   What is the ratio of zinc ions to iodine ions in zinc iodide? What is the ratio of magnesium ions to chlorine ions in magnesium chloride? Write the formulas for these compounds. (*Hint:* Compounds are neutral.)

Question 14–30   What do you predict will be the behavior of oxygen (atomic number 8) with respect to gain or loss of electrons? Write an equation to show this.

Question 14–31   Write a reaction for the formation of magnesium ions (atomic number 12) and combine this with the oxygen reaction to obtain the equation for the reaction of magnesium with oxygen.

Question 14–32   Is the formula for magnesium oxide, which was given without evidence in Experiment 13–2, consistent with our model of the electron transfer for formation of chemical compounds?

Now we can examine the reactions of Experiment 14–4

more carefully and attempt to write equations for them. Different ions and atoms have different strengths of attraction for electrons. As a result, there may be a transfer of electrons when some pairs are brought into contact, and no transfer for other pairs. Write equations for only those pairs that reacted.

Question 14–33 What happened to the wire made of lead in copper nitrate? Can you interpret this in terms of electron transfers? Both the lead and copper ions have a charge of $+2$ in these reactions.

Question 14–34 Write an equation for each of the two electron-transfer ionic reactions of Question 14–33. Multiply each by some number so that the number of electrons lost in one reaction will be equal to the number of electrons gained in the other reaction.

Let us examine another case of electron transfer, the reaction that took place with copper metal and silver nitrate. Here the copper dissolved to form blue ions, which we know are Cu^{+2} in solution; the silver ions, each of which has a single positive charge, Ag^+, formed silver metal. These reactions occur simultaneously, and since neither will proceed without the other, each is called a *half reaction*.

$$Cu \rightarrow Cu^{+2} + 2e^-$$
$$Ag^+ + e^- \rightarrow Ag$$

If the second half reaction is multiplied by two, then the number of electrons transferred from copper in the first half reaction and the number of electrons transfered to silver ions in the second one will be equal. We can add the two half reactions to get a balanced equation for the replacement reaction.

$$Cu \rightarrow Cu^{+2} + 2e^-$$
$$2Ag^+ + 2e^- \rightarrow 2Ag$$
$$Cu + 2Ag^+ \rightarrow Cu^{+2} + 2Ag$$

Special names are applied to the two half-reactions that go to make up the complete reaction. The half-reaction in which electrons are lost (for example, Cu forms Cu^{+2}) is called an *oxidation;* the half-reaction in which electrons are gained (for example, Ag^+ forms Ag) is called a *reduction*. In other words, oxidation is defined as a loss of electrons; reduction as a gain of electrons.

14-10   *Oxidation and reduction*

The definitions of oxidation and reduction in the preceding section are based on our current model of the structure of the atom, in which chemical reactions are brought about by changes in the number or arrangement of the electrons outside the nucleus of the atom. But the terms oxidation and reduction were used long before we had any idea of the nuclear atom. The original meaning of oxidation was "combination with oxygen." Almost all the elements combine with oxygen to form compounds, known as oxides. Some elements will form a series of compounds with oxygen in which the ratio of the number of atoms of the element to the number of oxygen atoms varies. Carbon, for example, forms both carbon monoxide (CO) and carbon dioxide (CO_2). Carbon is said to be oxidized to CO, which can be further oxidized to CO_2. Copper forms cuprous oxide (Cu_2O) and cupric oxide (CuO). The suffix *-ous* indicates the lower ratio of oxygen to the element in the oxide, while the suffix *-ic* indicates the higher ratio. Cuprous and cupric derive from the Latin *cuprum*, which means copper.

The reverse of the process of addition of oxygen, that is, the removal of oxygen from a compound, became known as reduction. The relationship between oxidation and reduction can be shown in a reaction that takes place at elevated temperatures, and which is described by the following equation:

CuO (solid) + H_2 (gas) → Cu (metal) + H_2O (steam)

In this equation we see that hydrogen removes oxygen from the cupric oxide. The oxide is reduced to the metal; this is reduction. During the process the hydrogen is oxidized to water; this is oxidation. Note that in the overall reaction, oxidation and reduction take place together. You can never have one without the other.

Let us extend our concept of oxidation. We have seen that copper forms two oxides with oxygen, Cu_2O and CuO. Similarly, chlorine forms two chlorides of copper, cuprous chloride ($CuCl$) and cupric chloride ($CuCl_2$). Evidently chlorine acts similarly to oxygen in combining with copper. We also call the combination of chlorine with copper an oxidation process even though there is no oxygen taking part in the reaction. Why?

In Table 13–3 we see that the magnesium atom readily loses two electrons, forming the magnesium ion Mg^{+2}. In Experiment 13–2 you prepared magnesium oxide, which has the formula MgO. We know that the compound MgO is electrically neutral, and if the magnesium is present as Mg^{+2} the oxygen must be present as O^{-2}. In fact, in all common oxides, oxygen has a charge of -2. In Section 14–2 we found that the chlorine atom readily adds an electron, forming the chlorine ion, which has a charge of -1.

Now let us look again at the oxides and chlorides of copper, keeping in mind that all substances in the elementary state are electrically neutral. In the successive stages of the oxidation of copper to the oxides, we find that elementary copper ($Cu°$, indicating the copper atom) is first oxidized to Cu_2O, in which each copper must have a charge of $+1$ (Cu^{+1}), and finally to CuO in which each copper must have a charge of $+2$ (Cu^{+2}). Similarly, in the combination of elementary copper with chlorine, cuprous chloride ($CuCl$), in which the copper has a charge of $+1$, is succeeded by cupric chloride ($CuCl_2$), in which the copper has a charge of $+2$. In both types of combination (with oxygen and with chlorine) the copper goes through the series

$$Cu° \rightarrow Cu^{+1} + e^- \text{ and } Cu^{+1} \rightarrow Cu^{+2} + e^-$$

In both cases we call the processes oxidation processes, and on this basis we define oxidation as a process that results in making the charge on the atoms or ions a higher positive value or a less negative value. Reduction is the converse of oxidation.

Now let us summarize the discussion about oxidation and reduction by stating their three definitions in the chronological order of their development.

1. Oxidation is combination with oxygen; reduction is removal of oxygen.

2. Oxidation is an increase in the charge on an atom (or ion); reduction is a decrease in the charge on an atom (or ion).

3. Oxidation is a loss of electrons; reduction is a gain of electrons.

14-11   *Cells*

In Experiment 14–4, when part of the lead wire dissolved and copper ions from the copper nitrate solution formed copper metal, there was an oxidation-reduction reaction involving a transfer of electrons. Let us review the equations for this process.

1. $Pb \rightarrow Pb^{+2} + 2e^-$: A lead atom loses 2 electrons to form the lead ion Pb^{+2}.

2. $Cu^{+2} + 2e^- \rightarrow Cu$: The copper ion Cu^{+2} accepts two electrons to form a copper atom.

3. $Pb + Cu^{+2} \rightarrow Pb^{+2} + Cu$: Half reactions 1 and 2 are added to obtain Equation 3 for the complete reaction.

Inspection of the two half reactions 1 and 2 indicates the transfer of two electrons as an atom of lead goes into solution and a copper ion is transformed to copper metal.

Could we use such a spontaneous change of electrons to produce an electric current? We will explore this possibility in the following experiment. However, magnesium will be used instead of lead because magnesium loses electrons more easily and more quickly than lead. Consequently the reaction in the cell will be more vigorous when magnesium is used.

EXPERIMENT 14-6   An electric current by electron transfer

We will have to set up this experiment so that a loss of electrons occurs in one place and a gain occurs in another. These two differently charged places are connected through a flashlight bulb that will glow if charges flow between them. The electrons given up by atoms at the negative electrode must travel through a wire if they are to supply electrons for changing positive ions to atoms at the surfaces of the positive electrode. One way to do this is shown in Figure 14–4.

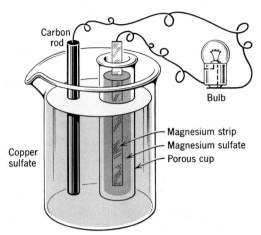

Carbon rod

Bulb

Magnesium strip
Magnesium sulfate
Porous cup

Copper sulfate

FIGURE 14-4
Electrochemical cell.

Place a porous porcelain cup inside a beaker. The porous cup is designed to permit the slow passage of ions through its wall, which will prevent the rapid mixing of the copper and magnesium ions. Fill the cup with a saturated solution of magnesium sulfate (Epsom salts). Put a saturated solution of copper sulfate into the beaker around the cup. Place a carbon rod in the copper sulfate solution and a piece of magnesium ribbon in the magnesium sulfate solution. Connect a low-current (60 milliampere = 0.06 amp) flashlight bulb between the carbon and magnesium strips. After making connections to close the circuit, and after observing the bulb, raise each electrode, one at a time, so that only a small portion of it is submerged in the solution. Do you see any change in the brightness of the bulb? Replace the electrodes and let the experiment run for about ten minutes. Open the circuit, remove each electrode, and examine it carefully.

Question 14–35 Why did the bulb light up?

Question 14–36 What is the source of the energy?

Question 14–37 Write half-reaction equations to describe what is happening at each electrode.

Question 14–38 In which direction are the electrons flowing through the wire, from magnesium to carbon or from carbon to magnesium?

Question 14–39 Why does one electrode appear to have partially dissolved while the other one has a deposit on it?

The moving electrons did light the lamp. The chemical energy released by the reactions that result from the transfer of electrons between atoms and ions has been converted into electrical energy and made to do work for us. In fact, we used just this kind of chemical change, going on inside the batteries, to provide the electric current for the electrolysis experiments. We used electrical energy there to separate compounds into their component elements. When those elements are allowed a combine again, their energy is released, usually in the form of heat.

Energy changes form in many ways. We have seen some of them. Although we have not always made quantitative determinations of the amount transferred, such measurements have often been made. These measurements reveal that within any isolated system the total energy remains the same. Energy is neither gained nor lost; energy is conserved.

QUESTIONS 14-40 Following is an electron-acquisition competition list such as you constructed earlier in this chapter.

Element	Symbol
Lithium	Li
Potassium	K
Sodium	Na
Magnesium	Mg
Zinc	Zn
Chromium	Cr

Iron	Fe
Cobalt	Co
Nickel	Ni
Tin	Sn
Hydrogen	H
Copper	Cu
Silver	Ag

Name two elements that will react with water to form hydrogen at room temperature. Name two elements that will not so react.

14-41 You observed a gas escaping from the magnesium electrode in Experiment 14-6. What was this gas? How can you prove this experimentally?

14-42 In an electrolysis experiment, at which electrode does oxidation take place? At which electrode does reduction take place?

14-43 Two chemical reactions you have seen or investigated yourself are the reaction of zinc with iodine and the reaction of a sodium chloride solution with silver nitrate. Does either or do both of these reactions involve oxidation and reduction? Explain.

14-44 Cobalt is a metallic element resembling iron. It is used in small amounts with iron in making some kinds of steel. Like iron, it combines with oxygen to form two different oxides, CoO and Co_2O_3. Give the names of these two oxides, and write an equation for the reaction of oxygen with cobalt to form each.

14-45 If electric charges were passed through a water solution of cupric chloride $CuCl_2$, what would be deposited at each of the electrodes?

References 1. Chem Study, *Chemistry, an Experimental Science*, W. H. Freeman & Co., 1963, Chapter 12, Section 12-1, pages 199 to 207.
2. Greenstone, A. W., F. X. Sutman, and L. B. Hollingworth, *Concepts in Chemistry*, Harcourt, Brace & World, New York, 1966, Chapter 7, pages 115 to 121

An explanation of the meaning and use of formulas and equations in chemistry. Also, Chapter 13, pages 201 to 208. The behavior of particles in solution.

3. PSSC, *College Physics,* Raytheon Educ. Co. (D. C. Heath), 1968, Chapter 25, Sections 25–2 and 25–3, pages 467 to 471.
   Ions in solution and their behavior at electrodes.

ALSO OF INTEREST

4. Sanderson, R. T., *Principles of Chemistry,* John Wiley & Sons, 1963, Chapter 20, pages 400 through 412.
   Electrolysis of liquids and solutions, with diagrams.

John Shelton

CHAPTER 15

THE NATURE OF AN IONIC CRYSTAL

In our study of the nature of solid matter, we have devoted considerable time to the crystalline solids. In Chapter 5 macroscopic observations provided us with a clue to their microscopic structure, and we assumed that the macroscopic regularity we observed results from some orderly microscopic structure. In Chapter 11 we proposed the hypothesis that this microscopic structure is a regular arrangement of very small particles held together by forces due to the electrical charge carried by the particles. In Chapter 13 we were led by many considerations to conclude that the smallest constituent particles of all matter are atoms, and that they themselves have a structure. All relevant experimental evidence indicates that each atom consists of a single positively charged nucleus whose mass is almost that of the atom, surrounded by a swarm of negatively charged electrons whose masses are negligible. The amount of positive charge on the nucleus and the number of negative electrons required to form a neutral atom is a characteristic property of all the atoms of a given element.

We found no completely satisfactory model for visualizing the arrangement of the electrons in an atom. One model, based on studies of ionization energies, classified the electrons into groups called shells. This shell model is quite useful in explaining the origin of atomic spectra, in grouping the elements to form the Periodic Table, and in explaining some chemical properties. Another model postulates that the electrons are not particles but clouds of negative charge arranged in a definite structure. This cloud model proves to be very useful in describing the interactions of large numbers of electrons within atoms and molecules.

In Chapter 14 we saw that the shell model could be used to describe the tendency of certain atoms to form ions by losing or gaining one or more electrons. On the basis of our experiments in Chapter 12 and Chapter 14 we concluded that in compounds which are combinations of a metal with fluorine, chlorine, or other elements in that family, the particles present are ions. This is true whether the material is in the solid state, is melted, or is dissolved in water.

The hypothesis proposed in Chapter 11—that the particles involved in a rock salt crystal carry electric charges—appears to be substantiated. Thus we have verified another hypothesis proposed in Chapter 11—that the force holding a rock salt crystal together is due to the electric charge carried by ions constituting the crystal. We are now ready to look for answers to the question that is the theme of this course: "What is the nature of solid matter?" At the moment we must restrict our answers to those solids in which the bonding is ionic; later we will see whether there are other types of bonding. To be more specific, the question may be restated in the following way: "What is the structure of ionic crystals, and how does that structure account for the observed properties of ionic materials?"

15-1 Preliminary steps in determining the structure of a crystal

You have already seen several examples of crystals that you now know to be composed of ions: sodium chloride, sodium acetate, and zinc chloride (used in Experiment 12–3, Part B); lead chloride (Experiment 14–1); copper bromide (Experiment 14–2); and lead nitrate (Experiment 14–3). We now add to this collection one other similar crystalline substance, cesium chloride (CsCl), for it is one we will use to take a close look at the way in which we determine the structure of a crystal—that is, the arrangement of the particles of which it is composed.

First we will examine the crystal shape and properties of cesium chloride, and then compare them with sodium chloride, with which we are more familiar.

Question 15-1 In Experiments 7–2 and 12–3, Part B, we used the following crystalline materials: copper bromide, iodine, lead chloride, lead nitrate, naphthalene, *para*-dichlorobenzene, sodium acetate, sodium chloride, sugar, and zinc chloride. Which of these substances are ionic crystals? What property do the ionic crystals have that enables you to distinguish them from nonionic crystals? To refresh your memory you may wish to refer to the properties of these materials as you observed them in these experiments.

EXPERIMENT 15-1   Crystals of sodium chloride and cesium chloride

Part A   Growing the Crystals

You will be given two small glass vials, one containing 2.5 ml of a solution of cesium chloride, and the other 2.5 ml of a solution of sodium chloride. Put two microscope slides and two plastic boxes in a place where they will not be disturbed, but where you will remember to look at them from time to time. Use a glass stirring rod to transfer four or five drops of one solution to a microscopic slide and the rest of the solution into one side of a plastic box. To insure that the solution remains in one blob, prop up one side of the plastic box with any thin, flat object. Repeat this procedure with the other solution, slide, and box. Be sure you apply an identifying label to each slide and box. It will require between five and twenty-four hours for the water to evaporate, depending on the relative humidity. Be sure to examine the solutions from time to time so that you can see the beginning of the crystal growth. Watch the crystals as they grow, using the magnifying glass as needed.

Question 15-2   For each of the two materials: (a) What is the form of the first crystal you saw? (b) Describe the pattern of growth of the first crystal. (c) Describe the appearance of the materials left on the microscope slides after all the water evaporated, noting any significant difference between NaCl and CsCl.

Part B   Examination of the Crystals

Carefully transfer some of the crystals of each substance from the plastic box to the clean end of a microscope slide. Using the strongest magnifying glass available, examine the crystals carefully to observe their shape. Using the point of a pin, try to cleave or break the crystals.

Question 15-3   Describe the shape of the crystals you have grown. Classify them according to geometrical form. By looking at them, can you guess the shape of the smallest possible crystal?

Question 15-4   Did the NaCl crystals break or cleave in the same way as the rock salt you used in Experiment 5-3? Do they appear to have the same structure as the rock salt crystals?

Question 15-5   Could you cleave the cesium chloride crystals? Were they too small to cleave or is it possible that they don't have cleavage planes?

Question 15-6   From the appearance of the crystals and their cleavage properties, what have you deduced about the structure of NaCl? about the structure of CsCl?

Having recognized that crystals of different materials have different shapes, we want to relate these shapes to the arrangement of ions within the crystal. Before we can determine a detailed arrangement of ions, however, we must identify the ions in the crystal and then establish their relative numbers.

The chemical analysis to determine the kinds of ions present may be an involved process. A simple flame test helps; for example, if a sodium salt is placed in a flame, a characteristic yellow flame is seen. A cesium salt produces a violet flame, but then so does a potassium salt. The uniqueness of the bright-line spectrum of each element, however, permits the spectroscopic test to produce reliable chemical identifications.

The silver nitrate you used in Experiment 3-1 serves as a good test for the presence of the chlorine ion if you are sure that no other ions in the chlorine family are present. You will recall from the experiment that when a few drops of silver nitrate solution were added to a water solution of sodium chloride, a finely divided white solid was formed that was insoluble in water and settled to the bottom. Such an insoluble product formed by the addition of two solutions is called a *precipitate*.

$$NaCl + AgNO_3 \rightarrow NaNO_3 + AgCl\downarrow$$

The downward arrow indicates that the silver chloride settles out as a precipitate. The formation of this precipitate upon the addition of a few drops of silver nitrate to an unknown solution verifies the presence of the chloride ion. Tests such as these are an important step in determining the structure of a crystal since they establish the identity of the ions present.

Next we must determine the relative numbers of the various types of ions in a given sample. We can find the relative number of each ion present if we first know the atomic mass of each element and the fractional part each contributes to the total mass of the sample. The percent by weight of each constituent of sodium chloride and cesium chloride has been determined by chemical analysis. The data for these crystals are given in Table 15–1.

Table 15–1   Chemical Analysis of Sodium Chloride and Cesium Chloride

	Atomic Mass	Sodium Chloride (Percent)	Cesium Chloride (Percent)
Sodium (Na)	23.0	39.3	
Cesium (Cs)	132.9		78.9
Chlorine (Cl)	35.5	60.7	21.1

The data of Table 15–1 tell us, for example, that 100 g of sodium chloride contain 39.3 g of sodium ions and 60.7 g of chlorine ions. Similarly, 100 g of cesium chloride contain 78.9 g of cesium and 21.1 g of chlorine ions. If we divide each of these masses by the mass of the appropriate ion we will obtain the number of each kind of ion in 100 g. What are the masses of the Na^+, Cl^-, and Cs^+ ions? The atomic masses in Table 15–1 are taken from the Periodic Table in Appendix C. Because the mass of the electron is so much less than the mass of the atom, these atomic masses can be considered equal to ionic masses.

To simplify the calculation of the mass of an ion, recall how we defined atomic mass: atomic mass is the ratio of the mass of a given atom to the mass of the hydrogen atom. Therefore, since we know that the mass of the hydrogen atom is 1.7×10^{-24} gram (from your calculations in Question 13–3), we can find the mass of any atom by simply multiplying its atomic mass by 1.7×10^{-24} gram or more precisely, 1.67×10^{-24} gram. Therefore, the masses of the sodium and chlorine ions are:

Mass of Na$^+$ ion $= 23.0 \times 1.67 \times 10^{-24}$ g $= 38.4 \times 10^{-24}$ g
Mass of Cl$^-$ ion $= 35.5 \times 1.67 \times 10^{-24}$ g $= 59.2 \times 10^{-24}$ g

Consequently the number of sodium ions and chlorine ions in 100 g of sodium chloride is

$$\text{Number of Na}^+ \text{ ions} = \frac{39.3 \text{ g}}{38.4 \times 10^{-24} \text{ g}} = 1.02 \times 10^{24}$$

$$\text{Number of Cl}^- \text{ ions} = \frac{60.7 \text{ g}}{59.2 \times 10^{-24} \text{ g}} = 1.02 \times 10^{24}$$

Obviously the number of sodium ions and the number of chlorine ions are equal. Although we used 100 grams in this calculation, if we had chosen any other mass we would still have found that there are equal numbers of sodium and chlorine ions in a chunk of sodium chloride.

Question 15–7 Use the data given for cesium chloride and the method used for sodium chloride to calculate the relative number of cesium ions and chlorine ions in cesium chloride.

Question 15–8 For water, 11.1% of the mass is due to hydrogen atoms and 88.9% is due to oxygen atoms. Use the method just outlined and the atomic masses from the Periodic Table to show that there are twice as many hydrogen atoms as oxygen atoms in any sample of water.

Question 15–9 Sodium chloride and cesium chloride crystals each contain two kinds of ions. As we have just seen there are equal numbers of the two kinds of ions in any sample of either crystal. Suggest a regular, repetitive, three-dimensional arrangement containing equal numbers of the two kinds of ions. Try to think of more than one such arrangement. The shape of sodium chloride crystals suggests that you limit the possible arrangements for that subtance to those which would form cubes.

Could you detect anything about the cleavage of cesium chloride in Experiment 15–1? Did it cleave neatly with straight faces at right angles to each other as sodium chloride did? No doubt you found that cesium chloride is a crystal that has no cleavage. But it has the same ratio of atoms of two kinds as sodium chloride does (1:1). When its crystals grew on the glass slide, did they look like those of sodium chloride? The different behavior in crystal growth and cleavage suggests that the ions in cesium chloride are arranged differently from those in sodium chloride. Their arrangement is determined from the results of x-ray diffraction work.

15-2   *Use of x-ray diffraction in determining the structure of a crystal*

In Chapter 6 we discussed the way in which water waves are diffracted by a regular array of posts. Figure 6–4 is modified slightly as Figure 15–1 to remind you of the geometrical relationships. The incident and diffracted wave crests are represented by the short parallel lines, separated by the wavelength λ; the spacing of the posts is d.

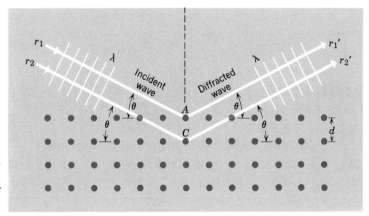

FIGURE 15-1
Waves diffracted by a rectangular array of objects.

If the incident waves travel in the direction indicated by r_1, and r_2, they will be scattered by the posts as a number of circular waves traveling outward in all directions from each post. These waves will interfere constructively to form rays r_1' and r_2' if two conditions are satisfied. First, the angles θ and θ' between the direction of motion of the incident and diffracted rays and the row of obstacles must be equal. Second, only those values of θ that satisfy the Bragg equation

$$n\lambda = 2d \sin \theta$$

will lead to constructive interference. (These conditions were discussed in Chapter 6.) It was stated in Chapter 6 that if we know the wavelength λ of the waves and observe the angles θ for which the diffracted wave has its largest amplitude, we can use these data in the Bragg equation to find the spacing d between layers.

It was also stated in Chapter 6 that the atoms or ions in a crystal are arranged in a regular three-dimensional array. It was suggested by von Laue that the spacing between the particles is of such a magnitude that it should be possible to observe diffraction if x rays were used as the waves. Friedrich and Knipping performed an experiment with x rays and observed an x-ray diffraction pattern. We will perform an imaginary x-ray diffraction experiment to illustrate the use of x rays in studying crystals. The imaginary experimental arrangement is shown in Figure 15–2.

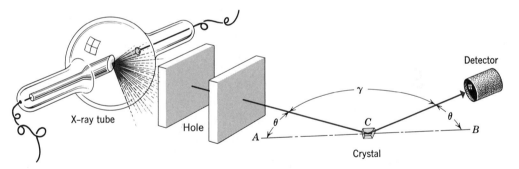

FIGURE 15-2   *X-ray diffraction experiment.*

The x-ray tube and the detector are each mounted on an arm that is free to pivot independently about the vertical axis through C. The crystal is placed at C and kept in a fixed position, and the x-ray tube is turned on. An x-ray tube has been selected that emits x rays of a single wavelength, so that λ is constant throughout the experiment. The arms carrying the tube and the detector are then rotated from the line $AB(\theta = 0°)$ through increasing angles of θ until an angle is found for which the detector records the existence of a diffracted x-ray beam. At this position the x-ray beam formed by the slit is diffracted by a set of ionic layers, received by the detector, and recorded as a maximum intensity.

The geometric relationship between the x-ray tube, the crystal, and the detector is given by the angle θ. That angle is related to the spacing of the layers of ions within the crystal by the Bragg equation. Let us consider this important relationship in some detail.

1. Determination of the Diffraction Angle

In making use of the Bragg equation, the first step is finding the angle θ. Remember that the layers in the crystal, whose spacing d we seek, are not obvious by looking at the crystal; consequently neither is the angle θ.

Suppose that with the arrangement shown in Figure 15–2, the detector records a maximum intensity of the diffracted beam. It follows, then, that the layers of ions responsible for the diffracted beam are parallel to the line AB and perpendicular to the plane of the page. These layers need not be parallel to the crystal faces. The angle θ is the angle between the incident x-ray beam and AB, or between the diffracted beam and AB. But even if this hypothetical experiment were performed, we could not see where the line AB would be; so how can we measure θ? To see how this angle is measured, look at the angles in Figure 15–2. We can measure the angle γ, the angle between the incident and diffracted beam, and since $\theta + \gamma + \theta = 180°$, we can calculate θ by solving this equation for θ. Having calculated θ we can use the Bragg equation to find d, as long as we know λ and n.

Question 15–10   In a hypothetical Bragg diffraction experiment such as that shown in Figure 15–2, a maximum intensity of the diffracted beam is found when the angle γ between it and the incident beam is 168°. What is the angle θ between the incident beam and the layers of ions?

2. Determination of n

Having found a value of θ for which the diffracted beam has a maximum intensity, we have found a value of θ that satisfies the Bragg equation. But other values of θ should also satisfy the equation even if we maintain the crystal in the same orientation so that the same ion planes are parallel to the line AB (and therefore we use the same spacing d), and even if we use the same wavelength λ, for n can take on various values 1, 2, 3, ... To find the particular angles θ_1, θ_2, θ_3, . . . that correspond to the respective values of n, we simply rotate the two arms simultaneously through ever increasing values of θ until other maxima are recorded in the detector. A curve showing the intensity of the output of the detector is shown in Figure 15–3. In this example

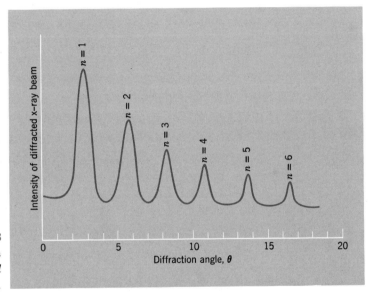

FIGURE 15–3

X-ray diffraction pattern from a hypothetical crystal composed of identical ions.

the scattering centers are all the same, but a curve like that in Figure 15–3 will also result if adjacent planes of atoms are the same, even though each plane contains more than one kind of atom. This curve, obtained from our hypothetical experiment, reveals that for $n = 1$, $\theta = 2.7$ degrees; for $n = 2$, $\theta = 5.4$ degrees; and for $n = 3$, $\theta = 8.2$ degrees. As we will see later the pattern does not always look like this, but it is always possible to deduce the values of θ and n from the observed pattern.

3. Determination of λ and d

If we wish to use the Bragg equation to find d from the diffraction pattern, we must know λ in addition to the values of θ and n. In Chapter 6 when the Bragg equation was introduced, no mention was made of the wavelength of x rays except that this wavelength must be very small to account for the penetrating power of x rays. The experimental results of von Laue were the first indication of how small they really are. When the experiments were first performed in 1912, neither λ nor d was known, but fortunately Ewald was able to give von Laue a rough idea of λ. Subsequently, experiments in x-ray diffraction and data from other kinds of experiments have given more precise values for λ, resulting in more exact values of d, and vice versa. As a result of this process of continuous improvement, the distance between the centers of Na^+ ions in NaCl is now known to be 5.627×10^{-10} meter; the wavelength of the strongest x-ray spectral line from an x-ray tube with a copper target, for example, is known to be 1.540×10^{-10} meter or 1.540Å.

Question 15–11   X rays having a wavelength of 1.5×10^{-10} meter are diffracted by a crystal for which the spacing of one set of layers of ions is 4.12×10^{-10} meter. What are all the values of $\sin \theta$ less than 1 for which there is a diffracted beam of x rays from these layers?

4.   Which d is Measured?

In Figure 15–1, d was used to represent the spacing between rows of atoms placed in a regular square array. However, these rows are not the only ones that can be found. If you have ever driven past an orchard of young trees planted in a regular pattern, such as that shown in Figure 15–4, you may have noticed that from different positions of the car you saw rows radiating outward in different directions. If x rays are incident on a crystal, a similar situation exists, except that we cannot see the trees—only the forest. We cannot see the ions—only the crystal. For this reason we will find many different values of d for various orientations of a particular crystal in an x-ray beam.

Each value of d reveals a different set of ion layers. In Figure 15–4 five of these sets are represented by dotted lines, and each can be described by a characteristic spacing d.

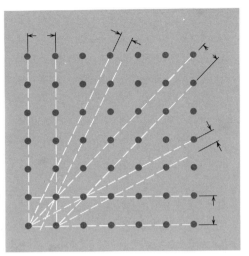

FIGURE 15-4
Regularly spaced objects form many different sets of parallel straight lines, each set with a different spacing. Hold the page with your eye near the lower right-hand corner of the figure, and look at the different lines of dots at a grazing angle.

You can see that a number of sets of planes are possible and that the values of d are different for each set. Hence, in determining the structure of a crystal, an x-ray crystallographer must make a series of measurements and from them determine first the values of d and then the arrangement of the ions.

5. The Height of the Peaks in the Diffraction Pattern

For each d corresponding to the spacing of a set of identical planes of ions in a crystal (and for a constant λ), there is a θ for every integer n that satisfies the Bragg equation. For this angle the waves scattered by these ions interfere constructively; for all other angles the scattered waves interfere destructively. Interference takes place in this way as long as all the scattered waves have the same amplitude; but the amplitude depends on the number of electrons in the scattering ions. If we can relate the diffraction amplitudes to the arrangement and kind of the scattering ions, we will have a means of finding out more about the arrangement of the ions in a crystal.

Suppose that we were doing a Bragg diffraction experiment, using equipment like that shown in Figure 15–2, on a hypothetical crystal consisting of a rectangular array of particles (Figure 15–5a). As in our previous example, all of the particles are the same, but their arrangement is differ-

(a)

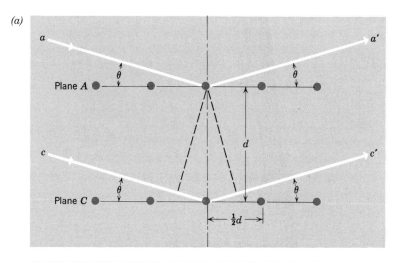

(b)

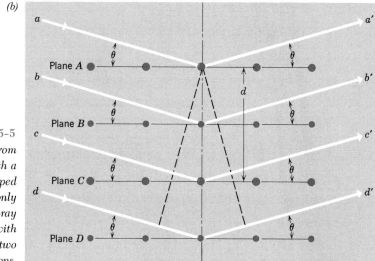

FIGURE 15-5
(a) X-ray diffraction from a crystal with a rectangular box-shaped unit cell composed of only one kind of ion. (b) X-ray diffraction by a crystal with alternating layers of two different kinds of ions.

ent. For reasons that will be apparent very soon, we will assume that the distance d between planes is twice the distance between the particles in any one plane. We will also assume that all of the particles are in their proper places, each at a corner of a rectangular-shaped unit cell. (Actually no crystal has such a structure. When crystals are made of just one kind of atom, the atoms are arranged in other ways, one of the commonest of which is shown in Figure 6–4. And of course a crystal could not be made of only one kind of *ion* since all of its constituent particles would repel each other.) Since the particles in the hypothetical structure shown in Figure 15–5a are all of one kind, the variation of intensity of the diffracted beam as the angle θ is varied would be like that shown in Figure 15–3.

The first peak ($n = 1$) occurs when θ is such that the path difference between rays a and c (Fig. 15–5a) is equal to one wavelength. The difference in the paths is equal to that part of the path of ray c that lies between the lines made with short dashes. The peak indicates that rays a' and c' interfere constructively. For some smaller angle at which the path difference is one-half, the rays a' and c' would interfere destructively, canceling each other completely since they have equal amplitudes.

Now suppose that we add planes of a different kind of ion from those in planes A and C of Figure 15–5a. The new ions form planes B and D as shown in Figure 15–5b. Our new crystal is composed of two kinds of ions appearing in parallel planes separated by a distance of $(\frac{1}{2})d$. The ions in any one plane are identical; the planes of different ions alternate. Therefore, the distance between planes of similar ions is still d. Furthermore, let us suppose that this new kind of ion differs from the original one in that each of these new ions has only half as many electrons as each of the original ones. The addition of the new ions will, of course, alter the diffraction of x rays.

The difference in the length of path traveled by rays b and d is the same as the difference in the length of path traveled by rays a and c (you can ascertain this by measuring), that is, it is 1λ and therefore rays b' and d' interfere constructively with each other just as rays a' and c' do. However, the difference between the length of path traveled by ray b and that traveled by ray a is only one-half λ so these rays interfere destructively. The same is true of rays d and c. The rays diffracted from the particles in planes A and C interfere constructively, those diffracted from planes B and D also interfere constructively. But the rays diffracted from planes A and B, and those diffracted from planes C and D, interfere destructively. So what happens?

If all of the particles had the same ability for scattering

x rays, the array would not diffract x rays at all at that same angle θ; the Bragg equation would not be satisfied since the spacing between the planes would be $(\frac{1}{2})$ d rather than d. However, the scattering ability of an ion depends on the number of electrons in the particle. And since the particles in planes B and D have only half the electrons that those in planes A and C have, they have only half the scattering ability. Therefore, the intensity of rays b' and d' is half that of rays a' and c'.

How does the difference in intensity of the two rays affect the destructive interference? What happens when a high crest in water waves meets a very shallow trough? By adding the amplitudes of these two waves, the result would be a crest, but not quite as high as the original one. So the rays a' and b' do not cancel each other completely, nor do the rays c' and d'. The greater intensity of rays a' and c' wins out, and the detector records some x rays but less than if the particles in planes B and D were missing as in Figure 15–5a.

The first peak of the diffraction record from such a crystal is therefore not the biggest peak, as shown in Figure 15–3, but is considerably reduced. For the first peak $n = 1$, because the path difference for rays diffracted from neighboring *identical* planes (such as A and C, or B and D) is 1λ.

For $n = 2$, at a larger angle θ, the path difference between rays a and c is 2λ. That between rays b and d is of course also 2λ. The path difference between rays a and b is half of this, as before, but now it is 1λ—and 1λ means constructive interference. So rays a', b', c', and d' interfere constructively when $n = 2$, and a high peak is recorded.

The reasoning that applies for the partial destructive interference at $n = 1$, and the resulting small peak in the diffraction record, also applies to $n = 3$, $n = 5$, etc. Therefore the peaks for odd numbered n's are weak.

Correspondingly, the reasoning that applies for $n = 2$ also applies for $n = 4$, $n = 6$, etc., and these peaks are higher than the odd-numbered peaks. The resulting plot of intensity versus angle θ would be like that shown in Figure 15–6.

By observing the relative intensities of x rays diffracted

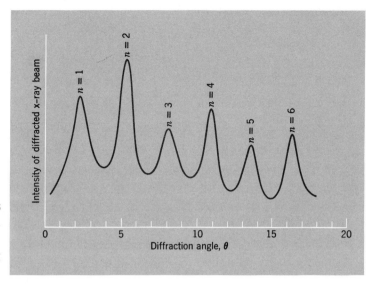

FIGURE 15-6
*X-ray diffraction pattern
from a crystal having two
kinds of ions in alternating
layers.*

from crystals as well as the angles θ of the diffracted beams,
an x-ray crystallographer is able to draw conclusions about
the nature of the atoms or ions and the way they are
arranged to form the crystal structure.

Ions in crystals are arranged in regular patterns. The unit
that contains the smallest number of ions, which if repeated
often enough in all directions produces an accurate model
of a large piece of crystal, is called the *unit cell*. The unit
cells of some crystals are rhombic-shaped, like those of
calcite (see Abbé Haüy's "molecules integrantes," discussed
in Chapter 5); some are brick-shaped units; and some are
cube-shaped. Very few other shapes are possible since the
cells of a given crystal must all be identical (if they are to
be repeat units of the pattern) and must be of such a shape
that they fit together to fill space solidly without any gaps.
(Imagine trying to fit together units shaped like a house with
a peaked roof and a chimney.)

Question 15–12 We know that sodium chloride has a cubic structure and
that along one edge of such a cube Na^+ ions and Cl^- ions
alternate, and are 2.81 Å apart. (a) How far apart are the
Na^+ ions? (b) The unit cell is cube-shaped and the arrangement of ions is the same along every edge. What are the
dimensions of the unit cell? (c) What is the volume of the
unit cell?

Question 15–13   Potassium chloride has the same structure as sodium chloride.
(a) Which of the three ions, K^+, Na^+, and Cl^- has the most
electrons? the least? (b) When oriented correctly, a sodium
chloride crystal yields an x-ray diffraction pattern with
weaker peaks for $n = 1, 3, 5, \ldots$ similar to those in Figure
15–6. If a potassium chloride crystal were oriented in the
same direction as the sodium chloride crystal, would it, too,
yield weaker peaks for values of $n = 1, 3, 5, \ldots$? (c) Justify
your answer in part *b*.

15-3   *The structure of cesium chloride*

When a crystal of cesium chloride is oriented in a particular direction on the x-ray diffraction equipment, the x-ray
diffraction record exhibits weaker peaks for values of $n = 1$,
$3, 5, \ldots$. The crystal might, therefore, contain parallel but
alternating planes of Cs^+ and Cl^- ions (shown horizontal in
Figure 15–7*a*). Calculations from these records reveal that
the spacing between planes of similar ions is 4.12 Å.

If that same crystal is now rotated through an angle of
90° so that those planes are perpendicular to their position

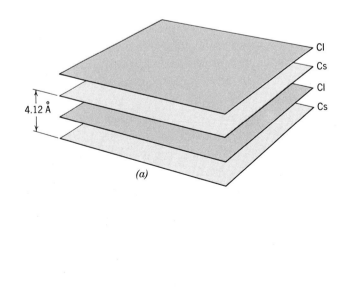

(a)

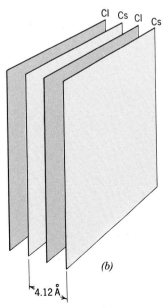

(b)

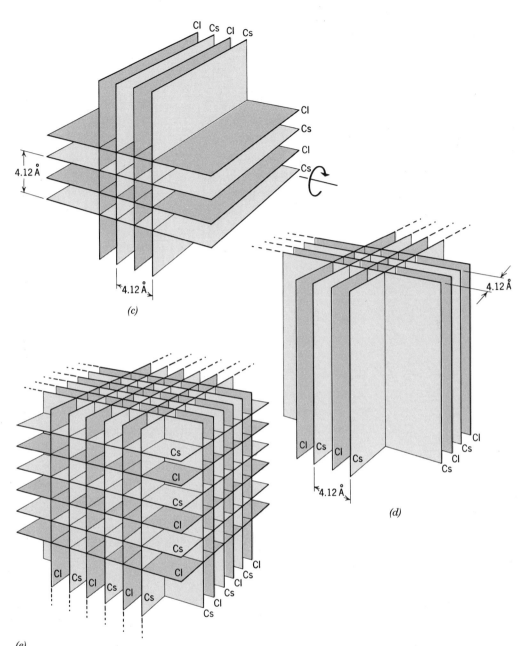

(e)

FIGURE 15-7 *(a) The first x-ray diffraction pattern (see Fig. 15-6) of cesium chloride reveals a set of parallel alternating planes of Cs⁺ and Cl⁻ ions. (b) The planes of a rotated so that they are at right angles to their previous orientation. (c) The second x-ray diffraction pattern reveals a second set of parallel alternating planes of Cs⁺ and Cl⁻. (d) The first and second sets of planes rotated so that each set is perpendicular to the first set of planes when oriented as shown in a. (e) The third x-ray diffraction pattern reveals a third set of parallel alternating planes of Cs⁺ and Cl⁻.*

in Figure 15–7*a* (see Fig. 15–7*b*), and an x-ray diffraction record is made again, the result will be the same as before: weaker peaks for $n = 1, 3, 5, \ldots$, and these peaks will be spaced the same as the first. Therefore, there might again be parallel but alternating planes of Cs^+ and Cl^- ions, with the planes of similar ions spaced 4.12 Å apart (Fig. 15–7*c*).

If that crystal is again rotated through an angle of 90°, so that all the planes that have been detected are perpendicular to the first set of planes detected as they were oriented in Figure 15–7*a* (see Fig. 15–7*d*), and a third x-ray diffraction record is made, the results will be the same as the previous two. Consequently, there might be a set of parallel but alternating planes of Cs^+ and Cl^- ions at right angles to both sets of the previously detected planes (see Fig. 15–7*e*).

The crystal can be rotated in various other angles to detect other sets of parallel planes that are oblique to the three sets of planes already detected. The results of such studies indicate clearly that the cesium and chloride ions are located in their respective planes as shown in Figure 15–8*a*. A cesium ion (shown in grey) is located in the very center of a cube formed of six chloride ions (shown in blue). Similarly, a chloride ion is located in the very center of a cube formed of six cesium ions. A unit cell of this crystal is either one of these cubes; the one with the cesium ion in the center is shown in Figure 15–8*b*.

In actual practice the crystallographer does not change

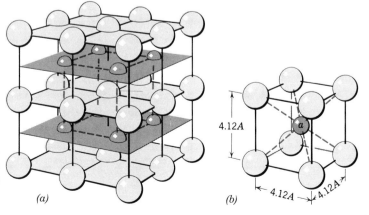

FIGURE 15–8
(a) The structure of cesium chloride. (b) The structure of a cesium chloride unit cell.

(a)

4.12A

a

4.12A   4.12A

(b)

the orientation of a single crystal and then rotate the arms containing the x-ray tube and detector to find peaks as we have described in our hypothetical experiment. The x-ray tube is not easily moved, so he moves only the crystal and detector.

For some studies the single crystal proves tedious; the crystal and detector must be moved through many different angles. The crystallographer, however, can obtain crystals in all orientations at the same time if he grinds the crystal into a fine powder and uses that instead of a single crystal. Each grain of powder is a tiny crystal; and since there are so many tiny crystals all mixed, they assume many different orientations. Consequently, there must be some tiny crystals oriented in just the right direction relative to the incident x-ray beam to satisfy the Bragg equation for each of the different spacings *d* within that kind of crystal. The resulting diffracted beams, therefore, assume many different angles with respect to the incident x-ray beam. Each of these diffracted beams can be located by using a curved photographic film covering a much larger area than the single detector in Figure 15–2. Each of the beams diffracted will expose the film in a particular spot.

EXPERIMENT 15–2 Construction of a model of cesium chloride

In this experiment (which is really an experiment only in that you study models of crystal structure) the class will build two different models of the cesium chloride structure. To save time, small groups of one-half the class can each build one kind of model, and small groups of the other half can each build the other kind. Both sets of groups will use polystyrene balls to represent the ions. Use the larger balls to represent the Cl^- ions, and the smaller ones to represent the Cs^+ ions. It is surprising to learn that even though the atomic number of cesium is larger than that of chlorine, the Cs^+ ion is smaller than the Cl^- ion. Their relative sizes have been measured by x-ray diffraction studies and will be discussed in Section 15–5.

The Open Model

The first model to be constructed is the "open model," in which the balls representing the ions are both held together and separated by toothpicks.

Use round full-length toothpicks sharpened on both ends to fasten the chloride ions into an arrangement as shown in Figure 15-9a, in which two cubes are joined. It will help if you use a small nail to make a hole in the ball before inserting the toothpicks. Take care in making this hole to aim for the center of the ball and to make the angles between the toothpicks as close as you can to 90 and 180 degrees. Then prepare a layer of six cesium ions, using full-length toothpicks, like that shown in Figure 15-9b, and use broken

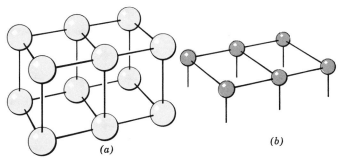

FIGURE 15-9
*Partially completed model
of the cesium chloride
structure.*

(a)

(b)

toothpicks to support this layer so that the cesium ions will lie halfway between the chlorine ions. Then insert the layer of cesium ions into the chlorine-ion cube by temporarily removing some of the toothpicks holding the layer together (see Fig. 15–10). This model indicates how the cubic lattice of chlorine ions interprenetrates the cubic lattice of cesium ions, and vice versa, to form the cesium chloride structure.

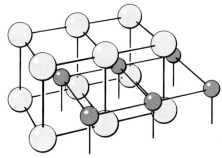

FIGURE 15-10
*Completed model of the
cesium chloride structure.*

The Packing Model

The second model, called the "packing" or "space-filling" model, can also be used to represent the cesium chloride structure. Although the packing model is easier to build than the open model, it does not reveal the structure of the unit cell as clearly. To build the packing model, use twenty-seven one-inch polystyrene spheres to represent chlorine ions, and eight $\frac{3}{4}$-inch spheres to represent cesium ions.

Use polystyrene cement to glue three chlorine ions in a straight line, using two blocks or other objects to hold them while they dry. Make up nine such groups. When they are thoroughly dry, take three sets of chlorine ions and glue them to form a square with three ions on a side. Repeat with the other groups of chlorine ions until you have prepared three square layers.

When the layers have dried completely, assemble the crystal model by placing four cesium ions in the depressions in one chlorine-ion layer, glue another chlorine-ion layer on top, place four more ceium ions, and finally glue the third chlorine-ion layer. After the glue is dry, the crystal model is sufficiently rigid to be picked up and moved around.

Question 15–14   Examine the diagram in Figure 15–8*a* and the open model of the CsCl crystal, and justify the statement that a unit cell composed of eight Cl⁻ ions at the corners of a cube with a Cs⁺ ion at the center is equivalent to a unit cell composed of eight Cs⁺ ions at the corners of a cube with a Cl⁻ at the center. In combining a large number of unit cells to form a model of a crystal, does it matter which of the two unit cells you use as long as you use the same one?

Question 15–15   (a) If one uses the model of the unit cell shown in Figure 15–8*b*, how many Cl⁻ ions are in each unit cell? how many Cs⁺ ions?
(b) In a large crystal composed of many unit cells, in how many of them does each Cl⁻ ion participate? In how many unit cells does each Cs⁺ ion participate? How do these statements agree with the calculation you made in Question 15–7? If they do not agree can you make them compatible?

Question 15-16 We have just seen that x-ray diffraction studies indicate that the unit cell of CsCl is a cube. Does this model correspond to the appearance of the crystals you grew in Experiment 15–1?

15-4 *The structure of sodium chloride*

Using the same experimental techniques and the same reasoning process as we used with cesium chloride, crystallographers have determined the structures of a great many other crystals. The only other example we will use, however, is sodium chloride.

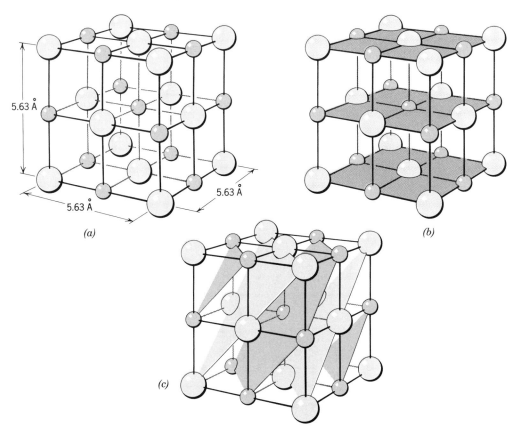

FIGURE 15-11 *(a)* NaCl *unit cell. Blue spheres represent* Cl⁻ *ions and grey spheres* Na⁺ *ions. (b)* NaCl *unit cell showing planes parallel to cell faces containing both* Cl⁻ *and* Na⁺ *ions. (c)* NaCl *unit cell showing diagonal planes containing either* Na⁺ *or* Cl⁻ *ions, but not both.*

Sodium chloride also has a cube-shaped unit cell, as you might have suspected from the way its crystals grow and from its mutually perpendicular cleavage planes. The unit cell found by x-ray diffraction measurements is shown in Figure 15–11*a*. The dimensions of this cube are 5.63 Å. Notice that this is the distance along an edge of the cube from one Cl⁻ ion to the next, even though other Cl⁻ ions (those along a diagonal of a face) are closer than this. It should be pointed out that the spacing *d* in the Bragg equation is the distance between layers that are exactly alike, such as the top and bottom layers of this cell (Fig. 15–11*b*). Figure 15–11*c* shows a series of diagonal planes, each of which contains either all Cl⁻ ions or all Na⁺ ions. X-ray beams diffracted from these diagonal planes will show patterns with the characteristic alteration of intensity pictured in Figure 15–6. It is on the basis of records such as these that the sodium chloride structure shown in Figures 15–11*a* – 15–11*c* was deduced.

As with cesium chloride, our understanding of the structure of sodium chloride will be aided if we construct and examine an actual model in which the ions are represented by polystyrene balls.

EXPERIMENT 15–3   Construction of a model of sodium chloride

In this experiment you will construct two models of sodium chloride similar to the cesium chloride models. The representation of relative ion sizes for these models is obtained by using 1-inch spheres for the chlorine ions and $\frac{1}{2}$-inch spheres for sodium ions. As in Experiment 15–2, half the class should make open models and the other half packing models.

The Open Model

To construct an open model, you will make up two kinds of layers (Fig. 15–12*a* and 15–12*b*). First construct two layers of the kind that has a chlorine ion at the center (Fig. 15–12*a*) by using toothpicks as connectors. Use a small nail to make the holes at right angles, and use glue to

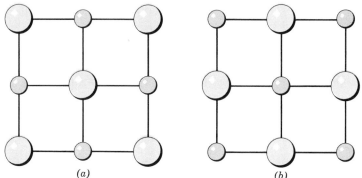

FIGURE 15–12

*Layers for sodium chloride
model.*

(a)   (b)

hold the toothpicks in place once you are sure they will not
have to be removed. Next, construct one layer with a sodium
ion at its center (Fig. 15–12b). Finally, use toothpicks
to fasten the three layers together. The best way to do this
is to put the layers flat on a table top, use the small nail to
make vertical holes in the four corner atoms in each layer,
insert toothpicks, join the layers, and then fasten it all
together with glue. Note that when the model is finally
assembled and placed on the table, the small spheres in
the bottom plane will not touch the table.

The Packing Model

The packing model of sodium chloride that you will construct will be somewhat larger (63 chlorine ions and 62
sodium ions) and is a little more difficult to construct, because one must glue together spheres of different diameter
in one layer. There are several ways of holding the spheres
in place while the glue is drying. One is to prepare small
supports, $\frac{1}{4}$-inch high and approximately $\frac{1}{2}$-inch long and
$\frac{1}{2}$-inch wide, which can fit under each $\frac{1}{2}$-inch diameter
sphere and hold it in place, as shown in Figure 15–13a
while the entire layer is held in place by blocks or other
supports from the side. A second way is to use short pieces
of toothpicks as connectors, as shown in Figure 15–13b.
Use one of these methods to make up five layers, three with
1-inch chlorine ions at the corners and two with $\frac{1}{2}$-inch
sodium ions at the corners. Then glue the layers together.
It is instructive to prepare the top layer by omitting the
glue that would hold the center chlorine ion in place. Then

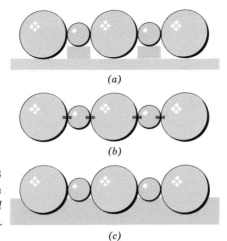

(a)

(b)

FIGURE 15-13
Methods for construction of a row of sodium and chlorine ions.

(c)

one can drop it in place in the finished crystal model, and remove it at will to show the symmetrical arrangement of sodium ions around any one chlorine ion.

Question 15-17 Consider the unit cell shown in Figure 15–11a and the open model you constructed. How many Cl^- ions are present in this unit cell? how many Na^+ ions? If these numbers are different could you construct an equivalent unit cell in which the numbers of Cl^- ions and Na^+ ions are interchanged?

Question 15-18 Again consider the unit cell shown in Figure 15–11a, but this time as if it were part of a large crystal. (a) In how many unit cells does each of the following ions participate: a Cl^- ion at the corner; a Cl^- ion at the center of a face; a Cl^- ion inside the unit cell; a Na^+ ion along an edge; a Na^+ ion inside the unit cell? (b) In Section 15–1 it was found that there are equal numbers of Na^+ and Cl^- ions in a sample of NaCl. Does this agree with your answer to Question 15–17? If not, use the first part of this question to eliminate the discrepancy.

Question 15-19 We have just seen that x-ray diffraction studies indicate that the unit cell of NaCl is cubic. Does this model correspond to the crystals you grew in Experiment 15–1?

Sodium chloride cleaves along cleavage planes that are parallel to the ion layers shown shaded in Figure 15–11*b*. There are just as many positive sodium ions as there are negative chlorine ions in these layers. The neighboring layer of ions have the same arrangement. A positive sodium ion of one layer lies directly above a negative chlorine ion of the next lower layer. Recall that an ion is attracted by the ion of the opposite charge but repelled by the ion of the same charge.

Does the crystal easily cleave along these planes because the exposed cleavage surfaces are electrically neutral? We don't know. Many people have tried to explain why crystals cleave in various ways, but no one explanation seems to fit all crystal structures. Many crystals have no cleavage at all, like alum, quartz, and cesium chloride. Some have poor cleavage in one or more directions; some have excellent cleavage, like sodium chloride in the three mutually perpendicular directions. Mica, which has the most excellent cleavage of any crystal, cleaves in only one direction because its structural units (ions and complex groups of atoms) are arranged in sheets with weak attraction between one sheet of structural units and the next, but strong attraction between neighboring units within a sheet.

The unit cells of sodium chloride, cesium chloride, and some other crystals are cubic-shaped, but this is not true of the unit cells of all crystals. Recall the shapes of the cleavage fragment of calcite. They are very different from those of sodium chloride. Calcite is not a cubic crystal. Mica, which cleaves in sheets, has a unit cell that is different from either calcite or cubic crystals. Patterns of different ions assume many different shapes.

15-5   *Sizes of ions*

In Figures 15–8 and 15–11 the crystal structures of CsCl and NaCl are shown with different size circles to represent Cs^+, Na^+, and Cl^- ions. In Experiments 15–2 and 15–3 you used different sizes of polystyrene balls to represent these same ions. This certainly implies that ions have different relative sizes. What do we mean by the size of an ion? How do we know these relative sizes? Does the open model or the packing model properly represent the sizes of the ions?

The distances we measure with x-ray diffraction are the equilibrium distances between ions. By this we mean that any effect that tends to decrease this distance will be opposed by repulsive forces, and any effect that tends to increase this distance will be opposed by attractive forces. The attractive forces are electrical and result from the opposite charges of neighboring ions. The repulsive forces result in part from the repulsion of the electron clouds of neighboring ions, and in part from like charges of more distant ions. We can think of the repulsion between electron clouds of neighboring ions as the result of forces that arise when clouds of two ions are pushed together so closely that they effectively overlap. Consequently, the packing model is useful when we are considering forces that determine the sizes of ions; the open model, however, better reveals the geometrical arrangement of ions.

You constructed an open model of sodium chloride, one plane of which is shown in Figure 15–12*b*. You also constructed a packing model for sodium chloride, one plane of which is shown in Figure 15–14. The distance between the centers of the Na^+ and Cl^- ions is 2.81 Å. This is half the size of the NaCl unit cell drawn in Figure 15–11*a* and is obtained from x-ray diffraction measurements. The distance between the centers of two adjacent ions is called the *interionic distance;* and although it tells us a great deal about the size of the unit cell, it does not directly tell us the sizes of all the ions in that cell. If the electron clouds of two adjacent ions touch, the interionic distance is the sum of the radii of those two ions. But we can find the radius of each ion?

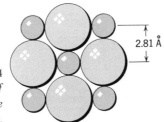

2.81 Å

FIGURE 15-14
Arrangement of a layer of sodium ions and chlorine ions in sodium chloride.

Table 15–2   Interionic Distances of Selected Ion Pairs in Crystals (Ångströms)

Positive Ion	Negative Ion			
	Fluorine F⁻	Chlorine Cl⁻	Bromine Br⁻	Iodine I⁻
Lithium Li⁺	2.01	2.57	2.75	3.00
Sodium Na⁺	2.31	2.81	2.97	3.23
Potassium K⁺	2.67	3.14	3.29	3.58
Rubidium Rb⁺	2.82	3.29	3.43	3.66
Cesium Cs⁺	3.01	4.11	4.29	4.56

In a search for an answer to this question, it may be helpful to compare the interionic distances of a number of crystals (see Table 15–2). The distances in this table were obtained by x-ray diffraction experiments and are given in Ångströms. But again, to know that the interionic distance between a sodium ion and a fluorine ion is 2.31 Å and the distance between a sodium ion and a chlorine ion is 2.81 Å, does not tell us the radii of the sodium, fluorine, or chlorine ions. However, if we can determine the radius of just one ion in this table, then by assuming that the radius of a given ion does not change much from one crystalline compound to another, we can learn the radii of all of them.

It is not at all easy to find the radius of one ion, but let us propose a method, try it, and then see how well it works by checking the result with values determined by crystallographers. If a crystal were formed of two kinds of ions, and if one of those kinds of ions were much bigger than the other kind, it might be that the smaller ions are small enough to be tucked away in the space between the bigger ions (see Fig. 15–15). Then assuming that the electron clouds of the bigger ions do indeed touch, the interionic distance of those ions is equal to two ionic radii, or one

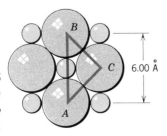

FIGURE 15–15
Proposed model of one plane of LiI crystal used to determine ionic radius.

6.00 Å

ionic diameter. Let us try this method from values for the ions given in Table 15–2.

Which of the ions from Table 15–2 should we try? We will need one positive ion and one negative ion. Suppose we choose the positive ion as the smaller of the two? Which of the positive ions is the smallest? It would appear to be lithium, for the interionic distance of lithium fluoride is smaller than that of any other fluoride compound; and the same applies to the compounds with chlorine, bromine, and iodine.

Which of the negative ions is the largest? Using the same reasoning that we used with the positive ion, we would select iodine.

But what is the structure of the unit cell of LiI? X-ray diffraction studies reveal that it is the same as that of sodium chloride (see Fig. 15–11). One layer of the sodium chloride unit cell is shown in Figure 15–14. The dimensions of the LiI unit cell, however, are 6.00 Å on each edge of the cube. If we assume that the electron clouds of the iodine ions are touching (but, since electron clouds are fuzzy things, it is reasonable to ask what is meant by "touching"), and if you refer to Figure 15–15, you can see how to estimate the diameter of the iodine ion.

The right triangle ABC has a hypotenuse equal to 6.00 Å. The two legs of this right triangle are equal, and we will call them each s. Each of these legs equals two radii (or one diameter) of an iodine ion. By applying the Pythagorean theorem

$$s^2 + s^2 = 6^2$$

we find that

$$2s^2 = 36$$
$$s^2 = 18$$
$$s = 4.24 \text{ Å}$$

(You can verify that the square root of 18 is 4.24 by multiplying 4.24 by 4.24.) This, then, is our best estimate of the diameter of the iodine ion. Its radius, therefore, is estimated to be 2.12 Å.

Using the radius of the iodine ion as 2.12 Å, we can now

estimate the diameter of the lithium ion. Side AB of the triangle ABC is composed of the radii of two iodine ions and the diameter D of one lithium ion. Assuming that the electron clouds of the lithium and iodine ions are touching, and recalling that the side AB is 6.00 Å long, we can write

$$2.12 + D + 2.12 = 6.00$$
$$D = 6.00 - 4.24$$
$$D = 1.76 \text{ Å}$$

The values for the diameter of the iodine ion (4.24 Å) and the lithium ion (1.76 Å) are comparable with the values determined by crystallographers using many different crystals. The best values for these diameters are 4.20 Å and 1.36 Å. Consequently, it would seem that the electron cloud of lithium does not touch the electron clouds of the iodine ions surrounding it.

The result of this investigation only indicates some of the difficulties in determining ionic sizes. Nevertheless, using many different ions, continued studies of this sort have resulted in estimates of the sizes of all of the common ions. These studies indicate that the radius of a particular ion is not fixed for all crystals of which it is a member. Apparently the ionic radius depends to some extent on the other ions present and on the structure of the unit cell.

Question 15–20   List the positive ions of Table 15–2 in order of increasing size. Do the same for the negative ions.

Question 15–21   How much larger is the chlorine ion than the fluorine ion? the potassium ion than the sodium ion? Do you get more than one answer for these differences? If you do, what does this tell you about the assumption that the ion size is the same for similar compounds?

Question 15–22   In Chapter 13 we described Rutherford's scattering experiment to determine the size of the nucleus. Why can't a similar experiment be performed to determine the size of an ion?

It is curious that the radius of the chlorine ion (atomic number 17) is about 1.81 Å, while that of the cesium ion (atomic number 55) is only 1.67 Å. Why should the ion with more electrons have a smaller radius? There seem to be several reasons.

The nucleus of the cesium atom has a much higher positive charge than that of the chlorine atom, so all the electrons are pulled a little closer to the nucleus. In addition, a chlorine atom must acquire an additional electron to form a chlorine ion. The cesium atom, on the other hand, loses an electron to become an ion. This outermost electron of the cesium atom is alone in its shell and is easily lost. The loss of this one electron means that the cesium ion has one electron shell less than the cesium atom. In contrast, during the formation of the chlorine ion, a chlorine atom merely fills up its outermost shell; the chlorine ion and atom each have the same number of shells. Consequently, the cesium ion is much smaller than the cesium atom, and the chlorine ion is just a little larger than the chlorine atom.

Still another factor affects ion size. When the cesium atom loses an electron to become an ion, its total negative charge is reduced. Therefore the effect of the positive nucleus on each electron is stronger and the whole thing shrinks. However, when the chlorine atom acquires an electron to become an ion, the effect of the positive nucleus on each electron is reduced, and the whole thing expands.

Could you have predicted the relative sizes of the cesium and chloride ions from the data in Table 5–2?

15-6 *The mass of an atom and Avogadro's number*

Now that we are able to determine the size of a unit cell of a given crystal as well as the arrangement of ions within that crystal, we can calculate the mass of an ion (or of an atom since the electron's mass is negligible). We can calculate the number of unit cells in a given chunk of crystal by measuring the volume of that chunk and by knowing the volume of one unit cell. Knowing how many unit cells exist in our chunk of crystal, and knowing how many of each kind of ion are in each unit cell, we can calculate the number of each kind of ion in our chunk. Knowing the number of each kind of ion and the relative mass of the ions permits us to calculate the mass of each of the ions present.

As an example, let us calculate the mass of the Cs^+ and Cl^- ions from our knowledge of the CsCl crystal. The cubic-shaped unit cell of CsCl is shown in Figure 15–8b, and how that unit cell repeats itself is shown in Figure 15–8a as well as in the open model you built in Experiment 15–2. The length of one edge of the cube in a unit cell of CsCl is 4.12 Å, and we can calculate the volume by taking the cube of this number. Let us perform this operation with units of centimeters rather than Ångströms (1 Å = 10^{-8} cm):

$$\text{Volume of unit cell} = (4.12 \times 10^{-8} \text{ cm})^3$$
$$= (4.12)^3 \times (10^{-8})^3 \text{ cm}^3$$
$$\text{Volume} = 70 \times 10^{-24} \text{ cm}^3$$
$$\text{Volume} = 7.0 \times 10^{-23} \text{ cm}^3$$

How many unit cells are there in our chunk of crystal? For convenience, let's choose a chunk of crystal with a volume of 1.0 cm³:

$$\text{Number of unit cells} = \frac{1.0 \text{ cm}^3}{7.0 \times 10^{-23} \text{ cm}^3}$$
$$\text{Number} = 0.14 \times 10^{23}$$
$$\text{Number} = 1.4 \times 10^{22}$$

Now, how many ions are there in each of these unit cells of CsCl? From Figure 15–8 and from your open model of the crystal, you see that the ion in the center of the unit cell is in that cell exclusively. However, each of the corner ions is shared by other cells. How many cells share one corner ion? Eight; count them. Consequently, only one-eighth of each corner ion is in our unit cell in Figure 15–8b. Therefore, since each unit cell has eight corners, there is the equivalent of one cesium ion and one chlorine ion in each unit cell. In our chunk of crystal, then, there must be 1.4×10^{22} cesium ions and 1.4×10^{22} chlorine ions.

We can now determine the mass of each ion. To start, we can either measure the mass of our crystalline chunk on an equal-arm balance or determine the mass from its experimentally determined density, which is 4.0 g/cm³. Its volume is 1.0 cm³, so its mass is clearly 4.0 g. The mass of the ions in one unit cell, that is, a pair of cesium and chlorine ions, is the mass of the whole chunk divided by the number of unit cells in it.

$$\text{Mass of pair of Cs and Cl ions} = \frac{4.0 \text{ g}}{1.4 \times 10^{22}}$$

$$\text{Mass of pair} = 2.9 \times 10^{-22} \text{ g}$$

From the Periodic Table we find that the atomic mass of cesium is 133 and that of chlorine is 35.5. The total mass of the pair is 168.5. Therefore, the part of the total mass due to each ion in that pair, expressed as the fractional part of the total mass, is:

$$\text{Fractional part of mass of Cl} = \frac{35.5}{168.5} = 0.21$$

$$\text{Fractional part of mass of Cs} = \frac{133}{168.5} = 0.79$$

Finally, since the mass of the pair is 2.9×10^{-22} g, the mass of each ion in the pair is:

Mass of Cl⁻ ion $= 0.21 \ (2.9 \times 10^{-22} \text{ g}) = 0.61 \times 10^{-22}$ g
Mass of Cs⁺ ion $= 0.79 \ (2.9 \times 10^{-22} \text{ g}) = 2.3 \times 10^{-22}$ g

Question 15–23 As a rough verification of these values, calculate the mass of the hydrogen atom from the mass of the chlorine ion and its atomic mass; from the mass of the cesium ion and its atomic mass.

Question 15–24 Potassium bromide has a cubic structure like that of sodium chloride. The density is 2.8 g/cm³ and the unit cell is 6.6 Å (6.6×10^{-8} cm) on each side. Find the masses of the potassium and bromine ions.

In Section 14–3 Avogadro's number (6.02×10^{23}) was thrown into the discussion for no apparent reason. The amount of material containing 6.02×10^{23} particles was called a mole. Why? Let us calculate the mass of one mole of chlorine and cesium ions, using the mass of the individual ion. To make this calculation more meaningful, we will use values for the masses of the individual ions derived from calculations using three significant figures rather than two.

On this basis the mass of the chlorine ion is 0.589×10^{-22} g*, the mass of the cesium ion is 2.23×10^{-22} g. Therefore, the mass of 1 mole of each ion is:

$$\text{Mass of 1 mole of Cl}^- = (6.02 \times 10^{23})(0.589 \times 10^{-22} \text{ g}$$
$$= 35.5 \text{ g}$$
$$\text{Mass of 1 mole of Cs}^+ = (6.02 \times 10^{23})(2.21 \times 10^{-22} \text{ g})$$
$$= 133 \text{ g}$$

The mass of one mole of each ion is numerically equal to the atomic mass of that substance, therefore a mole of Cl^- ions is 35.5 g; a mole of Cs^+ ions is 133 g.

Question 15–25   Use the results of Question 15–24 to find the number of potassium atoms in 39.1 g of potassium, and the number of bromine atoms in 79.9 g of bromine.

15-7   *Crystal forces and potential wells*

On the macroscopic level a crystal is indeed placid and static. On the microscopic level, however, the ions in a crystal are teeming with activity. The ions, apparently held in place by electrical forces of both attraction and repulsion, actually vibrate in a manner not unlike the two air carriages hooked by a spring in Chapter 9. We discussed the energy of those air carriages, and we will discuss the energy of ions in a crystal in a similar way. However, to simplify our discussion, we will want to concentrate our attention on only one ion. Therefore, let us alter our spring-carriage system a little by imagining a single carriage hooked up to fixed posts by two springs (see Fig. 15–16). This air carriage can vibrate back and forth if energy is given it by an outside agent.

If the carriage is displaced to the right the net force of the springs on the carriage is directed to the left and, if released, the carriage will respond by accelerating to the left. If

---
* The value 0.589×10^{-22} gram disagrees with the 0.61×10^{-22} gram in our calculation above because of rounding off within that calculation.

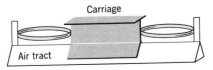

FIGURE 15-16
An air carriage mounted between two springs.

displaced to the left, the net force on the carriage is directed toward the right. The net force on the carriage is always directed toward the center, that point where the net force is zero. This point is called the *equilibrium position.* As the air carriage (assuming no friction) moves through its equilibrium position from one extreme position to the other, it maintains a constant energy, but there is a change in the division of that energy between kinetic and potential.

When the carriage starts from rest at one extreme, its kinetic energy is zero, so at this position all its energy is in the form of potential energy. When it is released, the forces exerted on the carriage by the springs accelerate it; its kinetic energy increases, and its potential energy decreases.

At the equilibrium position the oppositely directed forces exerted on it by the springs are equal. The speed of the carriage is maximum, and all its energy is in the form of kinetic energy. Moving on toward the other extreme position, the net force on the carriage opposes its motion, so its speed decreases. As a consequence, its kinetic energy decreases, and its potential energy increases.

It will be helpful in our discussion of ions vibrating in crystals if we plot the potential energy of the carriage as it moves from one extreme to the other. This graph is shown in Figure 15–17. In position *a*, at the extreme left, all the

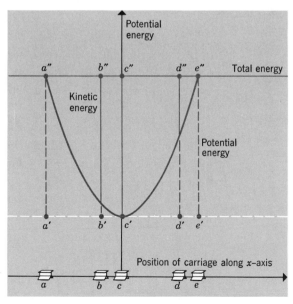

FIGURE 15-17
Potential energy graph.

energy is potential. The amount of potential energy is indicated by the dotted line $a'a''$. At position b, part of its energy is kinetic and the amount is indicated by the solid portion of line $b'b''$. The length of the dotted portion of this line represents the potential energy. At position c all of the energy is kinetic. The energies at positions d and e are similar to the energies at b and a.

The more energy we give the carriage, that is, the farther we displace it before releasing it, the more exaggerated will its motions be, and the greater will be the kinetic energy at various positions (see Fig. 15–18). When a potential

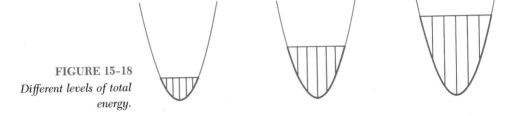

FIGURE 15-18
Different levels of total energy.

energy curve has a minimum we refer to the region around the minimum as a *potential energy well*. If the springs don't break, the carriage is trapped in the well. The potential energy curves are drawn to extend beyond the maximum potential energy of the system to indicate the overall shape of the energy curve—so long as the springs don't break.

The water in a lake, for example, Crater Lake shown in the photograph that opens this chapter, is in a gravitational potential energy well. Given enough energy the water would escape from its well by washing right up and over the mountain walls that contain it.

The vibration of the ion within a crystal lattice is similar to the motion of our carriage, but the ion can move in three dimensions. A model that would represent the motions of the ion more closely would be one in which a carriage is suspended in the center of a closed cubical box with six springs connecting it to each of the six sides. For simplicity, however, we will consider motion in only one dimension along the x axis (position axis).

The shape of the potential curves for an ion in a crystal

differs from the potential curves of the carriage-spring system. The electrical forces of attraction and repulsion acting on the ion are different from the forces exerted on the carriage by the spring. The ion does have an equilibrium position and a potential well, but the well is surrounded by "hills" that the ion can cross. The total energy of the ion can be increased by a transfer of energy from outside the crystal (e.g., when the crystal is heated); if the energy becomes large enough, the ion can escape from its well. The shape of a potential well with hills on the sides is shown in Figure 15–19. Beyond the hills the potential energy

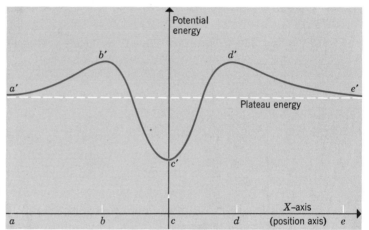

FIGURE 15–19
Potential energy well and hills.

approaches a constant value, the "plateau." See the line labeled *plateau energy* in the figure.

If we could move the ion along the x-axis from position *a* to *b*, *c*, *d*, and on to *e*, its potential energy would first increase to a maximum at *b'*, then fall to a minimum at *c'*, then increases to a maximum at *d'*, and finally drop to nearly the plateau energy at *e'*. An ion trapped in a potential well will escape if its total energy exceeds the potential energy at *d'* or at *b'*.

Suppose we have a particle trapped in such a well, and during one of its oscillations it is struck by another particle with a lot of energy. Initially, our particle has a total energy that might be represented by the line through *a'*, *b'*, *c'*, and *d'* in Figure 15–20. With the added amount of energy, the total energy of the particle is now represented by the line through e', f', g', h', and i'. With this amount of energy, the particle can escape from the potential well.

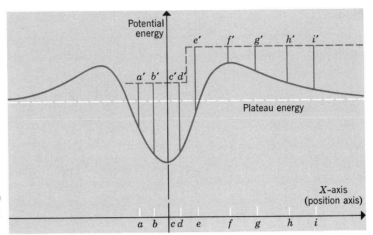

FIGURE 15-20

Escaping from potential energy well.

So far we have discussed only one ion, and considered how that ion is trapped if its energy is less than a specified amount and how that ion can be freed by receiving energy from the outside. What happens to an ion when it is freed? Where does it go? Does it become trapped again?

Suppose an ion is trapped in a potential well that keeps it confined to its unit cell in the crystal. The ion is nicely placed in a well-ordered crystal structure. It will oscillate back and forth about its equilibrium position as long as its energy is less than that required to free it. But suppose we heat the crystal and in so doing we give the ion more energy. If the ion is given enough energy, it will break free from its equilibrium position in the unit cell. It then has no choice; it must leave its unit cell, forming a vacancy, and wander about among the other ions in the crystal.

If the ion should find itself close to other ions that have an electric charge that is the same as that of our ion, it would be repelled by them. The ion would be forced away from this position. Such a position is represented on the energy diagram of Figure 15–21 as hills at a', c', e', and g'.

Should the ion find itself close to other ions whose charge is opposite, it would be attracted by those other ions.

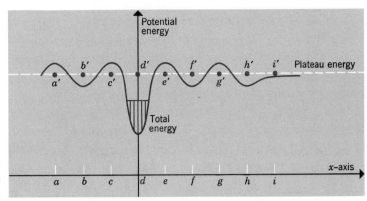

FIGURE 15-21
*Potential energy of an ion
in a crystal.*

However, if this new position is not equivalent to its original position in the unit cell, the forces holding it will be less than those which held it in its original unit cell. If the forces holding it are less, then the potential well is not as deep as the potential well at position c in the original unit cell. Three shallower potential wells are shown at b', f', and h' in Figure 15-21.

An ion in one of the shallow wells does not fit nicely into any unit cell; it is almost an interloper, and as such it is called a *defect* in the otherwise well-ordered crystal (see Fig. 15-22).

If the ion should move far enough away from position c (Fig. 15-19) before losing its energy, it will drop to the plateau and thus escape from the crystal altogether and become part of the liquid or vapor that surrounds the crystal. Most of the ions in a crystal at ordinary temperatures are trapped in their deep wells. There are very few ions at defect positions. However, some ions at the surface are always escaping from the crystal.

Let us now consider a crystal in contact with its melt, such as one of the crystals you melted in Experiment 12–3. We focus attention on a particular ion at the crystal surface and let the ion move to various positions along the x-axis, passing from the crystal into the liquid. The potential energy graph for our special ion would look something like the one given in Figure 15–23.

The ion has its primary crystalline well at A. To the left of A are the shallow wells of defect positions in the crystal.

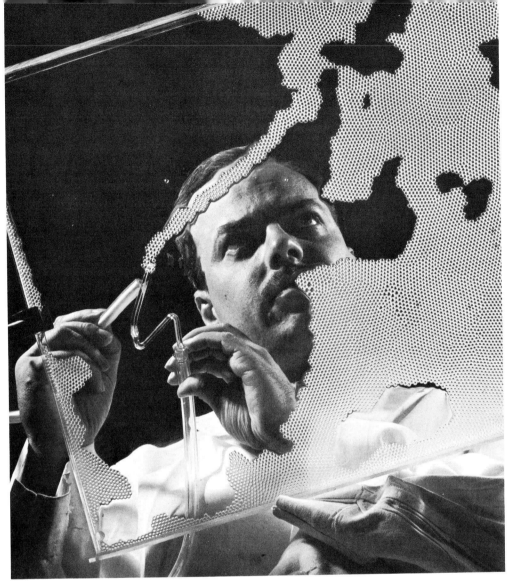

FIGURE 15-22   Can you find defects in the otherwise well-ordered
arrangement of bubbles? (Courtesy of the Westinghouse Corporation)

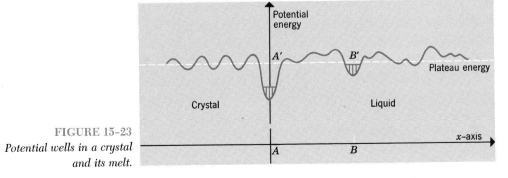

FIGURE 15-23
*Potential wells in a crystal
and its melt.*

To the right of A the graph looks very different. It represents the disorderly forces that would be encountered by our special ion at various positions in the liquid. That part of the graph varies chaotically in time, never looking the same from one moment to the next. If the ion moved to B and became trapped in the well there, it would become a typical ion in the liquid melt. The well at B must be due to an accumulation of the neighboring ions of charge opposite to our special ion. They tend to follow the special ion in its subsequent motions, and to the extent that they are successful they maintain a moving potential well in the liquid for our special ion. The liquid must have such wells or it would not hang together.

The depths of the wells in the liquid may vary, but let us say that the well shown in the figure at B is an average liquid well. It is not as deep as a primary crystalline well, and hence the average potential energy of an ion in the liquid is higher than the average potential energy of an ion in the crystal. If the average kinetic energies are the same, as when the crystal and its melt are at the same temperature, then the average total energy for an ion in the liquid is higher than for one in the crystal.

If a crystal is placed in a test tube and heated (as with those crystals you melted in Experiments 2–6 and 2–7), the initial energy you add goes into raising the temperature of the crystal. This energy simply increases the kinetic energy of the ions in the crystal.

As soon as the melting temperature is reached, the ions, in general, have enough energy to break free of the bonding forces that hold them in. With continued heating those ions do indeed break free, but the added energy does not increase the kinetic energy of the ions; it simply increases their potential energy. The potential energy of ions in the melt is greater than that of ions in the crystal because the bonding forces in the melt are weaker. Continued heating does not increase the temperature of either the crystal or the melt, it simply causes more ions to break free.

If the crystal is melted entirely, the kinetic energy of the ions in the melt can be increased by continued heating, and the temperature of the melt can be increased beyond the melting temperature.

If that melt is now allowed to cool, the kinetic energy of ions will decrease until some of them become locked into place to form a new crystal. It is at this stage that a seed crystal helps the process of crystallization. As each successive ion loses energy by becoming trapped in the crystal, it loses some of its potential energy—not its kinetic energy; the ion drops from a shallow potential well in the melt into a deeper potential of the crystal. Since the kinetic energy is not altered by this process, the temperature remains constant during the process of crystallization from a melt. You observed this as the plateau in the cooling curve of Experiment 2–6.

In Experiment 2–7 you observed that during the process of solidification, however, the temperature of the water surrounding the test tube increased. You concluded that heat was given off by the solidifying melt. This heat came from the ions losing potential energy as they became trapped in the crystal.

If the liquid surrounding the crystal is not its own melt, but a solvent such as water, the situation is quite different. Under these conditions, changing the solid to a liquid is not a matter of melting the solid but of dissolving it in a solvent. It is still true that the arrangement of ions in the liquid is disorderly, but the potential wells in the solvent may be deeper than the potential wells in the melt because of the attraction between ions torn free from the crystal and the molecules of the solvent.

The water molecule, for example, contains two hydrogen atoms bonded to one oxygen atom such that the angle between the two hydrogen-oxygen bonds is 105 degrees (see Fig. 15-24). Although the water molecule has no net charge, there is an excess of positive charge on the hydrogen atoms and an excess of negative charge on the side of the

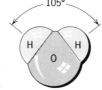

FIGURE 15-24
The water molecule.

oxygen atom. Thus, ions dissolved in water tend to surround themselves with water molecules oriented in the appropriate way, as indicated in Figure 15–25. Such ions are said to be *hydrated.*

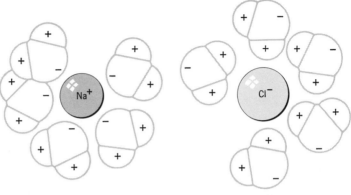

FIGURE 15-25
Hydration of ions.   (a) Hydrated sodium ion   (b) Hydrated chlorine ion

If the hydration deepens the potential wells of the liquid so much that they are deeper than the crystal wells, then the potential energy of an ion in the liquid is lower than that in the crystal. A crystal dissolving at a fixed temperature will give up energy that can be carried away as heat. If the energy is not carried away as heat, the kinetic energy of the ion will increase and the temperature of the solution will be higher than the temperature of the original crystal. If the liquid wells are shallower than the crystal wells, the dissolving crystal will absorb energy and the solution will cool down. The issue is decided by the charges and the sizes of the ions. Smaller ions with higher charges draw molecules into a tighter assembly and make deeper wells.

Question 15–26 Energy is required to melt NaCl, but energy is given out by NaCl dissolving in water. Explain.

Question 15–27 Reconsider Experiment 2–2. Explain why the sodium carbonate solution was warm, while the ammonium chloride solution was cold.

Question 15–28 Why does it require more energy to vaporize a crystal than to melt it?

If the water molecule has a charge distribution like that

shown in Figure 15–24, it resembles those particles discussed above in connection with Figure 11–5*b*. Since those particles were attracted to the charged plastic strip, even though the particles were electrically neutral, it would seem that the water molecules ought to be attracted to a charged plastic strip. Having made this prediction, you should now perform Experiment 15–4 to see whether our prediction is borne out. If it is, then our discussion of how water acts as a solvent to dissolve ionic crystals is supported.

EXPERIMENT 15–4   Water in an electric field

This experiment can be performed at home or in the laboratory, at any place where there is a water tap that can be regulated to let a thin stream fall through a distance of at least 6 inches, preferably more. You will also need a glass rod, a piece of cotton cloth, a plastic strip, and a piece of woolen cloth to prepare a positively charged object (the glass rod) and a negatively charged object (the plastic strip).

Adjust the water tap to deliver a stream of water that is as small as possible, but not discontinuous (broken into drops). Charge the glass rod positively, and bring it slowly toward the stream. Hold it as close as possible without touching the water, and watch the effect. Repeat the experiment with the negatively charged plastic strip.

Question 15–29   How does the deflection of the stream of water depend on the charge on the rod?

15–8   *Melting, dissolving, and equilibrium*

To better understand the melting process, we have to extend our discussion of potential wells of ions in a crystal. We must consider not just one ion but many ions at the surface between a crystal and its melt. Recall from Chapter 10 that the temperature is a measure of the mean kinetic energy of the particles in both the crystal and the melt.

Suppose you are melting a crystal. You add heat in such a way that the temperature of the melt is higher than the temperature of the crystal. On the average, the ions of the melt have a higher kinetic energy than the ions of the crystal. In a typical collision between a slow ion at the surface of the crystal and a fast ion in the melt, the fast particle will lose kinetic energy and the slow one will gain it. As the two materials continue to kick each other through collisions, the crystal is kicked harder and kinetic energy is transferred to the crystal. That is, heat flows into the crystal.

During the collision process many of the ions of the crystal escape and enter the melt, but some of the ions of the melt may get trapped in the crystal. However, since the crystal is melting, there are more ions escaping from the crystal to the melt than there are returning. The ions leaving the crystal become part of the disorderly tumble of ions in the melt.

If the source of heat is removed, and the system loses heat to its surroundings, the melt will no longer absorb energy and the ions of the melt will be trapped by the crystal more rapidly than ions of the crystal will escape into the melt. The crystal grows as the melt freezes.

It is possible, however, to maintain a temperature such that the rate at which the ions of the crystal escape into the melt equals the rate at which the ions of the melt are trapped by the crystal. The two actions are balanced and if the crystal melts a little bit here, it grows a little bit there; the overall effect is that the crystal as a whole neither melts nor grows.

The balance between melting and freezing may be better understood by using an analogy. Suppose you have a construction crew trying to put up a brick building. Theirs is exacting work, and they work more slowly at higher temperatures. At the same time there is a demolition crew knocking bricks out of place with sledge hammers. This crew becomes more excited at higher temperatures and the destruction proceeds more rapidly. If the temperature is such that the two crews work at the same rate, the building remains in a partially finished condition.

If, however, the temperature were raised slightly, demolition would proceed a little faster than construction. The building would be gradually reduced and ultimately disappear. The construction crew would make subsequent attempts to build anew, but they would not get off the ground. However, if the temperature were reduced slightly, the construction crew would have an edge and the construction would proceed to near completion; it would be easy to repair the minor damage done by the frustrated demolition crew. A small change in temperature can completely change the direction of the process.

In the melting process, ions of the melt collide with ions in the crystal and give those ions enough energy so that they escape from their potential wells; those ions enter the melt. Once in the liquid, an ion will move freely until it has the kinetic energy kicked out of it by a collision. Then it will drop into a liquid well. At the same time ions from the melt are being brought to the crystal, where they have the kinetic energy kicked out of them so they can drop into a crystal well. These events are analogous to what happens to a brick in the construction process. A brick mortared into surrounding ones is in a deep potential well. If it is hit hard enough by a sledge hammer, however, it can be raised out of its well and become a flying brick. Eventually it hits the ground, where it loses kinetic energy and becomes part of the disorderly pile of candidates for trapping in a mortared well.

When the escaping ions join a solvent rather than a melt, the disordering processes are favored; the crystal order can be destroyed at temperatures below the crystal's melting point. Escaped ions become hydrated, and they become less susceptible to trapping at the crystal surface. This slows down the ordering process. Some or all of the crystal will dissolve.

Returning to our building construction analogy, let us suppose that the temperature is such that construction is faster than destruction. Without changing the temperature, we introduce a pool of water for the bricks to fall into after they are knocked out of place. This makes it harder for the construction crew to retrieve the bricks, and it slows down the construction. In fact, construction is stopped completely at first, but as bricks are accumulated in the pool, retrieval becomes easier until the rate of construction once again balances the rate of destruction. From that time on, the building fluctuates around a partially finished condition as the number of bricks in the pool fluctuates around a constant value. A balance is reached. If the pool were big enough, it would contain all of the bricks before a balance was reached and the building would disappear altogether.

We have considered a number of different types of balance involving crystals. In every situation there is some property that is stationary in time because of a balance of forces or motions. In one situation the positions of the ions are fixed because of a balance of forces. In another situation the temperature of the crystal is constant because of a balance between transfers of energy into and out of the crystal. In still another the amount of crystal melted or dissolved is constant because of a balance between ordering and disordering processes. In each of these situations one property of the crystal remains essentially unchanged even though competing processes tend to cause small fluctuations. Under any one of these conditions the competing forces or processes are said to be in *equilibrium*. For example, if the temperature of the crystal remains constant we say that it is in thermal equilibrium.

In both melting and dissolving, however, our picture of opposing processes is not complete. It does not explain why a crystal becomes involved in such competitions. Why does a crystal melt at all? A melting crystal does not have so much energy that all the ions are forced to rise out of their potential wells. If it did, it would become a vapor, not a liquid. Evidently the distribution of energy among the ions is uneven. Some of the ions have enough energy to escape and some do not. The ions that escape may be carried away altogether or they may become involved in the competition of opposing processes. If the temperature were low enough, the number of escaping ions might be too small to form a liquid, but even then some of the crystal could evaporate.

15-9 *Entropy*

What assurance do we have that a crystal will have an uneven distribution of energy that will allow it to become involved in the competitions of melting and dissolving? There is nothing in the nature of the crystal forces to prevent an even distribution. Any distribution of the energy is possible, and it is conceivable that a crystal can be put together in such a way that all the ions have the same energy. How do we know that an observed crystal has an uneven distribution of energy?

The same question can be raised about the distribution of ion velocities. If the velocities were distributed evenly, in the sense that all the ions were moving in the same direction at a particular moment, then the crystal could fly up from its resting place and crash into the ceiling. Such behavior is not observed, and we are forced to conclude that the number of ions moving in one direction must be about the same as the number moving in the other direction. The velocities must be distributed unevenly if the crystal is to sit still long enough to be given a name, and for a long time thereafter.

In order to consider these questions about the distribution of ion properties further, we will first present a simpler problem, which will turn out to have a close connection to them. Suppose you are given a closed box containing a hundred pennies, each lying flat on the bottom. You are asked to predict whether the distribution is uniform with all heads, or uniform with all tails, or is nonuniform with both heads and tails. There is no way to make a reliable prediction with the information given, because whoever laid the coins out could have laid them out in any way. It is true that there is only one way to lay the coins out all heads, and there are many ways to lay them out with, say 30 heads and 70 tails, but this does not mean that the nonuniform distribution is more likely. Whoever laid them out may have preferred, for a joke, some unusual distribution. On the other hand, several hard shakes of the box would wash away the designs of the practical joker. It would then be possible to make some reliable predictions. Let us pursue this problem further for the simple case of two coins and then work our way back to the hundred coins.

Consider a pair of coins that have been flipped in a way that does not favor any particular arrangement of heads and tails. The possible results can be described as HH (both heads), HT (first coin heads, second coin tails), TH (first coin tails, second coin heads), and TT (both coins tails). We will call these arrangements the possible states, which are equally likely from our point of view. (If you do not believe this, flip a pair of coins many times and count the number of times you find each state.) In each state there is a distribution that can be described in terms of the number of heads. Let x be the number of heads, which can be 0, 1, or 2. For example, the state HH has the distribution $x = 2$; the state TT has the distribution $x = 0$. The number of states in a particular distribution is called the *entropy* of that distribution. The various distributions, states, and entropies for two coins are listed in Table 15–3.

Table 15–3

Distribution x = Number of Heads	States	Entropy
0	TT	1
1	HT and TH	2
2	HH	1

The entropy of a distribution is important if the coins have been flipped so that the states are equally likely. Then the probability, or likelihood, of a particular distribution is measured by its entropy. Because the distribution $x = 1$ with one head has the largest entropy (namely, 2), it is the most probable distribution, and the one we would predict if two coins have been flipped. The prediction is not very reliable, however, because the most probable distribution is not much more probable than other distributions. Nevertheless, it is the best prediction that can be made.

Question 15–30   What are the possible states and distributions for three coins? Find the entropy of each distribution.

If ten coins are flipped it is possible to make predictions that are more reliable than if two or three coins are flipped. The various distributions and their entropies are listed in Table 15–4. We see that there is only one state in which

Table 15–4

Distribution (Number of Heads)	Entropy	Distribution (Number of Heads)	Entropy
0	1	6	210
1	10	7	120
2	45	8	45
3	120	9	10
4	210	10	1
5	252		

there are no heads, 10 different states in which there is 1 head (each of the 10 coins can be the one that is heads), 45 different states with 2 heads, and so forth. The most probable number of heads is 5, which can occur in 252 different ways. The total number of possible states is the sum of the entropies, which is 1024. There is only one chance in 1024 of having no heads, but there are 252 chances of having 5 heads. A prediction that there will be some heads is extremely reliable. On the other hand, the prediction that the number of heads will be exactly five is not very reliable, even though 5 is the most probable number.

What is the chance that the distribution will be near the most probable one in the sense that the number of heads is between 3 and 7? Adding the entropies for those distributions, we find 912 chances out of 1024 (91%) for a number of heads between three and seven, which is pretty good. If we use 100 coins, the predominance of distributions near the most probable distribution becomes much more impressive. A prediction of between 45 and 55 heads for a flipped coin collection is quite reliable.

No matter what the initial state is, after flipping, the system is most likely to be in a state having high entropy. For a system with many nearly indistinguishable states, the entropy of the system rapidly approaches a maximum. After reaching the maximum, it fluctuates around that value without changing very much. The collection comes to equilibrium in the sense that the number of heads remains in the neighborhood of the value with maximum entropy. The tendency of the collection to reach and maintain maximum entropy is referred to as the *law of entropy:* Entropy tends to increase and, for a system with many interacting particles, does so rapidly if it can.

EXPERIMENT 15-5   The law of entropy

Place a hundred coins on the bottom of a box all with heads up. Shake the box gently to cause some flipping but not too much. Count the number of heads. Shake the box again, and count again. Repeat until the number of heads falls into the range between 45 and 55. This will happen immediately if the shaking is vigorous. Continue to shake and count to see whether the distribution will return to the initial distribution. What is happening to the entropy?

The law of entropy, which applies to a collection of coins that have been flipped, also applies to a collection of ions in a crystal. The entropy of a particular distribution of ion properties, such as energy, is the number of ways to divide up the energy to result in that distribution. There is only one way to divide the energy for a uniform distribution—one in which each ion has exactly the same energy every instant of time; and there are many ways to divide the energy for a nonuniform distribution. If the ions have interacted with each other so that their energies and other properties have changed (been "flipped") several times, then the various ways of dividing the energy (i.e., the various states) have become equally likely and the distribution becomes the distribution of maximum entropy. The entropy can be further increased if additional energy is put into the crystal, for there are many ways to divide up the added energy, but the entropy will reach a maximum value for any particular energy.

The process of bringing a collection of coins or a crystal to a distribution of maximum entropy can be called "aging." Aging washes away any peculiar distribution of low entropy that might have been established by our practical joker. In a crystal, aging is so rapid that it is difficult to conceive of a way to prepare an unaged crystal. The aging is much more rapid than the rate of growth in Experiment 1-1, and the growing crystals can be regarded as fully aged at each stage of their development.

It is the aged condition of crystals and other observable materials that causes them to behave in a way described by the law of entropy, and it is this law that describes the orderly and predictable behavior of the materials in their response to changing external conditions. For example, all observable sodium chloride crystals behave in the same way. They melt at the same temperature and dissolve to the same extent in the same solvents. They must become involved in the competitions of melting and dissolving because they are aged enough to have a nonuniform distribution of energies and other ion properties. If the aging of crystals and other observable materials were slow compared to the rate at which the materials are subjected to changes in the environment, then distributions of low entropy could become prevalent. The law of entropy would be violated and the world would become a highly chaotic place, full of nightmares ruled by demons.

QUESTIONS

15-31   Look at a pencil. What are you taking for granted about its history when you identify it as a particular thing? How do you know it will not suddenly rise into the air or separate into two parts flying in opposite directions?

15-32   What will happen when you puncture a filled balloon? How can you be so sure?

15-33   Predict what will happen if you drop a crystal of potassium permanganate into a glass of water. Refer to your notes for Experiment 1–3. Can the reverse process be observed? If not, why not?

15-34   The entropy of a crystal is reduced when the crystal is cooled by removing heat. Why doesn't this contradict the law of entropy?

15-35   What factors other than the distance between positive and negative ions might affect the attractive forces in crystals? Are these other factors the same in sodium fluoride, sodium chloride, sodium bromide, and sodium iodide? Are they the same in magnesium oxide and sodium chloride? in cesium chloride and sodium chloride?

15-36 An iron object can be plated with chromium metal by an electrolytic process, that is, by a process in which electric current is involved with a chemcial reaction. Without using a reference book, suggest the essential details of a process by which steel automobile bumpers can be chrome-plated.

15-37 With a sufficiently powerful microscope, would you expect to be able to see the individual sodium and chlorine ions in a sodium chloride crystal, and thus determine their positions?

15-38 If a mixture of sodium chloride and potassium chloride is dissolved in water, and the water is then permitted to evaporate slowly, sodium chloride crystals will form. It is likely that some potassium ions will be found in place of some sodium ions in these crystals? If the original mixture was composed of sodium chloride and cesium chloride, would you expect cesium ions to be found in place of sodium ions? Explain your answers.

15-39 Magnesium oxide crystals have the sodium chloride structure, that is, the pattern of arrangement of magnesium ions and oxide ions is like that of sodium ions and chlorine ions. But the magnesium ions have a charge of $+2$ and the oxygen ions a charge of -2. What prediction can you make about the melting point of magnesium oxide compared with sodium chloride? What can you predict about the relative solubilities of the two substances in water?

15-40 Any ionic crystal dissolves to some extent in water. Some crystals dissolve to a great extent and others to a small extent. Why?

15-41 The entropy of any crystal would be increased if the crystal were vaporized. Why don't all crystals vaporize?

15-42 If the molecules in the air in your room were all moving toward one wall, that wall would be blown out and you would be left sitting in a vacuum. Is such a catastrophe impossible? improbable? Why?

References 1. Campbell, J. A., *Why Do Chemical Reactions Occur?*, Prentice-Hall, Englewood Cliffs, New Jersey, 1965, pages 80 to 90.

A discussion of energy and randomness. This includes chemical systems, conservation of energy, and entropy and changes in isolated systems. The discussion is not quantitative.

2. Christiansen, G. S., and P. H. Garrett, *Structure and Change*, W. H. Freeman & Co., 1960, Chapter 19, pages 310 to 324.

A discussion of the theory of ions, and how this is related to properties of substances.

3. Sienko, M. J., and R. A. Plane, *Chemistry: Principles and Properties*, McGraw-Hill, 1966, Chapter 6, Sections 6.1 through 6.4, pages 131 to 142.

Structure of crystals. Drawings representing the packing of atoms in lattices.

4. Wood, E. A., *Crystals and Light*, D. Van Nostrand Co., Princeton, New Jersey, 1964.

Pages 10 to 12 discuss the unit cell and the structure of cesium chloride; pages 18 to 24 discuss x-ray diffraction by crystals.

ALSO OF INTEREST

5. Bassett, L. G., *et al.*, *Principles of Chemistry*, Prentice-Hall, 1966, pages 149 to 166.

Crystal structure from a more advanced point of view.

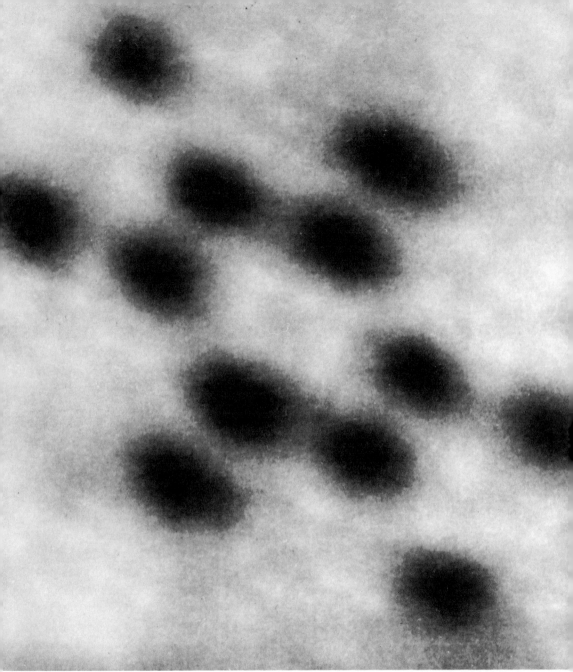

Courtesy of Dr. Maurice L. Huggins

CHAPTER 16

MOLECULES

You have now made a rather thorough investigation of certain crystalline solids, and you have seen convincing evidence that the nature of the forces holding these solids together is electrical. Furthermore, you have seen the development of a model of crystals in which the constituent particles have been identified as ions rather than atoms. The model exhibits a regularity in ionic structure that helps to account for some of the observed properties of crystals.

Are all substances composed of ions? Remember that the way you began the investigations that led to the idea of ions (in Chapter 12) was by examining the conductivity of a variety of substances. Some of the materials you tested did conduct an electric current, but others did not. Prominent among those that did not conduct, or at best conducted very poorly, was water. Because we have associated electrical conductivity of liquid materials with the existence of ions in both the liquid and the solid states of these materials, we must infer that neither water nor ice is composed of ions, or that at best there are very few ions in water and that most of its structure consists of some other kind of particles.

Of what kind of particles is water composed? You have been told that water is H_2O and we have used this formula several times without apology. Let us now propose that water is made not of charged ions but of electrically neutral molecules, each of which is an H_2O unit consisting of two atoms of hydrogen and one atom of oxygen somehow bonded together. Solid water, then, must consist of a great many H_2O molecules held to each other in a rigid way. Thus in addition to the ionic bonding in crystals we have already explained, we must explain two new kinds of bonding.

The first new problem is to discover what holds the two atoms of hydrogen and the one atom of oxygen together to form a water molecule. The second problem is to discover what holds the water molecules to each other in an ice crystal.

16-1 *An attempt to electrolyze water*

We have been successful at taking several substances apart by causing an electric current to flow through them in either the liquid or dissolved state. We know that pure water will not conduct an electric current; but it will conduct if we dissolve a small amount of sodium sulfate in it. We will electrolyze a dilute solution of sodium sulfate in the following experiment and see what we can learn from the results.

EXPERIMENT 16-1 Electrolysis of water (a demonstration)

A type of apparatus designed for the electrolysis of solutions when the products are gases is shown in Figure 16-1. The circuit is similar to that used in Experiments 14-1, 14-2, and 14-3, except that there is a milliammeter A, instead of a flashlight bulb, to detect a current in the circuit.

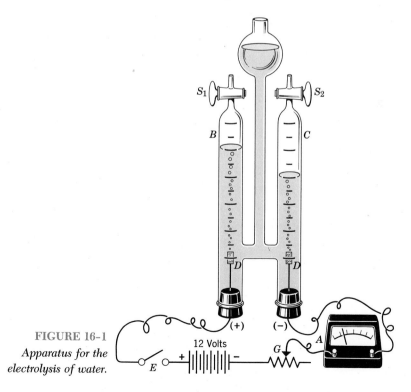

FIGURE 16-1

Apparatus for the electrolysis of water.

The circuit also includes a rheostat G, which is a device that enables us to change the conductance of the circuit, so that we can increase or decrease the current. The zigzag line with the arrow is the conventional symbol for a rheostat. The long and short lines are the conventional symbol for a battery.

At the start of the experiment, the dilute sodium sulfate solution fills both side arms and the central tube. Any gas produced by passing current through the solution will rise from the electrodes D and will collect in the side arms B and C. These tubes are graduated so that the volume of gas can be measured. To start the current we connect the battery with a clip.

We are going to observe the rate of gas evolution and determine the current by reading the milliammeter. You should make a table with the following headings: Time, Current, Volume of gas in B, Volume of gas in C, Total volume in B and C, and Ratio of the volume in B to the volume in C. You will be given the values of current, and volumes of gas in B and in C at specified times. Record these values. The current should be kept constant by adjusting the rheostat. When the volume of gas in tube C reaches about 20 cm^3 the circuit will be disconnected so that no further electrolysis takes place.

Do you know, or can you guess, what gas is in B? in C? We cannot be certain of our answers until we check properties. For gas identification we can observe color, odor, density, flammability, and the ability to support combustion. Before making any tests, however, we should consider what gases might possibly be present (as we did in Chapter 14). From water we could obtain hydrogen and oxygen. From the dissolved sodium sulfate we could obtain oxygen or some oxide of sulfur, such as sulfur dioxide. You can all observe the colors of the gases collected; the experimenters can check for odor while driving out the gases at the end of a run. (Sulfur dioxide has a distinct choking odor.) You tested for hydrogen in Experiment 14–3. A test for oxygen depends on the ability of the gas to support combustion more readily than air does. Some gas is collected in a test tube, and a glowing splint is inserted. If the test tube contains oxygen the splint will burst into flame. The gases in the two arms of the apparatus will be tested and identified.

Now you should be ready to answer some questions.

Question 16-1 List the properties of the gas collected over the negative electrode. What is the gas?

Question 16-2 List the properties of the gas collected over the positive electrode. What is the gas?

Question 16-3 Which gas was liberated in greater volume?

Question 16-4 What was the ratio of the volume of gas collected over the negative electrode to that collected over the positive electrode?

Question 16-5 Hydrogen has a very low density. Can you suggest a test other than combustibility for it?

We have observed that hydrogen and oxygen are produced in this electrolysis process and also that the volume of hydrogen is twice the volume of oxygen. Is water the only substance that has been changed during the electrolysis? How could you find out? To answer the question requires additional experimental work, which we will not do. We could test for the amount of water, the amount of sodium ion, and the amount of sulfate ion. Of course this extension of the experiment has been done by others, and it has been found that only the water changes in amount. All of the sodium sulfate added in the beginning can be recovered at the conclusion of the experiment.

Why does the oxygen come from the water molecule and not the sulfate ion SO_4^{-2}? We had a similar situation in Experiment 14-2, when a lead nitrate solution was electrolyzed. Oxygen gas was liberated at the positive electrode in that experiment too. The oxygen in ions like the nitrate ion NO_3^- and the sulfate ion SO_4^{-2} is held much more tightly than the oxygen in the water molecule. Therefore, oxygen is more easily liberated from the water molecule than from the sulfate or nitrate ions. In other words, it takes less energy to oxidize the oxygen in the water molecule than it does to oxidize the oxygen in the nitrate and sulfate ions.

Thus it seems appropriate to conclude that in this experiment we have separated the components of water. Assuming that this is so, we have found two very important pieces of information. First, the nature of the bonds holding the parts of the H_2O molecule together seems to be electrical. Let us consider the evidence for this.

We described Experiments 14–1, 14–2, and 14–3 in terms of an exchange of electrons at the electrodes in the cell. Similarly, we may interpret the results of the present experiment as follows:

at the positive electrode

water $\rightarrow$ oxygen + electrons

at the negative electrode

water + electrons $\rightarrow$ hydrogen

An exchange of electrons appears to be involved in these reactions, so it seems reasonable to assume that the bonds within the H_2O molecule do involve electrical forces.

The second piece of information is that the volumes of these gases are in the ratio of $2:1$—two parts of hydrogen to one part of oxygen. Furthermore, were we to measure the mass of each of the gases evolved in this experiment, we would find that the ratio of the mass of oxygen to that of hydrogen is $8:1$. But the ratio of the mass of the oxygen atom to that of the hydrogen atom is $16:1$. To relate these two ratios, we point out that a ratio of $8:1$ is the same as a ratio of $16:2$. Consequently, we can conclude that there must be 2 atoms of hydrogen evolved for every atom of oxygen.

We can draw still another conclusion from this argument. Because the ratio of the volumes of the hydrogen to that of the oxygen is $2:1$, and since there are twice as many atoms of hydrogen as oxygen, it can be concluded that at equal pressures and temperatures, equal volumes of these gases contain equal numbers of atoms. Assuming that a molecule of oxygen contains the same number of atoms as a molecule of hydrogen, we can propose that at equal pressures and temperatures, equal volumes of these gases contain equal numbers of molecules. This proposal will be discussed again in Section 16–4. First, however, let us attempt to confirm our findings regarding the volumes of hydrogen and oxygen obtained from water.

16-2 *Combining hydrogen and oxygen*

If hydrogen and oxygen gases can be obtained by electrolyzing water, we would expect to obtain water again if these gases were recombined. Furthermore, we would expect exactly two volumes of hydrogen to combine with one volume of oxygen to produce water with none of either gas left over.

EXPERIMENT 16-2   Combining hydrogen and oxygen (a demonstration)

The apparatus shown in Figure 16–2 is used to study conditions under which oxygen and hydrogen can be combined to give water. It is similar to the apparatus used in the experiment we have just completed except that both elec-

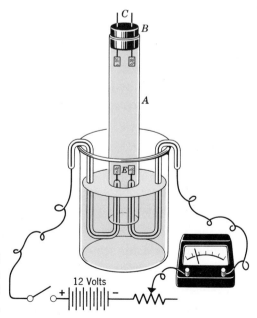

FIGURE 16-2

Electrolysis apparatus to collect hydrogen and oxygen in an ignition tube.

trodes are under the same tube so that hydrogen and oxygen are mixed as they evolve. A long tube is securely stoppered at one end with a two-hole rubber stopper *B*. A wire electrode *C* is sealed in each hole of the stopper. The tube is filled with a dilute solution of sodium sulfate. Two platinum electrodes *E* are inserted in the tube and connected with wires to the battery, as shown.

Gas will bubble from both electrodes and collect in the tube when the electrical circuit is closed. After about 20 cm^3 of gas is collected, note the level of the solution. The circuit is then opened to stop the electrolysis. A spark coil is brought up to the electrodes *C*, so that a spark jumps from one to the other in the mixture of hydrogen and oxygen. The effect of this is to heat the gases in the vicinity of the electrodes. A violent explosion will take place and the level of solution in the tube will change. Read the level again. What is your interpretation of the change of level of the liquid?

16-3   *Combining volumes*

If oxygen and hydrogen are available, they can be introduced into the tube in Experiment 16–2 in any desired proportions. Can you devise a simple experiment to determine quantitatively what would happen if oxygen and hydrogen were mixed in various proportions before being exploded? The results of such an experiment are described below.

Oxygen and hydrogen are mixed in different tubes in different proportions. Water can be used to confine these gases because neither one is soluble in water. Volumes of the two gases are mixed in the ratios of 1 oxygen to 1 hydrogen; 1 oxygen to 2 hydrogen; 1 oxygen to 3 hydrogen; and 2 oxyen to 1 hydrogen. These proportions of the gases are indicated by the lines drawn on the tubes in Figure 16–3*a*. After ignition by a spark, residue gases are detected in the amounts shown in Figure 16–3*b*. Can you draw a general conclusion from these results about the combining volumes of hydrogen and oxygen?

Regardless of the proportions in which they are mixed, oxygen and hydrogen always combine in the ratio of two

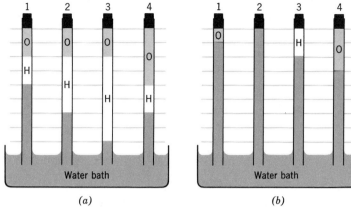

FIGURE 16-3
*Combining proportions of
hydrogen and oxygen.*

parts hydrogen (by volume) to one part oxygen to form
water. If that water remains in the gaseous state, its volume
will equal the volume of the hydrogen. Usually, however,
the water condenses to a liquid and its volume essentially
disappears.

Do such simple volume ratios appear when other gases
combine? Experiments can be devised to answer this question. With time, patience, and suitable equipment, one can
readily cause many gases to react with other gases. Results
such as those summarized in the next few paragraphs are
obtained from studies of reacting gases.

When sparked, hydrogen and chlorine react explosively
as do hydrogen and oxygen. When equal volumes of these
gases are placed in an ignition tube over water and ignited, no
gas remains. The reacting gases, hydrogen and chlorine,
are only very slightly soluble in water. Hydrogen chloride,
the product of the reaction, is very highly soluble in water.
Hydrogen combines with an equal volume of chlorine. The
combining ratio, therefore, is 1 to 1. If this same experiment
is performed using mercury instead of water to confine the
gases, there remains a gaseous product that has a volume
equal to the volume of hydrogen plus the volume of
chlorine.

In still another experiment, which is somewhat more
difficult to perform, nitrogen can be made to react with
hydrogen to produce ammonia. If the volumes of the react-
ant gases and of the product gas are carefully determined
as before, the ratio is found to be 1 nitrogen to 3 hydrogen

to 2 ammonia, when all are compared at standard conditions of temperature and pressure. The proportions of volumes of reacting gases and gaseous products in the three experiments just described are summarized in Table 16–1.

Table 16–1   Combining Volumes of Selected Gases

Gases	Ratios
Hydrogen and oxygen → water	2:1:2
Hydrogen and chlorine → hydrogen chloride	1:1:2
Nitrogen and hydrogen → ammonia	1:3:2

These results were first summarized by Gay-Lussac: *When gases react chemically, the volumes of the gases are to each other as the ratio of small whole numbers* (assuming equal temperatures and pressures).

16-4   *Avogadro's hypothesis*

It is important to recognize that 2 volumes of hydrogen combine with 1 volume of oxygen to yield 2 (not 3) volumes of water vapor. The number of atoms is not altered by the chemical reaction, so we certainly would not suggest that, equal numbers of atoms always occupy equal volumes.

But since a chemical reaction is involved, why should we consider atoms? In 1811, Amadeo Avogadro hypothesized that under the same conditions of temperature and pressure, equal volumes of gases contain equal numbers of molecules. This hypothesis is very simple, but it is very bold. Avogadro was the first to propose that the basic building block of gases such as hydrogen and oxygen is not the atom but the molecule. A liter of oxygen contains the same number of molecules as does a liter of hydrogen, chlorine, hydrogen chloride, ammonia, or any other gas, provided that the temperature and pressure are the same for all gases.

As has happened so often in the development of science, Avogadro was acting on a hunch, that is, he was making a guess. He relied a great deal on experimental evidence, but had to guess that molecules of some gases are composed of two or more atoms.

In Chapter 15 we determined that a mole of ions in a crystal (i.e., 6.02×10^{23} ions) has a mass in grams equal to the atomic mass of the ion. An equal number of molecules of a compound make up a mole of that compound. A mole of any compound has a mass equal to the sum of the atomic masses of the atoms making up the molecules of that compound. This number, as you may recall, is known as Avogadro's number in honor of his brilliant suggestion.

16-5 *Diatomic molecules*

Avogadro's insight makes understandable the ratios of gas volumes, and it leads to a very important conclusion about the molecules of hydrogen, oxygen, nitrogen, chlorine, and certain other gases.

Remember that *one volume* of hydrogen combines with *one volume* of chlorine to produce *two volumes* of hydrogen chloride. "Volume" could mean liter, quart, etc., but whatever volume is meant we must assume (with Avogadro) that it has the same number of molecules in it whether the gas is hydrogen, chlorine, or hydrogen chloride. The smallest volume of hydrogen chloride we can imagine is one molecule. But from the combining ratio this one molecule must have been produced from half a molecule each of hydrogen and chlorine, i.e., an atom of a diatomic molecule.

Both halves of a molecule of hydrogen surely are the same; therefore a single molecule must consist of *two atoms* of hydrogen, and the same is true for chlorine. This is the least number possible. Hydrogen could be H_4 or H_6, etc., but we will assume the simplest explanation and abandon it only if it does not agree with later experimental results.

The chemical equation for the formation of hydrogen chloride is

$$H_2 + Cl_2 \rightarrow 2HCl$$

Question 16-6   The word equation for the formation of water is

2 volumes of hydrogen + 1 volume of oxygen →
2 volumes of water

Write the chemical equation for the formation of water.

Reasoning similar to that above, coupled with additional information, suggests that oxygen, nitrogen, hydrogen, chloride, bromine, and iodine are all composed of diatomic molecules. We can now consider the nature of the bonds that hold these atoms together in pairs as well as the bonds that hold the atoms together in a water molecule.

16-6   *Bonds in molecules*

What is the nature of the force that holds a molecule such as H_2 together? The molecule contains two nuclei (positive) and two electrons (negative). Perhaps these four charged particles are arranged in some stable array, as in a crystal. This is doubtful, however, because there are not enough particles to produce a crystalline environment for each particle. Furthermore, the electrons have very little mass compared to the nuclei: their resistance to acceleration resulting from an unbalanced force is very small. Making a crystal out of these particles is like making a crystal out of two positively charged horses and two negatively charged flies. In the molecule, as in the atom (see Section 13–6), the electrons form a cloud while rapidly buzzing around the nuclei, which move relatively slowly. We may regard the nuclei as being essentially fixed and use the cloud model for the electrons.

The cloud model as applied to electrons in atoms and molecules is very useful. The model is appropriate here because of the tremendous difference between the motions of the massive nuclei and mobile electrons. When we use the electron cloud model we might as well forget about a moving particle; we are committed to the picture of an electron as a motionless fog hovering near motionless nuclei. The cloud is dense near the nuclei, and it thins out rapidly at larger distances from the nuclei. The dense part has a characteristic size and shape that resembles a sphere. For example, we may think of a hydrogen atom as a negatively charged spherical cloud with a positive nucleus embedded in the cloud at the center.

Let us consider what happens when two hydrogen atoms approach each other (see Fig. 16–4a). The electron clouds

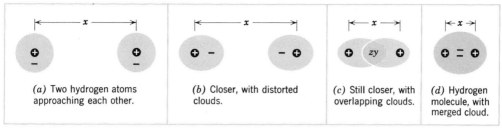

(a) Two hydrogen atoms approaching each other.

(b) Closer, with distorted clouds.

(c) Still closer, with overlapping clouds.

(d) Hydrogen molecule, with merged cloud.

FIGURE 16-4 *Formation of hydrogen molecule.*

repel each other, and the nuclei repel each other, but the cloud of each atom is attracted to the nucleus of the other atom. At large distances these forces cancel, leaving no unbalanced force between the atoms. When the atoms come closer together the cloud of each atom is distorted so that it is drawn toward the other atom (Fig. 16–4b).

The distortion of the clouds occurs because the attraction of the cloud of one atom to the nucleus of the other atom is stronger than the repulsion between the clouds; the nuclear charge is more concentrated and is more effective than the spread-out charge of the clouds. In fact, the repulsion between the clouds is actually reduced by overlapping (Fig. 16–4c). This reduction can be explained as follows.

Any piece of one electron cloud will repel any piece of another electron cloud. Therefore, if we let y in Figure 16–4c represents a piece of the left-hand electron cloud, and z a piece of the right-hand cloud, those two pieces will repel each other and in so doing push the two clouds together. The overall repulsion of the negatively charged clouds is reduced by the overlap. This is not to say that the clouds do not repel each other, for there are always more pieces repelling in such a way as to drive the clouds apart than there are pieces repelling in such a way as to push the clouds together. The clouds do repel each other, and they come together only because of their attraction to the positively charged nuclei. Nevertheless, the repulsion of the clouds is reduced by the overlapping, and the balance of attractive and repulsive forces is upset so that the atoms are attracted to each other.

As the atoms come still closer, the repulsion between the concentrated nuclear charges becomes too large to be compensated by forces involving clouds, and the net force between the atoms changes from attraction to repulsion. There is a distance between the nuclei at which the forces are balanced. This distance D is called the *bond length* for the H_2 molecule. Hydrogen has some measurable properties that depend on the size of individual molecules, so it has been possible to determine from experimental data that the bond length is about 0.8 Å, which is a little less than twice the radius of the hydrogen atom. This bond length D is the x in Figure 16–4*d*.

The competition between attractive and repulsive forces just described can be represented in a simple way by a potential-energy graph similar to those used in Section 15–7. If we were able to move the two nuclei relative to each other, the potential energy of the molecule would change correspondingly.

The curves in Figure 16–5 indicate the potential energy

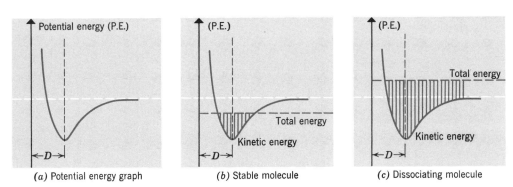

| (a) Potential energy graph | (b) Stable molecule | (c) Dissociating molecule |

FIGURE 16-5   *Energy of two hydrogen atoms.*

of the molecule for distances x (see Fig. 16–4) separating the two nuclei. There is a potential well centered around the separation distance $x = D$ (see Fig. 16–5*b*). The atoms can be trapped in the well if they do not have enough kinetic energy to escape. A hydrogen molecule is a pair of trapped atoms. Figure 16–5*c* shows a pair of atoms that have too much total energy to hang together as a molecule. The molecule in *b* can become the dissociating molecule in *c* if sufficient energy is received from a collision with atoms or other molecules. Alternatively, if a pair of atoms with the energy shown in *c* loses some kinetic energy before separation, the two atoms can drop into the well and become a molecule, as in *b*. Once in the well, the nuclei vibrate and the separation distance *x* oscillates around the bond length *D*.

Atoms are held together by forces to form molecules. How do these forces vary with the separation distance *x* between the nuclei of the atoms? We can answer this question by using the potential energy curve in Figure 16–5*b*. Suppose the atoms are moved together so that the nuclei are separated by a distance *x* less than *D*; this corresponds to moving left from the bottom of the potential well. The curve rises rapidly; this means that the potential energy increases rapidly. An external agent forcing the nuclei together would have to do a great deal of work to move the nuclei a short distance in order to supply this energy. Therefore the external agent would have to exert a very large force. This results from the fact that the atoms repel each other strongly when they are closer together than the distance *D*.

On the other hand, we can imagine moving the atoms apart so that the nuclei are separated by a distance *x* larger than *D*; this corresponds to moving right from the bottom of the well in Figure 16–5*b*. Again the potential energy of the molecule increases, and the external agent must do work to supply this energy. It follows that the external agent must exert a force to pull the atoms apart; this results from the fact that the atoms attract each other when they are farther apart than the distance *D*.

However, as the atoms are much farther separated we see that the potential energy no longer changes with increased distance. This means that the forces between the atoms must approach zero at large distances. At these distances the forces of attraction and repulsion between the two atoms cancel, leaving no unbalanced force. Our picture of the chemical bond is based on Figures 16–4 and 16–5. It is the overlapping in Figure 16–4 that causes the minimum in the potential energy graph of Figure 16–5. We can say that the bond exists whenever there is a well in a potential energy graph and insufficient kinetic energy to escape from the well. We encountered the bond in the last chapter, but there one ion in a crystal was bonded to others. That bonding is called *ionic bonding*. The bonding illustrated in Figure 16–4 is called *covalent bonding*.

Question 16–7   Suppose two hydrogen atoms are very close to each other. Do they necessarily form a hydrogen molecule?

Question 16–8   Why was the spark necessary to initiate the combination of hydrogen and oxygen gas in Experiment 16–2?

Question 16–9   In what way are bonds in molecules and bonds in ionic crystals alike? How are they different?

We have seen that the formation of a bond between two hydrogen atoms can be explained in terms of the overlapping of two electron clouds. Now we ask whether or not the resulting diatomic molecule can form additional bonds with other atoms. It is known that hydrogen molecules composed of more than two hydrogen atoms do not form in sufficient numbers to be easily detected. Evidently the electron cloud in the H_2 molecule is already so crowded because of the overlapping of two electron clouds that further overlapping with a third is not very likely.

The behavior of electron clouds as just described leads us to a reliable rule about overlapping. Two electron clouds can substantially overlap with each other, but three or more clouds cannot. This rule will be easy to use if we form the concept of "live" and "dead" electrons. An electron is alive if its cloud is not substantially overlapped with the cloud of another electron. An electron is dead if its cloud is overlapped. Two hydrogen atoms with live electron clouds come together to form one hydrogen molecule with a pair of dead electrons. The electrons are chemically dead in the sense that they will not participate in the formation of new bonds. Generally we can say that any collection of atoms with live electrons will form bonds through the overlapping of live electron clouds until all the electrons in the collection have become dead. Each bond that is formed can be regarded as a pair of electrons that have killed each other by overlapping. Bonding of this nature is called covalent bonding.

Let us now consider molecules composed of atoms

larger than hydrogen. The two electrons within one helium atom can overlap with each other and become dead. Therefore, it is not possible for the electron cloud of a helium atom to overlap and form bonds with clouds of other atoms. And indeed, the behavior of helium supports this model; that atom does not participate in bonding with other atoms. Helium is chemically dead.

A lithium atom has three electrons. Two of them can overlap to form a dead cloud, which is concentrated in a sphere around the nucleus. We will call this dead cloud plus the nucleus the "core" of the atom. The third electron cannot overlap substantially with the core; it must form a cloud that is concentrated outside the core. The cloud could be spherical if the sphere were larger than the core and the charge were concentrated outside it. Such a cloud would be alive to the possibility of bonding with other atoms, as illustrated with LiH in Figure 16–6.

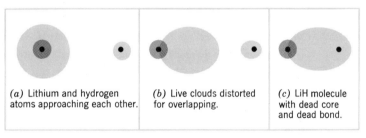

FIGURE 16–6
Formation of LiH
molecule.

(a) Lithium and hydrogen atoms approaching each other.

(b) Live clouds distorted for overlapping.

(c) LiH molecule with dead core and dead bond.

The live clouds of lithium and hydrogen in Figure 16–6 are shown in blue. The dark central dots represent the nuclei; the dark grey circles represent the dead core of the lithium atom. The cloud of electrons bonding the two atoms together in c is light grey to indicate that the electrons are overlapping, leaving the LiH molecule chemically dead.

Question 16–10 In a gas composed only of lithium atoms, the atoms combine to form diatomic Li_2 molecules. Make a sketch like the one in Figure 16–6 for the formation of Li_2.

A beryllium atom has four electrons. Two of them overlap to form the dead inner core just mentioned. They are the two electrons assigned to the shell of lowest energy in the shell model described in Section 13–6. The outer two electrons belong to the second shell. If their clouds are both like the cloud we described for the third electron in lithium, then they too are dead. If this were so, the beryllium atom would be as dead as the helium atom, and it would not form bonds. However, beryllium does form molecules, such as BeH_2, so evidently the two electrons in the second shell can form clouds that do not merge within the atom.

The bonding power of beryllium can be explained in terms of the cloud model if we assume that the two electrons in the second shell have different clouds. These could be two lobes poking out of the dead core in opposite directions, as indicated in Figure 16–7. These clouds would undoubtedly point in opposite directions in order to get as far apart as possible. The electrons within an atom can overlap, but they will not do so if they have a choice of avoiding it by pointing in different directions.

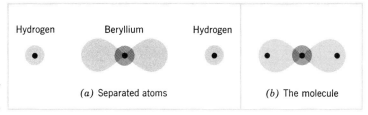

FIGURE 16–7
Formation of BeH_2.

Up to this point we have treated the electrons as clouds that can take on any shape, and we have guessed how the clouds behave. We are beginning to wonder how far this guessing can be carried, and how valuable a model full of guesswork is. Actually the results we have obtained can be derived from the modern mathematical theory of atomic and molecular structure. We will not consider the details of that theory, but will continue guessing, with assurance that the results are supported elsewhere by a systematic theory. In fact, the advantage of the cloud picture is that it can be developed intuitively. In some applications of the model, it actually serves as a guide for the development of the abstract mathematical model. However, we will find that we can deduce reasonable rules from the cloud model without resorting to the mathematical model.

Our model for beryllium in Figure 16–7 agrees with the experimental fact that beryllium forms two bonds in opposite directions. It also agrees with the mathematical theory about the shapes and directions of the clouds as they participate in bonding. However, the model does require modification for some applications. Consider, for example, a gas composed only of beryllium atoms. Most of the atoms persist as atoms, so the two outer electron clouds in beryllium atoms usually merge to form dead clouds. However, occasionally a diatomic beryllium molecule Be_2 does form. Apparently the outer electrons of the two beryllium atoms can, on some occasions, protrude as shown in Figure 16–8*a*. As the two atoms approach one another, one cloud from one of the atoms merges with one cloud from the other atom to form a covalent bond. But this merger does not represent the end of the story, for it leaves a molecule with two live clouds.

In order for the beryllium molecule pictured in Figure 16–8 to become dead, it is necessary for the two live clouds

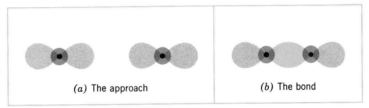

FIGURE 16-8
Beryllium atoms forming one bond.

(a) The approach   (b) The bond

to overlap, which means that they must bend around and approach each other. There is good reason for such a distortion. In the atom, the two live clouds point in opposite directions because they are not attracted to each other as much as to another atom. In the molecule, however, the live clouds on one atom are attracted to the nucleus of the other. This line of reasoning leads to the sequence of events in Figure 16–9.

The live clouds cannot exist as in *a*, so they must bend as in *b*. They cannot bend so far that they come together

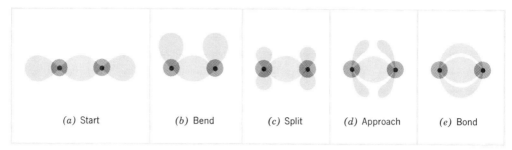

FIGURE 16-9   *Beryllium atoms forming second bond.*

between the atoms, because that region is already occupied by the dead pair that forms the first bond. Furthermore, there is no reason why the clouds should bend in such a way as to put each entire cloud on one side of the molecular axis, as in *b*. The split clouds in *c* give a more accurate picture of the distorted clouds. The additional distortion in *d* and *e* deepens the potential-energy well of the molecule, and the molecule can be trapped in the deeper well as a dead molecule of Be_2 with two bonds. As shown in *e*, one of the bonds is an ordinary one between the atoms and is said to be localized. The second bond has two parts that are strung in such a way as to be removed from the region between the atoms. The electrons in the second bond are said to be partially *delocalized*. They contribute to the bonding, but they are not localized in the region between the atoms. We will encounter other examples of such delocalization, and in each example it is used to explain the stability and behavior of various materials.

Less detailed descriptions of the bonding are given by formulas such as Be = Be or Be::Be. A stick or a pair of dots represents one bond. In these formulas there is no indication of the distinction between the two types of bonds. It is also possible to develop the more detailed cloud picture in such a way as to remove the distinction. When this is done, the bonds are alike and each electron is somewhat delocalized.

Question 16–11   Suggest a sequence of pictures to replace Figure 16–9 to show the formation of two identical bonds representing four partially delocalized electrons, rather than the localized and the delocalized electrons as in Figure 16–9*e*.

Let us now consider the other atoms in the second row of the Periodic Table. Each atom has a dead core formed by the two electrons in the first shell. In addition, each atom has a number of electrons in the second shell.

Atom	B	C	N	O	F	Ne
Number of second shell electrons	3	4	5	6	7	8

If the three electrons in the second shell of boron are alive, they will take the shapes of three lobes poking out of the core in three different directions that are as far apart as possible. Carbon has four lobes poking out in four directions that are as far apart as possible. We might expect the clouds for boron to be directed toward the vertices of a triangle and the clouds for carbon to be directed toward the vertices of a tetrahedron,* as indicated in Figure 16–10.

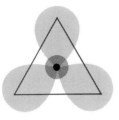
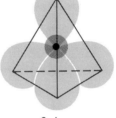

FIGURE 16-10
Live clouds for boron and carbon.

Boron   Carbon

The trigonal arrangement of clouds predicted for boron is easy to visualize, but the tetrahedral arrangement for carbon is not. If you have difficulty visualizing a tetrahedron, a three-dimensional model can easily be constructed by cutting equilateral triangles out of cardboard. Make four triangles of the same size and tape them together at their edges. Look at the tetrahedron and imagine a carbon atom inside with four live lobes each pointing toward one of the vertices.

* The prefix *tetra* means "four," and *hedron* means "side." A *polyhedron* is a three-dimensional solid with many sides. If the sides are all alike, the polyhedron is *regular*. The tetrahedron is a regular polyhedron with four sides, each of which is an equilateral triangle. A cube is a regular hexahedron. Two-dimensional figures such as triangles and squares are *polygons* (*gon* means "line"). Regular polygons include the trigon (triangle), tetragon (square), pentagon, hexagon, and so forth. We will also use the adjective forms *trigonal* and *tetrahedral*.

The purpose of the following experiment is to make a model of the clouds themselves.

EXPERIMENT 16-3   Model for carbon bonding (a demonstration)

Four balloons, each representing one electron cloud, are inflated and the necks tied closely together. The four-balloon combination is flipped into the air and allowed to bounce a bit. The ballons will point in four different directions that are as far apart as possible. Look at the resulting configuration and try to visualize the corresponding tetrahedron. The balloons repel each other by mutual contact and thus serve as a model for the electron clouds that repel each other because of the similar electric charge.

Before we continue with the other elements in the second row of the Periodic Table, let us pause to ask whether our models for boron and carbon are supported by experimental evidence. We have proposed a model of boron which has three live clouds, and a model of carbon which has four live clouds. From this we predict that these elements combine with hydrogen to form the compounds BH_3 and CH_4. In fact CH_4, which is called methane, is a very familiar compound. It is the main component of natural gas, which is burned as a fuel. The other compound BH_3 also exists but it is very unstable. It burns so explosively that it would make a very good rocket fuel.

The existence of CH_4 is evidence that carbon has four live electron clouds, but we need further evidence to support the tetrahedral structure. The four clouds all have equal negative charges and are similar in shape, so we would expect them to assume a symmetrical configuration. The two configurations that have the expected symmetry are the square and the tetrahedron.

If the four clouds of carbon all lie in the same plane and form the square configuration, there can be several arrangements of atoms in some molecules that contain carbon. For example, CH_2Cl_2 could form two distinguishable arrangements (see Fig. 16–11); consequently we would expect two substances, each with different chemical properties. The fact is, however, that we find only one compound with the formula CH_2Cl_2. Therefore, since observations do not support the planar model of carbon, we are left with the tetrahedral model.

FIGURE 16-11
Different arrangements for
CH_2Cl_2 in a square.

Question 16–12 Sketch a tetrahedron or use a model for a tetrahedron to convince yourself that there can be only one type of molecule CH_2Cl_2 if the bonding is tetrahedral.

Up to this point the cloud model has given us reliable results that agree with both observations and mathematical theory. When we consider nitrogen, however, we encounter a new problem. Because nitrogen has five electrons in the second shell we are tempted to picture the atom as a core with five live clouds poking out in five different directions. However, some compounds of nitrogen, such as NH_3 (ammonia), suggest three live clouds. Moreover, the properties of ammonia indicate that the bonds are not as crowded as they would be if there were five clouds poking out of the nitrogen core. In fact, the angles between the bonds in ammonia can be explained fairly well by assuming that nitrogen, like carbon, has a tetrahedral arrangement of clouds, with one of the four clouds dead. Evidently, two of the electrons in the second shell overlap to form a dead cloud that points to one of the vertices of the tetrahedron, and the other three electrons remain alive, pointing to the other three vertices. It would appear that there is no room to form five live clouds.

If our cloud model of nitrogen is correct, then we would expect oxygen, which has six electrons in the second shell, to have a tetrahedral structure with two live and two dead clouds, each composed of two overlapping electron clouds. This model of the oxygen atom is consistent with the known structure of the water molecule. If we were to ignore the dead clouds in the oxygen atom, we would expect the two live clouds to spread out as in beryllium, so that the two bonds in the water molecule would be in a straight line. However, because of the presence of the two dead clouds (see Fig. 16-12), the structure of the water molecule is not linear.

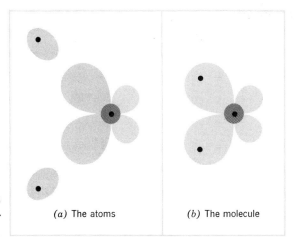

FIGURE 16-12
Formation of water molecule.

(a) The atoms   (b) The molecule

By continuing this line of argument, we could build still more cloud models of the atoms. Cloud models for the first ten elements of the Periodic Table are shown in Figure 16-13. The live clouds are shown in blue; a pair of electrons in a dead cloud is shown as light grey; the dark grey represents the core; the black is the nucleus.

Question 16-13   Explain why helium and neon are chemically inactive.

Question 16-14   Count the number of live electron clouds in each of the atoms in Figure 16-13 and write formulas for the molecule formed by each atom with hydrogen.

Question 16-15   Predict the number of bonds in each of the diatomic molecules N_2, O_2, and F_2. Sketch cloud pictures to show the delocalization of bonding electrons in N_2 and O_2.

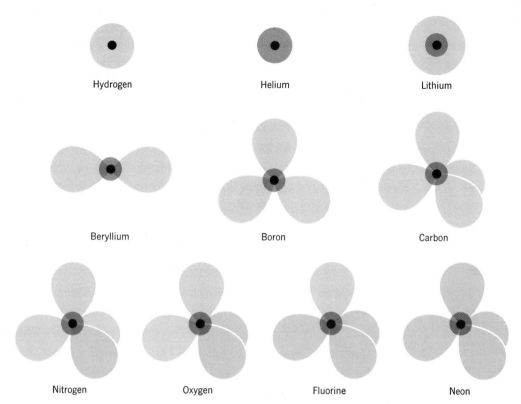

Hydrogen   Helium   Lithium

Beryllium   Boron   Carbon

Nitrogen   Oxygen   Fluorine   Neon

FIGURE 16-13   *Cloud models for atoms.*

We have considered the correspondence between experimental evidence and the electron cloud model. Let us review the variety of ways in which electron clouds can overlap to form bonds. They can merge head-on as shown in Figures 16-4, 16-6, 16-7, and 16-8, and thus remain localized between the participating atoms. They can also bend around to form a partially delocalized bond (consisting of delocalized electrons) as shown in Figure 16-9.

Our picture of covalent bonding is not complete, however, because we have not considered all the ways in which electron clouds can overlap. There are other types of delocalized bonds in which the delocalization is carried to such extremes that the bonding electrons are no longer associated with a pair of atoms.

The most extreme case of delocalization occurs in metals, which are considered in the next chapter. The atoms in a metal gather together so closely that the live electrons on each atom tend to overlap in all directions and merge to form one huge cloud that spreads throughout the entire metal. A metal can be pictured as a huge electron cloud with atomic cores embedded at various points in the cloud. This picture seems to contradict our rule that no more than two electron clouds can overlap substantially, but the rule really applies only to electrons that are trying to crowd into the space of one electron. In a metal, each electron spreads out so much that the cloud becomes very thin, and the merger is permitted.

An interesting example of delocalization that is less extreme than the metal delocalization is provided by the benzene molecule (C_6H_6). Notice in Figure 16–14 that

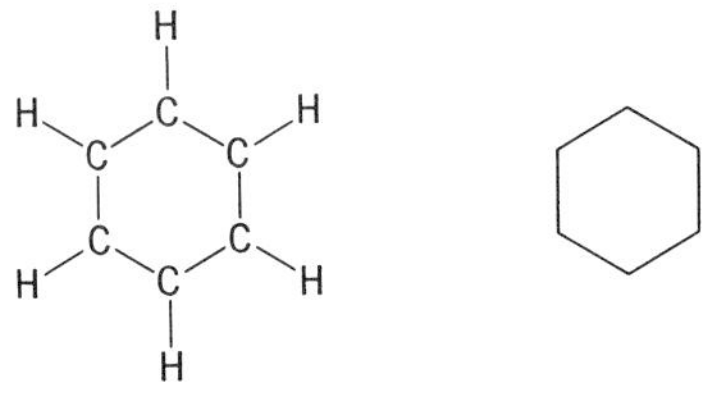

FIGURE 16-14   *(a)* Structure   *(b)* Abbreviation
Benzene.   for the benzene
                                          structure

each carbon atom is bonded to three other atoms, two carbon and one hydrogen. This leaves one live electron unaccounted for. Evidently the clouds for carbon in the benzene environment become distorted and assume a configuration like that of the electron clouds in the boron atom (see Fig. 16–10). The fourth cloud divides, as in Figure 16–9c, so that part of the cloud is above the molecular plane and part is below. These clouds are shown in Figure 16–15a, using the abbreviation in Figure 16–14b, which is called the *benzene ring*. The electrons that participate are delocalized in that they cannot be associated with any particular pair of atoms; they merge to form two doughnut-shaped clouds, one above and the other below the ring. These delocalized clouds produce a drastic lowering of the potential energy that makes the molecule stable. Without assuming this complete delocalization, the observed stability

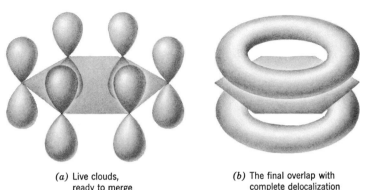

FIGURE 16-15
Cloud model for benzene.

(a) Live clouds,
ready to merge

(b) The final overlap with
complete delocalization

of benzene cannot be accounted for by the mathematical theory.

Many of the molecules we have discussed so far, such as CH_4 and C_6H_6, have a balanced distribution of electric charge. If such a molecule were placed between two oppositely charged parallel plates (see Fig. 16–16a), the

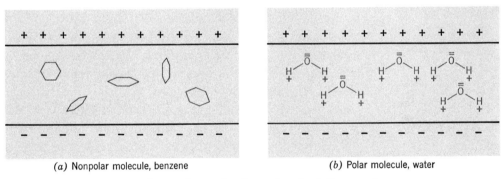

(a) Nonpolar molecule, benzene

(b) Polar molecule, water

FIGURE 16-16   *Polar and nonpolar molecules under the influence of electrical forces.*

molecule would not tend to rotate as a result of electrical forces. However, some molecules, such as water (see Fig. 15–24) have an unbalanced distribution of electric charge. If a water molecule is placed between two oppositely charged plates it will rotate as a result of electrical forces. The molecule will orient itself so that its more positive side is nearer to the negative plate than to the positive (see Fig. 16–16b). Recall what happened to the stream of water in Experiment 15–4. Any molecule that has a tendency to rotate as a result of electrical forces is said to be *polarized*.

The *polarity* of a bond between two atoms is a measure of the extent to which the bonding cloud is more concentrated near one atom than the other. If the atoms are the same kind, as in H_2, the attraction of the bonding cloud is the same toward both atoms, and the bond has no polarity —it is not polarized. On the other hand, if the atoms are different, as in HF, the bond may be strongly polarized. In the molecule of HF, the bonding cloud is more concentrated near the fluorine atom than near the hydrogen atom. The cloud is drawn toward fluorine because of the higher positive charge on the fluorine nucleus. It is true that fluorine also has other electrons, which cancel part of the nuclear charge, but the cancelation is not very effective. To understand why, we have to consider the fluorine atom in more detail.

The fluorine nucleus has a positive charge of nine units. If there were no electrons in fluorine except the one that forms a bond, the polarization would become complete, that is, the bonding cloud would be so strongly drawn toward the fluorine nucleus that it would be torn away from the hydrogen nucleus altogether. There is, however, an electron core between the fluorine nucleus and the bonding cloud. This core, which contains two electrons, is quite concentrated near the nucleus, so it effectively cancels two units of nuclear charge. The bonding cloud responds as if the combined core and nucleus were an enlarged nucleus with a positive charge of seven units. Even that is a tremendous charge, enough to pull the bonding cloud away from the hydrogen nucleus.

In addition to the core, however, there are six other electrons in fluorine that form three dead pairs in lobes directed away from the bonding cloud. If these six electrons were crowded into the core, the bonding cloud would be shielded from the nucleus by eight electrons; the net charge of the core would be one positive charge. But if the net charge were only one, it would have the same attraction for the bonding cloud as the hydrogen nucleus, and the hydrogen fluoride bond would hardly be polarized. However, it is polarized, so the three dead clouds simply cannot be in the core, they must spread out in a tetrahedral shape. Consequently, these dead clouds are less effective at shielding the nucleus than the core electrons, and the effective positive charge at the fluorine end of the bonding clouds is greater than one unit; perhaps it amounts to two or three.

Polarized molecules are strongly attracted to each other. The positively charged part of one molecule is attracted to the negatively charged part of another, and the molecules tend to become oriented in such a way that the attractive forces can hold them together. For example, hydrogen fluoride (HF) molecules tend to align themselves in chains (see Fig. 16–17). The forces holding the molecules together

FIGURE 16–17 $\ldots \overset{\oplus}{H}—\overset{\ominus}{F}$ $\ldots \overset{\oplus}{H}—\overset{\ominus}{F}$ $\ldots \overset{\oplus}{H}—\overset{\ominus}{F} \ldots$

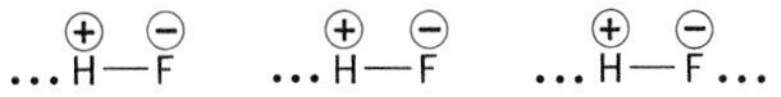

are not nearly so strong as the bonds within molecules, but they are strong enough to prevent the melting of crystalline HF at temperatures where the unpolarized molecules H_2 and F_2 cannot form crystals. These effects of polarization are illustrated in Table 16–2. The compound LiF is included in the table to show the drastic change in properties that result from complete polarization, which for solids results in ionic crystals.

Table 16–2

Substance (Formula)	Melting Point (Degrees Celsius)	Boiling Point (Degrees Celsius)
H_2	-259	-253
F_2	-223	-187
O_2	-219	-183
HF	-83	$+20$
H_2O	0	$+100$
LiF	$+870$	$+1676$

Question 16–16 Table 16–2 shows that H_2, F_2, and O_2 melt at much lower temperatures than HF and H_2O, which, in turn, melt at much lower temperatures than LiF. Explain why there should be such a sharp division of these compounds into three classes.

16-7   *Summary*

We began the consideration of bonding (in Chapter 11) with the hypothesis that certain bonds are electrical. Since then we have found evidence that not only substantiates this but also suggests that perhaps all bonds are electrical. We have constructed a model for bonding in molecules that is based on the idea of shared electron clouds. When single (live) electron clouds from different atoms overlap (become dead), a covalent bond is formed that holds the two atoms together.

Within this basic model we have found that in some molecules there may be more than one bond between atoms. However, each bond is formed by the overlapping of two electron clouds. In the formation of these multiple bonds, some of the electron clouds become delocalized. Several delocalized clouds can join together to form a larger cloud that is even more delocalized, as in benzene. This enlarged cloud covers the entire molecule.

We have seen that this is an extension of our earlier idea of ionic bonding. In most molecules one nucleus has a greater attraction for the bonding electrons than the other nucleus does. The result is that the dead electron pair is closer to that nucleus. If the attraction of one nucleus is great enough, the electron pair may stick to it exclusively, leaving the other nucleus with only the electrons remaining from its original complement. Thus a pair of ions is formed.

We have mentioned that metal crystals are held together by one huge electron cloud with the metal ions arranged regularly inside it. However, we have not seen how crystals of nonionic substances are formed. We have a model for the bonding within molecules but only a very incomplete idea of the bonding between molecules. This will be taken up in Chapter 17.

QUESTIONS   16-17   If we had a porous cup or barrier separating the two electrodes in Experiment 16–2, would we have a way of making sodium hydroxide and sulfuric acid? Explain.

16-18 Refer to the data collected in Experiment 16-1. From the current and the time, calculate the total charge that was transferred in the electrolysis. Calculate the number of grams of hydrogen that were produced by this charge transfer, and calculate from this the volume of hydrogen that was produced. How did the experimentally measured amount of hydrogen produced compare with this calculated amount?

16-19 Substances other than sodium sulfate can be used to make water conduct electricity when it is being decomposed by electrolysis. Name several. What properties do they have in common?

16-20 Your lungs hold about 4 liters (roughly a gallon) of air. If 6.02×10^{23} molecules are present in 22.4 liters of air, how many molecules can your lungs hold? (Give only an order of magnitude answer.)

16-21 When water is decomposed of electrolysis, the volume of hydrogen produced is twice the volume of oxygen. The density of hydrogen is about 0.09 grams per liter; the density of oxygen is about 1.43 grams per liter. What is the mass ratio of the gases formed when water is electrolyzed?

16-22 Use Avogadro's hypothesis and the densities in Question 16-21 to calculate the relative mass of molecules of oxygen and hydrogen.

16-23 The compound ammonium chloride (NH_4Cl) is bonded both ionically and covalently. Explain how this is possible and suggest a possible structure for the compound.

16-24 Some of the formulas that follow represent molecules that have been prepared experimentally. Guess which ones. In each case, indicate which atoms are bonded to each other, and suggest a likely shape for the molecule.

$$CCl_4, O_2F, OF_2, HCN, CBNF, PSNS$$

References   1. CBA, *Chemical Systems,* McGraw-Hill and Earlham College Press, 1964, Chapter 7, pages 229 to 278; Summary, pages 450 to 455 (this section is highly packed); and Chapter 11, Sections 11–1, 11–4 through 11–13, pages 456 to 458, 464 to 478.

Structure of atoms, molecules, and solids. Note that the word "orbital" is related (but not synonymous) to our term "charge cloud." Some of these sections will be more appropriate following Chapter 17, so you should return to these sections then.

2. Chem Study, *Chemistry, an Experimental Science*, W. H. Freeman & Co., 1963, Chapter 16, pages 274 to 297.

Bonding and the shapes of molecules.

3. Christiansen, G. S., and P. H. Garrett, *Structure and Change*, W. H. Freeman & Co., 1960, pages 81 to 85.

A discussion of the experimental data leading to the law of combining volumes and Avogadro's hypothesis.

4. Greenstone, A. W., F. X. Sutman, and L. G. Hollingsworth, *Concepts in Chemistry*, Harcourt, Brace & World, Inc., New York, 1966, pages 98 to 100.

Bonding of hydrogen molecules and other diatomic gases.

ALSO OF INTEREST

5. Herz, W., *The Shape of Carbon Compounds*, W. A. Benjamin, Inc., New York, 1963.

Pages 15 to 29 present the relationship between cloud shapes (the clouds are called "orbitals") and shapes of molecules; pages 30 to 47 discuss double and triple bonds; pages 92 to 106 discuss electron delocalization. More advanced than our treatment.

Satour

CHAPTER 17

NONIONIC MATERIALS

We *have found that a detailed study of the* structures of certain ionic solids, such as sodium chloride, has enabled us to understand, in terms of structure, why they behave as they do. But there are many solids that have properties very different from those of ionic solids. In this chapter we propose to consider a variety of nonionic solids and to see how a knowledge of their structures enables us to account for their properties. We can expect that the structures of all the nonionic solids we choose to study will include at least one common feature: there must be forces of some kind that hold the particles of the solid in place and keep them from escaping from the solid boundary to form a liquid or a gas.

17-1   *The structure of diamond*

Diamond is a solid whose properties are markedly different from those of the ionic solids we have studied. You probably know that diamond is the hardest substance known; no other substance can scratch it. Scratching a substance involves separting small bits of it from the remainder, which in turn requires separating one particle from another and overcoming the forces between them. Why is diamond so resistant to breaking? Let us study this question.

The following facts are known about diamonds.

1. When you burn a diamond in the presence of oxygen, the sole product is carbon dioxide (CO_2). Diamond must be pure carbon.

2. A diamond does not melt, but vaporizes above $3500°C$. This is a temperature at which almost every other known solid has already liquefied or vaporized.

3. A diamond will not conduct an electric current.

4. In a diamond crystal each carbon atom has four nearest neighbors that are also carbon atoms. We know this from x-ray diffraction evidence.

The observed properties of diamond must certainly depend on the tetrahedral structure of the carbon atom; but how? To help answer this question, let us first consider the nature of the bond joining a carbon atom to one of its neighbors. Clearly, the bond must be a strong one, since diamond is very hard and remains solid at very high temperatures. Moreover, the bonding electrons must be equally shared between the carbon atoms, since identical atoms have the same affinity for electrons. Since diamond does not conduct an electric current, we know that these electrons are not free to move randomly throughout the crystal. These properties strongly support the idea that the carbon atoms are bound to each other by covalent bonds.

As we have seen in Chapters 13 and 16, carbon has four electrons in its outer shell and can form four covalent bonds. We know that a tetrahedral arrangement of these bonds is the most stable one (recall the balloon model in Experiment 16–3). The bonding of carbon in diamond can

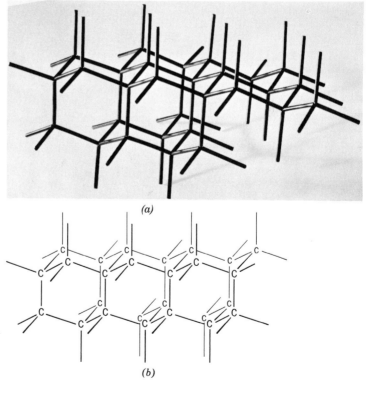

(a)

(b)

FIGURE 17-1
A model of bonding in diamond.

be explained similarly. Each carbon atom is bonded to four neighboring carbon atoms, and the bonds are tetrahedrally oriented. Figure 17–1 shows a photograph of a model and a drawing of the bond arrangement.

In *a* the tubes represent the bonds between atoms, so that the location of each atom is given by the intersection of four tubes. In *b* the bonds are represented by lines and each carbon atom by the letter "*C*"; each line represent a pair of bonding electrons.

Our model suggests that a diamond crystal is just one big molecule in which each carbon atom is tetrahedrally bonded to four other carbon atoms. X-ray diffraction evidence supports this model, which is also in accord with the properties we mentioned: hardness, high vaporization point, and lack of electrical conductivity.

FIGURE 17–2
A cut and polished diamond. (Burt Owen, D.P.I.)

Raw diamonds are usually formed during the rapid solidification of magma moving up through the vent of a volcano. The diamond may later be released from the rock in which it is formed by the processes of erosion and then be carried into streams. The photograph that opens this chapter shows two raw diamonds; they look something like common quartz and have none of the brilliance of a diamond set in a ring.

To shape a diamond into a gem prized in jewelry (Fig. 17-2) takes hours of laborious grinding and polishing. The facets (faces) on the front of the diamond must have just the right orientation to catch the light that enters the diamond. However, the shape of the back side determines where that light leaves the diamond. If the back side is shaped correctly, nearly all of the light that enters the diamond from the front and back leaves it through the front face, producing the brilliance you see. The diamond also acts like a prism in that it breaks white light into little spectra.

Question 17–1 Describe in terms of positions of atoms what you think may happen when a diamond is heated above 3500°C.

Question 17–2 Explain why diamond is a poor conductor of electricity.

Question 17–3 A diamond cleaves smoothly along four planes which are at right angles to the four covalent-bond directions. Can you suggest why this should be so?

17-2 *Other carbon compounds*

The carbon atom's ability to form covalent bonds with as many as four other carbon atoms, as in diamond, results in the possible existence of an almost limitless number of compounds containing carbon. Hundreds of thousands of such compounds are known, and many new ones are being prepared and investigated each day. We shall comment very briefly on a few of the compounds of carbon that have structures similar in some respects to that of diamond.

The simplest carbon compound is the gas methane (CH_4). As you know, the bonds in methane are tetrahedrally oriented. The shape of a methane molecule can be represented in various ways, some of which are shown in Fig-

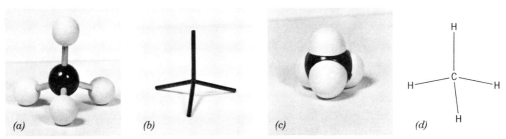

(a) *(b)* *(c)* *(d)*

FIGURE 17-3 *Models showing the arrangement of atoms in methane.*

ure 17–3. A photograph of a ball and stick model is shown in *a* in which the sticks show the direction of the bonds and the balls represent the atoms; each stick represents a pair of electrons. A framework model, such as was used to represent the diamond structure (see Fig. 17–1), is shown in *b*. Here you must imagine that a carbon atom is located at the intersection of the four plastic tubes used to represent bonds, and that a hydrogen atom is located at the outer end of each tube. In *c* bonding electrons are not represented and you must imagine bonds between the objects representing the carbon atom and the hydrogen atoms. In *d* each line represents a pair of electrons. We will find it helpful to be able to make use of these and other ways of representing structures.

A series of compounds containing carbon, hydrogen, and chlorine is formed when chlorine is allowed to react with methane. These are compounds in which one or more of the covalent carbon-hydrogen bonds has been broken and a carbon-chlorine bond formed instead. As shown in the following equations, it is possible to form four compounds by replacing hydrogen atoms in methane with chlorine atoms. The equations and the names of the compounds formed are as follows:

$$CH_4 + Cl_2 \rightarrow HCl + CH_3Cl \quad \text{(chloromethane)}$$
$$CH_3Cl + Cl_2 \rightarrow HCl + CH_2Cl_2 \quad \text{(dichloromethane)}$$
$$CH_2Cl_2 + Cl_2 \rightarrow HCl + CHCl_3 \quad \text{(trichloromethane)}$$
$$CHCl_3 + Cl_2 \rightarrow HCl + CCl_4 \quad \text{(tetrachloromethane)}$$

Trichloromethane is commonly called chloroform, and tetrachloromethane is usually referred to as carbon tetrachloride. Several of these compounds are useful as solvents in paint removers, dry cleaners, and spot removers. They should be handled with care because their vapors are toxic. You know from previous study that the bonds in methane are tetrahedrally oriented. Here is an experiment to help you decide what the bond arrangement is in the chlorine-substituted methanes.

EXPERIMENT 17-1 Electrical properties of two fluids

In Experiment 15–4 a charged rod caused a marked deflection of a stream of water as it flowed from a tap. We shall examine the behavior of carbon tetrachloride and dichloromethane in this way. Two burets will be set up so that the liquid will fall about 20 cm into catch beakers. Place dichloromethane in one buret and carbon tetrachloride in the other.

Before testing to see whether the charged rod will deflect either of the two liquids, try to predict what should happen if the bonds on the carbon atoms are tetrahedrally arranged. First consider the electron distribution in a single carbon-chlorine bond. Chlorine, like oxygen, attracts electrons, so electrons in a carbon-chlorine bond will be closer to chlorine and the bond will be polar. Now consider the shape of a carbon tetrachloride molecule and the shape of a dichloromethane molecule, assuming that the bonds in each are tetrahedrally oriented. Models of these molecules are shown in Figure 17–4. Would you expect carbon tetrachloride to be polar, like a water molecule (see Section 16-6)? Would you expect dichloromethane to be polar?

Test both substances and record your results. Is CH_2Cl_2 polar? Is CCl_4 polar? Do these experimental results support the hypothesis that the bonds on carbon in these compounds are tetrahedrally oriented?

Carbon forms many compounds with hydrogen. The number of possibilities is almost unlimited. Included among some of the simpler *hydrocarbons* (compounds containing

FIGURE 17-4   *Models of* CH_2Cl_2 *and* CCl_4 *molecules.*

only carbon and hydrogen) are the following:

C_2H_6 (ethane)
C_3H_8 (propane)
C_4H_{10} (butane)
C_5H_{12} (pentane)

Draw structures for each of these hydrocarbons. (You know that each carbon atom forms four tetrahedrally oriented covalent bonds, and that hydrogen can form only one covalent bond.) Is there any resemblance to the diamond structure? Is it possible to draw more than one arrangement of atoms for ethane with tetrahedral carbon bonds? for butane? Molecules that have the same molecular formula but different arrangements of atoms are called *isomers*. The determination of the structure of compounds containing carbon is often complicated by the existence of isomers. As you may guess from what we have said thus far, a study of the compounds of carbon is a lengthy subject; it is treated in detail in courses in organic chemistry.

Question 17-4   How many different compounds with the molecular formula $C_2H_4Cl_2$ are possible? Draw diagrams representing their structure.

Question 17-5   Knowing that carbon forms four covalent bonds, hydrogen one, and oxygen two, draw all the possible isomers for C_2H_6O.

17-3   *Graphite*

Diamond is not the only crystalline form of carbon. There is another form found in nature, called graphite, which has many properties different from those of diamond.

The density of diamond is 3.5 g/cm³; that of graphite is 2.2 g/cm³. What does this suggest about the closeness of packing in diamond as compared to graphite?

From x-ray diffraction evidence we know that each carbon atom in graphite has three nearest neighbors that lie in a plane with it. The distance between adjacent planes of atoms is greater than the distance between neighboring atoms within a plane. See if you can figure out a planar arrangement of atoms (pennies are convenient to use) in which each atom has three nearest neighbors. What geometrical shape do you find in your structure?

There are four excellent cleavage planes in diamond. Graphite, however, has only one, just as mica has. You can write with graphite (this is the "lead" in pencils), which means that some particles rub off onto the paper. It is also used to lubricate locks and door-latch mechanisms.

Diamond does not conduct an electric current. Does graphite? You could find out by a simple experiment; in fact, you have already done it. The carbon electrodes, which you used in several experiments in Chapter 14, were largely graphite. Does graphite conduct an electric current?

The electrode you used was made of many little crystals of graphite oriented in all directions. If you had one large single crystal of graphite you would find that its conductivity would depend on the direction in the crystal along which the charge flowed. If you place the electrodes on a single crystal so that the charge flows at right angles to the cleavage planes, the graphite crystal conducts very poorly. On the other hand, if you place the electrodes so that the charge flows parallel to the cleavage planes, you will find that the crystal conducts very well. All these properties suggest that graphite contains electrons that are free to move along particular directions in the solid and that it has at least some interatomic bonds that can be broken easily.

Figure 17–5 represents a section of a graphite crystal.

Layer

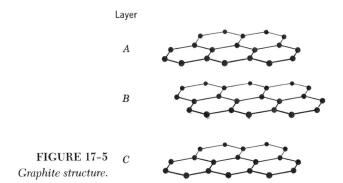

A

B

FIGURE 17-5
Graphite structure.   C

Note that within each layer every carbon atom has *three* nearest neighbors. Each carbon atom can form *four* covalent bonds, so this leaves one electron on each carbon atom unaccounted for.

This left-over electron is involved not with bonding within a layer, but with longer-range, weaker bonding between layers. Half of these electrons from any one layer lie above the plane of that layer and half below it. The attractive force between these electrons and the positively charged layers is somewhat weaker than the force holding atoms together within layers. In fact, these are the electrons that are free to move along between the layers and that are responsible for the electrical conductivity of graphite in directions parallel to the atom layers.

To help you visualize the structure of graphite, you can construct a model with styrofoam spheres and toothpicks, or with three pieces of chicken wire. If you use spheres, first make up three separate planar layers, using half-inch styrofoam balls and half-length toothpicks. All the angles between the toothpick bonds must be 120 degrees. Arrange the layers one above the other, as in Figure 17–5, using full-length toothpicks between adjacent layers. Note that layer *A* is directly above layer *C*, so that every atom in *A* has one directly below it in *C*. Also note that the atoms in layer *B* line up with the centers of the hexagonal spaces in layers *A* and *C*. Figure 17–5 does not show the electrons between the layers of atoms.

Question 17–6 Graphite is used as a lubricant at temperatures high enough to destroy an ordinary oil or grease. Suggest how it may function as a lubricant.

17–4 *Loosely bonded crystals*

In Experiment 7–2 you heated a number of different substances to learn how each reacted. What happened to the iodine when it was heated? Do you conclude that it is easy or difficult to break apart the particles in an iodine crystal?

Experiments show that one volume of iodine vapor combines with one volume of hydrogen gas to produce two volumes of hydrogen iodide gas. What is the nature of a molecule of iodine? (If necessary, see Section 16–5 on the formation of HCl.)

Even if you look at newly formed iodine crystals with a lens, you cannot determine the arrangement of the units in the crystal. This requires x-ray diffraction analysis. A representation of the structure of iodine determined in this way is shown in Figure 17–6.

Iodine crystals consist of I_2 molecules in which the atoms are covalently bonded to each other. When iodine is gently heated it forms a vapor without first forming a liquid. The formation of a vapor directly from a solid is called *sublimation*. This process is characteristic of substances that have little intermolecular attraction, that is, very shallow potential wells in the liquid state.

What is the nature of the force that holds iodine molecules in a fixed arrangement in an iodine crystal? We might expect that it is electrical, though not as strong as the electrical force between charged ions in an ionic crystal, or between two atoms in a covalent bond. There is no simple experimental evidence of the electrical nature of the bonding force in iodine crystals, but a theoretical model accounts

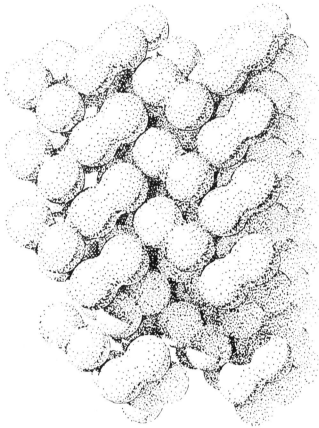

(From *College Chemistry* by Linus Pauling. W. H. Freeman and Company © 1950)

FIGURE 17-6
Iodine crystal.

moderately well for it in terms of fluctuating electric charge.

Consider an imaginary molecule, X, shaped like a sphere, in which the electrons are uniformly distributed so that the molecule is not polar (see Section 16–6); that is, it is more like a molecule of chlorine than of hydrogen chloride. We know that electrons in atoms and molecules (or electrons in motion in any situation) cannot have their positions specified exactly. But we can represent the average position of the electrons during a period of time. In Figure 17–7 two different instantaneous distributions of charge in our model spherical molecule are shown.

The distribution shown in a is the most likely one for the isolated, spherical molecule X, since the center of negative charge coincides with the center of positive charge. But at any one instant a charge distribution such as that shown in b is also possible.

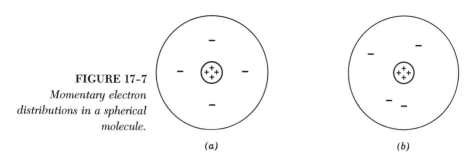

FIGURE 17-7
Momentary electron distributions in a spherical molecule.

(a)   (b)

How would distribution *b* of molecule *X* affect some other neighboring molecule *Y*? Temporarily, *X* has an uneven charge distribution, so that one side of it is temporarily negative and the other side temporarily positive. The positive side of *X* would distort the electron cloud of a neighboring molecule *Y*, so that for an instant the positive side of *X* would be attracted to the negative side of *Y* (Fig. 17-8).

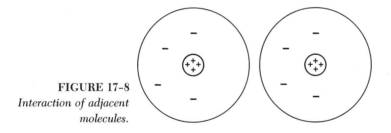

FIGURE 17-8
Interaction of adjacent molecules.

Such a shift in the charge distribution of *Y* would be temporary; at the next instant *Y* might respond to a different shift in *X* or it might respond to an uneven charge distribution in some other neighboring molecule. These instantaneous shifts are the basis for weak attractive forces between molecules. The forces depend little on the orientation of molecules as long as they are near each other; thus, these forces operate between molecules in the liquid state nearly as well as between those in solids.

This bonding force is electrical, just like the other bonding forces we have discussed, but it is weaker because of the nature of its formation, and it is different enough to have a special name. It is commonly called the *van der Waals force*, after a Dutch physicist.

Would you expect van der Waals forces to depend on molecular size? Consider the ease with which the electron cloud in an iodine molecule could shift, compared with the electron cloud in a hydrogen molecule. In the former, there are 106 electrons. The charged clouds extend out from the center of each iodine atom a distance of about 2.2 Å, at which point the cloud has thinned considerably, and the iodine atom centers are about 2.7 Å apart. A rough representation of this "molecular shape" is given in Figure 17–9a. By contrast, a hydrogen molecule is drawn to the same scale in Figure 17–9b.

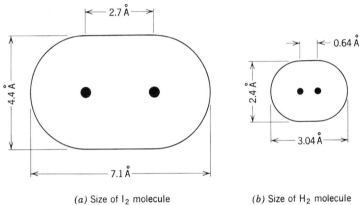

FIGURE 17-9
Molecular sizes.

(a) Size of I₂ molecule

(b) Size of H₂ molecule

The larger size and number of electrons in the iodine molecule increase its polarizability, that is, its tendency for charge fluctuations and for distortions of the electron cloud shape. You might, therefore, predict that the van der Waals forces are larger in larger molecules with more electrons. Iodine changes from a liquid to a gas at $183°C$; hydrogen at $-252.8°C$. Iodine melts at $114°C$. Are these data consistent with your predictions?

Question 17-7   Would you expect methane CH_4 and tetrachloromethane CCl_4 to have similar melting points? Explain.

17-5   *Hydrogen bonding*

In crystals in which the units of structure are molecules, there are bonding forces that are stronger than the weak van der Waals forces but weaker than the covalent bonds discussed in Chapter 16. Sugar, ice (which has a remarkably high melting point for such a small molecule), and aspirin are examples. Let us discuss the way water molecules are held together in an ice crystal.

In Figure 15–24 we saw that bonds in a water molecule are at an angle of 105 degrees to each other. This is close to the tetrahedral angle of 109 degrees. In addition, recall from our discussion in Chapter 16 that the water molecule is polar. The two vertices of the tetrahedron where hydrogen atoms are bonded to the oxygen atom have a net positive charge; the two vertices where there are only dead electron clouds are negative (see Fig. 17–10).

(a)

FIGURE 17–10
Water molecules.
(a) Clouds and bonding in a water molecule.
(b) Hydrogen bonds cause association in water molecules.

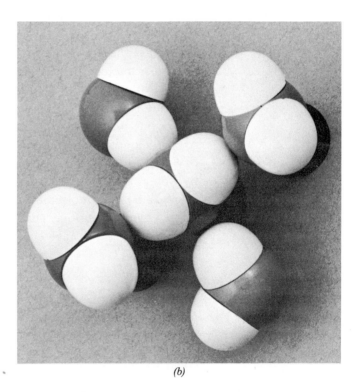

(b)

Because of the charge distribution around the water molecule, two water molecules will attract each other if placed close to one another. There are two ways that this attractive force can arise. First, there is an overall attraction of one molecule to another because of their polarity. This force may act over distances equal to or greater than the diameter of the molecule.

Second, a force arises when the negatively charged cloud of one molecule is touching or nearly touching the cloud that contains the positive hydrogen nucleus of another molecule. It is the latter kind of attraction that is favored by the small size of hydrogen atoms and the location of negatively charged dead clouds on the surface of oxygen atoms. The result is a kind of attraction between water molecules that is stronger than the simple polar attraction of one molecule for another. This kind of attraction is called *hydrogen bonding*. In general a hydrogen bond is only about one-tenth as strong as a covalent bond, for example, as within the water molecule itself. There are many compounds beside water in which hydrogen bonding is important.

What is the maximum number of water molecules that can be attracted to one water molecule? Figure 17–10*b* shows water molecules associated by hydrogen bonding. By referring to Figure 17–10*a*, we can see that each hydrogen end of one water molecule can attract one of the two dead oxygen clouds of another water molecule. And each dead oxygen cloud of our original molecule can attract one hydrogen end of still another water molecule. Therefore, since each water molecule has four clouds projecting from it, two with a positive charge and two with a negative charge, each water molecule can attract four other water molecules.

If every molecule in a group exerts this same attraction for others, how will the molecules be arranged in a crystal? Because the bonds between water molecules are arranged tetrahedrally, like the bonds of carbon, ice has an open diamond-like structure. When ice melts, the open structure collapses and allows the oxygen atoms to adopt a more nearly closed-packed arrangement, occupying a smaller volume.

We have studied a crystal with a tetrahedral arrangement of bonds (diamond) and found it very difficult to melt. Is ice difficult to melt? What are the differences between the bonds in diamond and the bonds in ice? Would you expect ice or diamond to be more closely packed?

Question 17–8 Would you expect hydrogen fluoride HF and fluorine F_2 to have similar melting points? Explain.

Question 17–9 What observation can you recall from everyday experience concerning the relative densities of ice and water? Is our discussion of the structural differences between ice and water consistent with this experience?

Question 17–10 Water has an unusually high boiling point for such a small molecule, as well as an unusually high melting point. How is this explained?

17-6 *Metals*

Lead, iron, copper, and silver are metals that were used in some earlier experiments. Summarize the physical properties of metals, either from your general experience or from earlier experimental work. Are metals crystalline? In what ways are they different from covalent crystalline substances such as diamond? In what ways are they different from crystals bonded by van der Waals forces such as iodine, or by hydrogen bonding such as ice? One way in which metals are different from other solids is that they are able to conduct an electric current in the solid state.

EXPERIMENT 17-2 Mechanical properties of metals

What happens to a crystal of sodium chloride if you strike it with a hammer? Do metal crystals behave in the same way? Let's find out. You will need a hammer, some wire (copper or iron, whatever kind you can find), and an anvil. The anvil may be another hammer or just a flat piece of steel or iron.

Lay the wire on the anvil and pound it with the hammer. What happens? Does it shatter?

Can you stretch a metal wire? Fasten the ends of a long

piece of copper wire to two pieces of broom handle, and
have a friend pull on one broom handle while you pull on
the other.

Can you stretch a salt crystal? The properties of metals
that are called *malleability* (you can pound them flat without breaking them) and *ductility* (you can stretch them)
certainly make metals different from the other crystals we
have studied.

Yet metals are crystalline. They have sharply defined
melting points; x-ray diffraction data tell us that the metal
atoms are arranged in very regular patterns. Nevertherless,
metals are malleable and ductile, and they do conduct an
electric current in both liquid and solid states. These are
the three properties that we must account for in a metal.

Because metals in the solid state are good conductors of
electric current, there must be many electrons in the metal
crystal that are free to move. How do we account for them?
Consider the metals included in Table 13–2: lithium, beryllium, sodium, magnesium, aluminum, potassium, and calcium. Do they have ionization energies that distinguish
them from the nonmetals? More extensive tables would be
required to answer this question for other metals. In Table
13–3 note that the energy required to remove *second* electrons in several metals (such as calcium and magnesium) is
also relatively low.

A model for a metal crystal is shown in Figure 17–11. It
consists of an orderly array of ions, each of which is made
up of an atomic nucleus plus all but the outer (one, two, or
three) electrons. The ions in the metals, however, differ
from the ions in an ionic crystal in that they frequently lose
and gain their charge. These positive ions are set in a "sea"
of electrons, which are those not firmly bound to the nuclei
and therefore relatively free to move through the crystal.
The example shown in the figure represents sodium with
eleven charge units in the nucleus.

The free electrons may be located in any of the total
space in the whole metal crystal not occupied by ions.

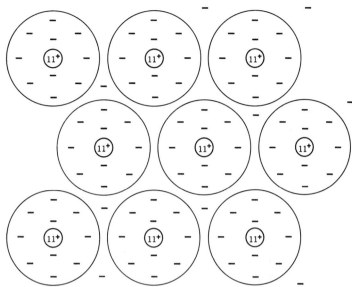

FIGURE 17-11
Model for a metal crystal.

These electrons are delocalized, that is, not associated with any specific ion. In a metal crystal not influenced by outside electrical forces, however, the electrons are partially localized between ions by electric attractions. As a matter of fact, it is this mutual attraction between positive ions and negative electrons that holds the metal crystal together.

We conclude, then, that the force that holds metal crystals together is also electric, but does not consist of electron-pair bonds as diamond does, ion-ion attraction bonds as cesium chloride does, or van der Waals bonds as iodine does. It arises from the attraction between positive ions and delocalized electrons.

17-7 *Packing in metal crystals*

What reasons can there be for a particular packing arrangement in a metal crystal? The structures of diamond and graphite are determined by the direction and number of the covalent bonds. In iodine the forces holding molecules in the crystal are not directional; they operate equally between neighboring molecules at the same distance regardless of position. The crystal structure of iodine is determined by the best packing together of dumbbell-shaped molecules. In ice the directional character of the hydrogen bonding requires a structure with hydrogens on a line between the oxygen centers and with oxygens arranged so that each has four neighbors.

We would expect no directional character to the bonds in metals. Moreover, the units that pack together are spheres, so the structure of a metal crystal generally results from packing the greatest number of spheres into a given space. Recall that you observed the close packing of spheres in Experiment 5–2.

17-8 *Applying the metal model*

Does the model for the structure of a metal explain the characteristics of metals that we noted earlier? First, it does conform with the arrangements of atoms found by x-ray diffraction. Second, it has electrons that are free to move. Under no external influence, these electrons are partially localized between ions; but when there is a difference of electrical potential, the electrons drift in the direction of the electrical force, and therefore the crystal conducts a current. To maintain a current in a metal wire, electrons must be added at one end and a like number removed from the other.

Recall that graphite also conducts an electric current in the solid state, and note that conductivity is explained in both cases by delocalized electrons. These electrons contribute to the bonds between layers as well as to the electrical conductivity. The bonding between layers in graphite, then, is partly metallic in nature.

The third characteristic that must be explained by our model is the deformability of metals. The experiments you performed showed that metals are malleable and ductile. Can the atom layers easily be moved over each other? If the top row in Figure 17–11 is moved a distance of one atomic diameter to one side, does the new structure differ from the original one? If you attempted to move one layer in an ionic crystal (say NaCl) in the same way, what would be the result? Such a move would bring negative ions close to negative ions and positive ions close to positive ions. Any attempt to do this causes the crystal to come apart. Why?

17-9 *A comparison of crystal bond types*

We have been using melting points of solids as a rough guide to the strength of bonding in crystals. A better measure of bonding strength is the energy required to cause sublimation, that is, to change the solid to a gas. By considering sublimation we avoid the comparison of the different bonding forces in liquids. Molecules in gases are sufficiently separated for the bonding forces to be nearly zero. The heats of sublimation and the melting points of the substances we have been discussing are given in Table 17–1.

Table 17–1   Melting Points and Heats of Sublimation of Some Crystalline Substances

Bond Type	Substance	Melting Point (°C)	Heat of Sublimation (Calories/Gram)
Covalent	Diamond	>3500	14,150
Metallic	Magnesium	651	1,500
Ionic	Sodium chloride	801	815
Hydrogen bond	Water	0	623
Van der Waals	Iodine	114	58

The covalently bonded crystals are indeed those with strongest bonding. From this table it would appear that metals have somewhat stronger bonding than salt crystals, but it should be realized that bonding strength is not uniform for all metals or for all kinds of salts. Hydrogen bonds in ice give bonding strength nearly equal to that of sodium chloride, but ice, with its strong three-dimensional hydrogen bonding, is unusual. Most hydrogen-bonded solids have crystal energies more like the energies of crystals bonded by van der Waals forces.

Question 17–11   What is meant by the statement, "There are no directed bonds in a metal crystal?"

Question 17–12   Suppose a metal has a metallic impurity whose atoms are larger than those of the host metal. What effect will this have on the properties of the metal?

17-10   *Sulfur—a solid that is sometimes noncrystalline*

Sulfur is an element that is mined either in volcanic deposits (the brimstone of the Bible and the classics) or in deposits deep below the surface of the Earth. Now it is used little as a household substance, but it is a source material for the production of chemicals in industry. Our present interest in sulfur comes from some unusual physical properties that are closely related to its structure.

Although sulfur is found as an element (see Fig. 17–12) it readily reacts with other elements. Its atomic number is 16, and its first ionization energy is 10.3 eV. How many electrons does it have in its outer shell? Try to predict how it combines with some other element.

Question 17–13   Give formulas for the compound of sulfur with hydrogen and for the compound of sulfur with sodium.

FIGURE 17-12   *Sulfur, which is brought up from under ground in the molten state, is allowed to solidify for bulk shipment. (Courtesy of Texas Gulf Sulfur)*

Question 17-14   Would you expect hydrogen sulfide to be ionic? Would you expect sodium sulfide to be ionic?

EXPERIMENT 17-3   Physical properties of sulfur

First examine some sulfur with a hand lens. Sketch the shape of the crystals. Then, before beginning the next part, have ready a funnel fitted with filter paper and a 250-ml beaker half-filled with water. Pour enough sulfur into a 150 by 16 mm test tube so that it is a little more than half full, and heat it carefully over an alcohol burner until the sulfur is all melted. The melted sulfur should fill at least one-third of the test tube. If it does not, remove the tube from the flame and add more sulfur. The reason for heating the sulfur carefully is that you do not want it to start burning, because the sulfur dioxide that would be produced is very irritating.

When the sulfur melts, insert a 360° thermometer into it and continue heating. Stir the melt with the thermometer, and observe the viscosity of the liquid as the temperature rises. (Viscosity means "thickness" in the sense of resistance to flow. Fingernail polish is very viscous; polish remover is not. SAE 30 motor oil is viscous; gasoline is not.)

Record your observations about viscosity at different temperatures. At what temperature is the melt the most viscous? What happens to the liquid as you continue to heat it beyond this temperature? When the sulfur is at as high a temperature as can be achieved with the alcohol burner, quickly pour only a small portion into the beaker of water. Remove the sulfur from the water and examine it. Record its properties.

Allow the sulfur remaining in the test tube to cool to about 200°C, and then pour it into the funnel containing the filter paper. As soon as a crust forms on the sulfur, carefully remove the filter paper from the funnel and open the paper to expose the inner liquid. Allow the liquid to cool, and sketch the shape of the crystals. If possible, save some of these crystals and examine them again after a few days.

How can the same substance exist not only as two kinds of crystal but also as a plastic material? What are the units of sulfur that are present and how are they held in the solid state? X-ray diffraction provides evidence that the primary unit in both kinds of crystallized sulfur is an S_8 molecule with angles of about 105 degrees between sulfur-sulfur bonds. In the next experiment you will construct a model of such a unit.

EXPERIMENT 17-4   A model of sulfur

Construct a small model of the S_8 molecule by using eight half-inch polystyrene spheres, two needles or pins, and eight small pieces of toothpick for connectors.

Use a protractor to mark on paper an angle of 105 degrees. Pierce each sphere with two holes directed toward the center at 105-degree angles to each other. The best way to do this is to place the center of the ball over the intersection of the 105-degree lines and insert two pins parallel to the lines.

Use the connectors to join eight sulfur atoms with spheres touching to form a puckered ring, that is, a ring in which the atoms are not all in the same plane. The connections can later be made secure by separating the spheres a bit, introducing a small amount of glue, and then pressing them together to make the unit. What holds the S_8 molecules to each other in a sulfur crystal?

Each of the two crystal forms of sulfur consists of S_8 molecules; but since they have different melting points, they must have a different packing of the molecules. Only one form, rhombic sulfur, is stable at room temperature. The monoclinic form that was obtained when the melt was poured into filter paper is unstable and changes within a few days into the rhombic form.

Now let us try to explain what happened to the structure of sulfur when you heated it. Recall that once all the sulfur was melted, it became less viscous as it was heated to higher temperatures. This behavior is similar to the usual behavior of liquids. But sulfur is unusual in that as the temperature of the liquid is increased still more, it becomes more viscous, and eventually less viscous as it nears the boiling point. It is curious that when hot sulfur is cooled suddenly, it solidifies into amorphous sulfur rather than into a crystalline structure. *Amorphous* means without form.

As sulfur is heated initially, it liquefies, and the S_8 molecules become free to move past each other. As the temperature is increased, some sulfur-sulfur bonds in the S_8 rings break open, and the resulting segments join end to end to form chains of different lengths. These may coil up, as shown in Figure 17–13, entangling other chains of sulfur atoms. Chains, both coiled and straight, move past each other much less easily than do the compact S_8 ring mole-

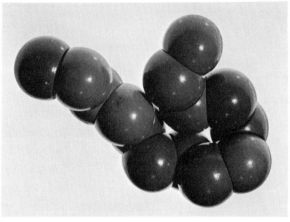

(a)

(b)

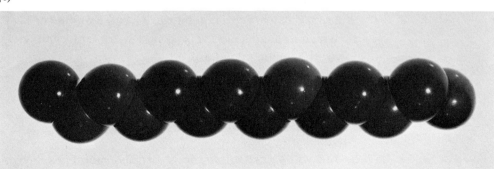

FIGURE 17-13 *Sulfur chains. (a) Folded. (b) Extended.*

cules. The viscosity increases as chains are formed from rings. As the sulfur is heated to still higher temperatures, the viscosity decreases again because the chains themselves begin to move more rapidly. This rapid motion results in less entanglement.

What happens when the liquid chiefly containing tangled chains is suddenly cooled? It becomes amorphous; the tangled chains lose their freedom of motion. They are only partially fixed in place by a combination of tangling and van der Waals forces; the chains have some freedom to move when a force is applied. Hence the amorphous sulfur is soft and pliable, and can be stretched to some extent.

Sulfur is not the most common example of an amorphous solid, but it is one in which the change from the crystalline to the amorphous state can be readily observed in the laboratory. In the following sections we shall be concerned with the properties of three more common examples of amorphous solids: glass, polyethylene, and rubber.

17-11   *Glass*

Glass for windows, containers, tubes, and rods is a man-made substance. It can have any one of a very wide range of compositions; its properties depend on its particular composition. In most glasses, silica (SiO_2) is a major constituent. Sand, composed of grains of quartz (an abundant form of silica), is the essential raw material in the making of glass. Glass made of pure silica has to be heated to a high temperature before it will soften enough to flow. For most uses, therefore, other substances are added so that the glass will soften at a lower temperature. (Recall Experiment 14–1, in which sodium chloride was added to lead chloride to lower the melting point.)

EXPERIMENT 17-5   Experiments with glass

Part A   Manipulating Glass

You will be given two pieces of glass tubing, each eight to twelve inches long. One piece has a high content of silica and is hard (a commonly available brand is Pyrex); the other has a lower content of silica (like common window glass).

Hold the tube of common glass by both ends, and place the center over the tip of the alcohol flame. Rotate the tube in your fingers so that the chosen spot gets heated all around the circumference of the tube. How long do you have to heat it before that spot is soft enough so that you can bend it a little? When it is soft, pull on the ends. What happens? Let the tube cool a bit and examine the results. Do you still have a tube? Play with various pieces of the tube, heating them and manipulating them in various ways. Can you make a circle of glass tubing? A square?

> *Caution.* Glass stays hot for a long time, and it looks just like cold glass.

Repeat your experiments with the hard glass. Can you make it soften in the alcohol flame? In a Bunsen burner?

Compare the way glass melts and solidifies with the way salol melts and solidifies. Which state of the glass would you call molten: when it just started to soften a little or when it became more fluid? How fluid would it have to be to be considered molten? Did this same question arise when the salol was melted? If necessary, repeat the melting of salol.

When the glass was soft enough to be bent, did you notice a characteristic color of the flame? Can you identify the element responsible for the color?

Part B   Testing Conductivity (A Demonstration)

Attach the leads of a 110-volt line to each of two carbon electrodes. Using rubber tubing for insulation, mount these electrodes in clips on a pegboard so that the electrodes point toward each other with the tips $\frac{1}{2}$ to 1 inch apart. Place a 2-inch piece of 7- or 8-mm common glass tubing so that it fits like a sleeve over both electrodes. Include in the circuit a 200-watt (no less) standard light bulb in series and a switch. When all is ready, turn on the switch. What happens? Use a Bunsen burner to heat the glass until it is soft where it touches the electrodes. Observe carefully. Be sure to turn the switch off after the glass has melted. Try to account for your observations.

Glass conducts electricity only when melted, but even then it is not as good a conductor as a melted ionic crystal. As the glass softens, ions are freed and they act as the charge carriers. The charge carriers are ions of sodium, calcium, and silicate.

Because glass has a melting range of temperature rather than a melting point, you may suspect that it is not crystalline. In crystalline substances the strength of the bonds in one unit cell of the structure is the same as that in the next because the orderly arrangement of the atoms is the same. Atoms of particular kinds maintain the same average distances from each other, as they oscillate about their average positions. As energy is added in the form of heat, the ions oscillate more rapidly. When just enough energy is added, the bonds break as the ions are freed from their potential wells. This happens at a definite temperature, called the melting point of the crystal.

In Experiment 17–5 you found that you could not say precisely when the glass changed from solid to liquid on heating or from liquid to solid on cooling. As the glass cools, it becomes stiffer and stiffer and at room temperature it is just a very stiff liquid. Its ions can no longer move past each other, but apparently they have not become arranged in a regular, orderly crystalline structure. The bonds between them weaken over a wide range of temperature, so there must be a wide range of bonding strengths between the ions in the glass.

Both glass and sulfur are amorphous, noncrystalline substances. Given enough time at a moderately elevated temperature, the molecules in sulfur will become arranged in a more orderly way; that is, sulfur will crystallize. Some glasses will also crystallize over a long period of time.

Would you expect amorphous substances to give orderly x-ray diffraction patterns as crystals do?

17-12   *Properties of a linear polymer*

Polyethylene is a synthetic plastic. The word *synthetic* is used to indicate that a substance is not a naturally occurring material, such as cotton, but is man made. The word *plastic* is used to designate certain physical properties that we shall mention in Section 17–13.

EXPERIMENT 17–6   Experiments with polyethylene

Put enough granular polyethylene in a six-inch test tube so that it fills one quarter of the tube after it is melted. Heat it carefully with an alcohol burner and note the behavior. Stir with a thermometer as you continue heating, and note the change of viscosity as the temperature increases.

After the maximum temperature obtainable with the alcohol burner is reached, allow the melted polyethylene to cool slightly, then draw threads from it by the following procedure. Withdraw the thermometer with a drop of the melted polyethylene, touch it to a cool surface, and then pull the thermometer away. You should be able to produce threads, whose diameter you can control by the amount of cooling and the rate of pulling. Save a number of them. Test the electrical conductivity of the threads and of the melted polyethylene. Examine the threads carefully. Are they flexible? stretchable? breakable? Hold a small portion of thread on a glass rod, and touch it to the flame of the alcohol burner. Observe carefully.

Question 17–15   Did you expect polyethylene to conduct electricity as a solid? Why?

Question 17–16   Polyethylene does not have a distinct melting point. Try to predict how this would affect the shape of its cooling curve.

Polyethylene is prepared commercially from ethylene C_2H_4. The prefix *poly* means many; a polyethylene molecule is composed of many ethylene units. There is good evidence that the synthetic polyethylene consists of long chains of carbon atoms of various lengths, and that the carbon atoms are joined by covalent bonds. The only other kind of atom present in polyethylene is hydrogen.

Let us try to visualize the structure of polyethylene and learn how it might be formed. The structure of ethylene can be represented as in Figure 17–14. Note in *a* that two

FIGURE 17-14
Ethylene structure.

(a)   (b)

pairs of electrons are shared between the two carbon atoms. One pair of electrons shared between two atoms constitutes a single bond; two pairs constitute a double bond. The carbon atoms in ethylene are bound to each other by a double bond. Each hydrogen atom is bound to a carbon atom by a single bond. By convention, a straight line is often used to represent a pair of electrons, so that the structure of ethylene might also be shown as in Figure 17–14*b*.

The formation of polyethylene from ethylene can be described in the following way. Suppose that one of the two bonds joining one carbon atom to another breaks open. Then each carbon atom would have one electron available to form a new bond, possibly with another ethylene unit. In this way ethylene units can react with each other to form long chains of variable lengths. Long chain molecules formed in this way are called *polymers*. The formation of a portion of a polyethylene chain from three ethylene units is represented in Figure 17–15. Note especially the single electron on each carbon atom of the intermediate molecule.

FIGURE 17-15
Polyethylene formation.

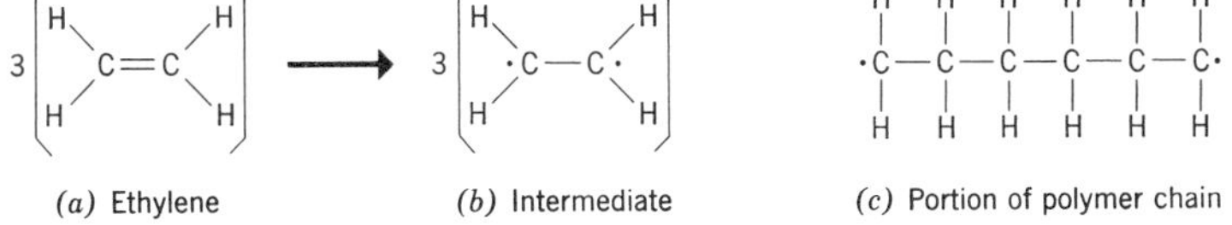

(a) Ethylene   (b) Intermediate   (c) Portion of polymer chain

Figure 17–15 does not attempt to picture the three-dimensional shape of a polyethylene chain. In the carbon compounds we studied, CH_4, CH_2Cl_2, CCl_4, etc., we saw that the bond orientation of carbon atoms is tetrahedral; the same is true in polyethylene. Because of this, a better way of visualizing a polyethylene chain is as follows. Imagine selecting one chain out of the diamond structure shown in Figure 17–1. Let all the bonds that join carbon atoms to each other in this one chain remain intact, but imagine that all other carbon-carbon bonds are broken and replaced with carbon-hydrogen bonds. The chain could take many turns. Figure 17–16 shows some possible structures for short polyethylene chains. Do you see how these arrangements can come from the basic diamond structure?

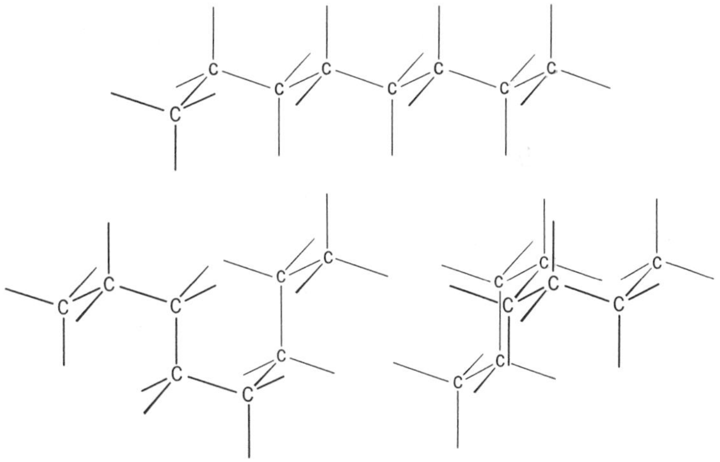

FIGURE 17-16
Portions of polyethylene chains.

For ease of showing structures, we have imagined only short carbon-carbon chains, but the chains in polyethylene are very long and contain thousands of carbon atoms. The polyethylene chains in molten polyethylene have sufficient thermal agitation to exist in flexible, randomly coiled arrangements. On cooling below the melting point (about 135°C), the chains can crystallize, and in this form they are extended, rodlike chains packed together in a somewhat regular array.

17-13   *Rubber—network polymers*

What physical properties do you think might distinguish a *rubber* (technically an elastomer) from a plastic (technically a thermoplastic) material? A *plastic* is a substance that will be deformed by a prolonged gentle force but will shatter under a sudden below. Both are solids and both are largely amorphous. The one property that is most characteristic of rubber is the ability to be stretched easily and to recover its original shape afterward. Some solid polymers, such as polyethylene and polystyrene, can be deformed easily only when heated to temperatures near their softening points. If an external force is applied to a plastic object at these temperatures, there will be sufficient motion and rearrangement of the polymer chains to produce a permanent change in the shape of an object. Rubber, however, is capable of deformation under a force, but it will restore itself to its original shape when that force is removed. Can you think of a model of molecular structure that would explain this?

EXPERIMENT 17-7   An experiment with rubber

Place some pieces of rubber and a whole rubber band in a test tube. Cover them with toluene, stopper the test tube, and allow it to stand overnight. (Toluene is highly flammable.) What changes do you observe in the rubber? Does it dissolve? Is it still elastic? Has it changed in shape?

The model we propose should agree with your experimental observations and with your experience in stretching rubber bands.

You may already know that natural rubber does not have the resilience that makes rubber so useful; it must be *vulcanized* (heated with sulfur) to give it this property. Before vulcanization, natural rubber is a polymer somewhat like polyethylene, but softer, more easily deformed, and less definitely crystalline. It is known that its structure consists of long chains randomly coiled and entangled. But there is a structural feature in unvulcanized rubber that is not present in polyethylene. Regularly spaced along the chains are carbon-carbon double bonds. These double bonds make the chains less flexible because carbon atoms joined by a double bond cannot rotate about the bond axis between them, whereas they can rotate about a single bond axis. The double bonds also provide sites that are active for building short links between polymer chains, as with ethylene. Vulcanization results in the formation of short bridges between chains, as shown in Figure 17–17.

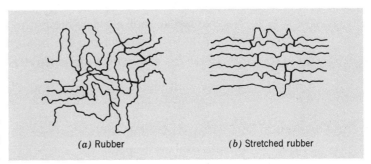

(a) Rubber   (b) Stretched rubber

FIGURE 17-17
Model of vulcanized rubber.

When vulcanized rubber is stretched, these links are not broken. Long sections of randomly coiled polymer chain between links can be straightened out as shown in Figure 17–17. When these are fully extended, the interactions between chains become stronger and the rubber is strengthened; that is, it cannot easily be stretched farther. When stretched, the chains are ordered; x-ray diffraction shows some degree of crystallinity. But because the polymer chains are linked together, each chain itself does not move *as a whole* relative to the others. The stretched position is not maintained when the force is released. The portions of the chains between cross links move back into their more random orientations.

Suppose that instead of a few cross links between long chains, there were many cross links extending in three dimensions and linking all chains so that a tight network of covalent bonds extended throughout the material. What physical properties would you expect to observe? One can make such a structure in many ways. One is to use a relatively large amount of sulfur when vulcanizing rubber, in order to obtain hard rubber. Another way is to use larger amounts of curing (cross-linking) agents mixed with linear polymers. This is the way epoxy resins harden after mixing. Still another way is to use mixtures of organic compounds, such as phenol and formaldehyde, which form a network polymer (bakelite) on heating under pressure. Electric switch plates, handles for cooking pots, and radio cabinets are made of bakelite.

Question 17–17   How are the properties of paraffin wax different from those of polyethylene? Paraffin has a structure of hydrocarbon chains like polyethylene, but the chain lengths are only 1/100 to 1/1000 of the chain lengths in polyethylene. Explain the differences in properties in terms of structure.

Question 17–18   Do you think a hardened phenol-formaldehyde resin would melt? Explain.

Question 17–19   Does stretching rubber cause a change in the entropy of the rubber? Explain.

Question 17–20   Would a network polymer such as a hardened epoxy resin yield a regular x-ray diffraction pattern? Explain.

17-14   *Mixtures*

Very few solids are completely amorphous, with their atoms and molecules in jumbled disorder. In nearly all solids the atoms, ions, or molecules have some degree of orderliness in their arrangement. Even a slight orderliness can be discovered by the use of x-ray diffraction. In wood, for example, the cellulose molecules have quite a regular arrangement and strong bonding in the same direction as the wood fibers, but comparatively random arrangement and weak bonding at right angles to the fibers. It is not surprising, therefore, that wood comes apart in splinters.

Many substances in which we cannot see crystals with the unaided eye are in fact made up of very good crystals with irregular boundaries between them. Some of the clusters of crystals grown in this course had such irregular grain boundaries. Figure 17–18 is a photomicrograph of the polished surface of a piece of magnesium. It has been slightly etched with an acid that attacked the boundaries between the grains themselves and thus caused the boundaries to look darker in the picture. Each grain is a crystal of magnesium. All pieces of metal show individual crystal

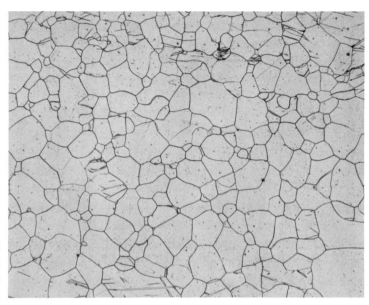

FIGURE 17-18
Section of an annealed magnesium rod (magnified 100 times).

grains under the microscope and diffract x rays in an orderly way.

In contrast to magnesium, which is a polycrystalline substance, let us consider granite, which is a mixture of different crystals. Figure 17–19 is a photomicrograph of a thin slice of granite 0.03 mm thick. Unlike magnesium, granite

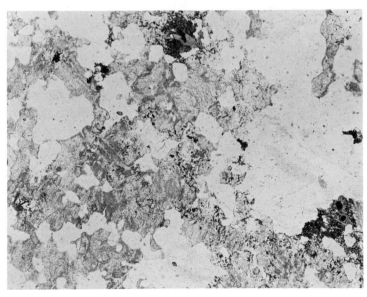

FIGURE 17-19
Photomicrograph of granite (magnified 12 times).

is made up of several different kinds of crystals. The two major components are quartz (SiO_2) and feldspar ($NaAlO_2 \cdot 3SiO_2$). In the photograph, the white (clear) areas with few specks in them are quartz grains, and the grayish areas with many small specks in them are feldspar grains that have particles of impurities. The dark grains are probably hornblende ($NaAlO_2 \cdot 2CaO \cdot 4FeO \cdot Al_2O_3 \cdot 6SiO_2 \cdot H_2O$), and the very black grains are magnetite ($FeO \cdot Fe_2O_3$), the mineral from which the magnet got its name.

Polished granite is widely used for decorating buildings. If its grains are big enough, you may be able to identify them. The quartz will look gray and glassy; the feldspar will be pinkish or creamy white and may show cleavage planes that will reflect the sunlight brightly; the black grains that are usually present may be hornblende or they may be flakes of the black mica known as biotite.

All rocks except lava are crystalline. Their intergrown grains can be identified in thin sections under the microscope, and they give excellent x-ray diffraction patterns.

Your teeth and bones are also made up of atoms in orderly arrays and would give good x-ray diffraction patterns. Note that the shadow pictures of bones and teeth used by dentists and doctors are very different from the diffraction pattern resulting from the constructive interference of x rays scattered by a crystal. The x-ray machines used for shadow pictures send out a broad beam of x rays of a wide range of wavelengths, whereas in the diffraction studies of crystals, a very thin (0.5 mm diameter) beam is used which consists primarily of a very narrow range of wavelengths. The wavelengths of the x rays used in crystallography are much longer than the x rays used for medical shadow pictures.

Wood, metals, rocks, fibers, bones, nearly all of the solid things in the world around us are more or less crystalline. Even clays, which are used to make such common objects as pottery, flower pots, and bricks are made of crystals, although they do not appear to be. Clays certainly do not have a cleavage plane. The crystals that make up clays are microscopic in size, as the electron micrograph in Figure 17–20 indicates. The magnification in this photograph is

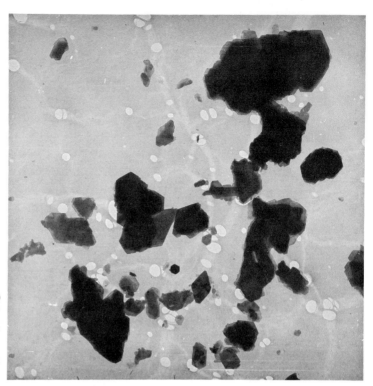

FIGURE 17-20
*Electron micrograph of
particles of clay. (College
of Earth and Mineral
Sciences, The Pennsylvania
State University)*

nearly 36,000 times. The largest piece of crystal in the upper right-hand corner is about 6×10^{-4} centimeter across.

17-15   *Conclusion*

By examining the properties and behavior of many different kinds of substances in this chapter and then trying to explain them, we have extended the process of observation and model building into new areas we had not explored before. In the process you have used old ideas and discovered some new ones: directed covalent bonds, van der Waals bonds, metallic bonds, long chain molecules, and cross linking of chains. We have also seen how properties of solids depend on order or crystallinity.

We should now ask if our model system is internally consistent and whether it is broad enough to be applied in new situations as well as in the specific examples we have considered.

Certainly the model system is consistent in that all bonds are explainable by one kind of force: the attraction between opposite electric charges and the repulsion of like charges. Bonds are called by different names, depending on the nature of the charged particles involved: ionic bond, ions; covalent bond, electron pairs and atomic nuclei; metallic bond, free electrons and ions; van der Waals bond, electrically neutral but polarizable particles. In each and every chemical bond the nature of the force is electrical.

The way a model has been used to help us understand behavior has been related to the strength and direction of bonds, as well as to the ease of changing, breaking, or forming new bonds. We should be able to apply the model system we have developed in a great variety of new situations, perhaps adding to it some kinds of bonds which may be new, but new only in detail, not in the basic attractive force. Indeed, *you* should be able to make some applications of the concepts we have developed in order to explain or predict the behavior of physical substances.

Figure 17–21 gives the names or structures of several substances we have not discussed. Which is likely to be a solid? which crystalline? What are the important bonding forces in each substance that is solid? Which has the highest melting point? Which conducts an electric current as a solid? as a melt? Can you propose a model for the structure of each substance in the solid state?

(*a*) Silicon carbide

(*b*)

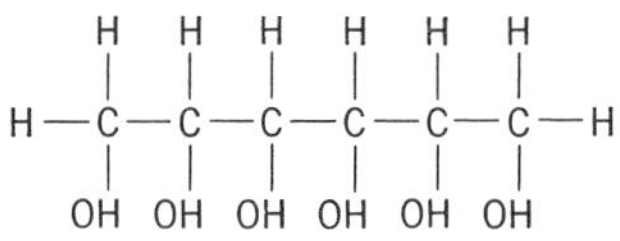

$$H-\underset{\underset{\displaystyle OH}{|}}{\overset{\overset{\displaystyle H}{|}}{C}}-\underset{\underset{\displaystyle OH}{|}}{\overset{\overset{\displaystyle H}{|}}{C}}-\underset{\underset{\displaystyle OH}{|}}{\overset{\overset{\displaystyle H}{|}}{C}}-\underset{\underset{\displaystyle OH}{|}}{\overset{\overset{\displaystyle H}{|}}{C}}-\underset{\underset{\displaystyle OH}{|}}{\overset{\overset{\displaystyle H}{|}}{C}}-\underset{\underset{\displaystyle OH}{|}}{\overset{\overset{\displaystyle H}{|}}{C}}-H$$

(*c*) Potassium fluoride

FIGURE 17-21

(*e*) Iron

(*d*)

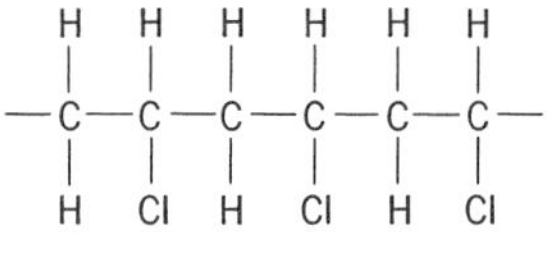

$$-\underset{\underset{\displaystyle H}{|}}{\overset{\overset{\displaystyle H}{|}}{C}}-\underset{\underset{\displaystyle Cl}{|}}{\overset{\overset{\displaystyle H}{|}}{C}}-\underset{\underset{\displaystyle H}{|}}{\overset{\overset{\displaystyle H}{|}}{C}}-\underset{\underset{\displaystyle Cl}{|}}{\overset{\overset{\displaystyle H}{|}}{C}}-\underset{\underset{\displaystyle H}{|}}{\overset{\overset{\displaystyle H}{|}}{C}}-\underset{\underset{\displaystyle Cl}{|}}{\overset{\overset{\displaystyle H}{|}}{C}}-$$ (A portion of a long chain)

QUESTIONS   **17-21**   (a) We know that an atom consists largely of empty space. The nuclei are very small indeed. What keeps the nuclei from coming much closer together than the distance of the sum of the radii of the two atoms?

(b) When you push on a table (made of atoms) with your finger (made of atoms), what keeps your finger from sinking into the table?

17-22 Heptane is one of the components in gasoline that causes knocking in automobile engines. It is a straight chain (i.e., unbranched) hydrocarbon and has seven carbon atoms and sixteen hydrogen atoms. Draw the structure of this hydrocarbon and give its formula.

17-23 Draw diagrams of models of the chloromethane (CH_3Cl) and trichloromethane ($CHCl_3$) molecules. Trichloromethane (called chloroform) is a liquid at room temperature. Predict whether it would be deflected by a charged rod as is water. Explain the basis of your prediction.

17-24 Suppose the structure of ice were more closely packed than that of water. How would this affect what happens in lakes in the winter? in the summer?

17-25 If a fish bowl were filled with water and placed outdoors when the temperature was below $0°C$, the bowl would probably break. Why?

17-26 When the whole weight of a person is concentrated on the narrow blade of an ice skate, the open structure of the ice beneath the blade is subjected to a very high pressure. What would you suppose might happen as a result? How is this related to the almost complete lack of friction between the skate blade and the ice?

17-27 Consider some substances not mentioned in this chapter. Describe a property of each that gives some clue as to its structure. What is the clue and what does it tell you?

17-28 Would you expect diamond to become a conductor at very high pressures? The answer is not known at the present time. If it does become a conductor, what has happened to the bonds?

References 1. CBA, *Chemical Systems,* McGraw-Hill and Earlham College Press, 1964, Chapter 7, pages 229 to 278; Summary, pages 450 to 455 (this section is highly packed); and Chapter 11, Sections 11–1, 11–4 through 11–13, pages 456 to 458, 464 to 478.

2. Sienko, M. J., and R. A. Plane, *Chemistry: Principles and Properties,* McGraw-Hill, 1966, Chapter 3, pages 49 to 74.
Bonds of various types. The latter half of the chapter is more advanced than our approach.

ALSO OF INTEREST

3. Hart, H., and R. D. Schuetz, *A Short Course in Organic Chemistry,* Houghton Mifflin Co., Boston, 1966.

4. Leffler, J. E., *A Short Course in Organic Chemistry,* Macmillan, New York, 1959.
This small book is useful in providing information on the basic structural units in many organic substances.

5. Sanderson, R. T., *Principles of Chemistry,* John Wiley and Sons, 1963, Chapter 4, pages 58 to 76, Chapter 11, pages 210 to 219.
Bonding of various types; structures of various types.

Mount Wilson and Palomar Observatories

CHAPTER 18

WHAT IT IS ALL ABOUT

"When dining, I had often observed that some particular dishes retained their Heat much longer than others; and that apple pies, and apples and almonds mixed (a dish in great repute in England) remained hot a surprising length of time. Much struck with this extraordinary quality of retaining Heat, which apples appeared to possess, it frequently occurred to my recollection; and I never burnt my mouth with them, or saw others meet with the same misfortune, without endeavoring, but in vain, to find out some way of accounting, in a satisfactory manner, for this surprising phenomenon."

So wrote Benjamin Thompson, Count Rumford, near the end of the eighteenth century. Busy though his social and political life was, his puzzlement led him to perform experiments on mechanisms by which heat may be transferred from one place to another. For most people, the difficulties of living from day to day are not interrupted by a concern for how the world is made. But for some, as for Rumford, that concern has seemed unavoidable—for a few, even joyful—and they have constantly remade the world of the others.

The preceding course of study has introduced you to the method characteristically employed by these people: observing nature with puzzled curiosity, perceiving relations between the observations, constructing hypotheses that are consistent with them, and contriving experiments, when possible, to test the range over which a hypothesis is applicable. That "scientific" procedure takes many forms, varying with the nature of both the question and the questioner. No one can perform controlled experiments on the stars; we must observe them and wonder about them as they are, not as they can be made to be. On the other hand, many chemical and physical questions can be pursued in a laboratory, and often only there.

As for the questioners, their tastes are quite diverse; scientists, like artists, come in all kinds. There are those who enjoy tangling with nature directly—gripping nature literally with their hands. Others prefer to invent or perfect or apply theories of natural behavior. In turn, the experimentalists and the theorists are divided among many different temperaments. Some prefer to work in teams, others to work alone. Some prefer exactness in experiment or rigor in theory; others choose quick but suggestive observations, or approximate but novel theoretical pictures. All these attitudes have respectable parts to play in the scientific enterprise.

Hence "science" is not easily defined. Surely it falls somewhere between an impersonal body of knowledge and a personal way of life. Science is made by people. It is an accumulation of their efforts, and it reflects its human origins in many conspicuous ways. It is subject to fashion, for example. And the more scientists there are, the more science there is; and it flies off in various directions, in obedience to the whims of the various scientists who make it.

But in all this variety, the scientific way of finding order in the world has an important common property that distinguishes it from the mathematical way. In mathematics we envisage sets of axioms and then deduce their consequences, confident in the worth of the internally consistent structures that result. By contrast, in science we encounter a detailed responsibility to what actually happens in nature, outside the mind.

Hence a well-known definition of science by the Nobel Prize-winning physicist, Percy Bridgman—"doing your damndest with your mind, and no holds barred"—tells only half the story. A mathematician could say the same of his discipline, and yet he need not concern himself at all with nature. For that reason, mathematics, in principle, is "easier" than science, and it is no accident that the character of scientific inquiry as we know it today is only about three hundred years old. The Greeks believed that scientific knowledge, like mathematical knowledge, could be gained by thought alone. Francis Bacon was stating the canon of a radically new philosophy when he said, "Man, who is the servant and interpreter of nature, can act and understand no further than he has observed, either in operation or in contemplation, of the method and order of nature."

The success of the "new philosophy," in spurring the construction of pictures of natural phenomena, is now conspicuous. Man's constantly increased command over nature has blossomed in rational medical practice and in the numberless technologies we now enjoy. From the preoccupation of a few amateurs, scientific inquiry has grown into the fulltime profession of several hundred thousand persons. Their cumulative labor has provided a body of knowledge about nature's ways that is far too large to be mastered by any one mind.

Facing this accumulation today, many people have unfortunately concluded that it is hopeless for them to attempt to understand science. But if you have understood this course, you know that they are wrong. The professional scientist gains his knowledge by the same methods that you have pursued, and by pursuing his methods you too have gained some of his knowledge. You can ask questions of his sort and seek answers in his ways.

Sir Charles Boys has described those ways charmingly in a book about soap bubbles. "I would remind you then that when we want to find out anything that we do not know, there are two ways of proceeding. We may either ask somebody else who does know, or read what the most learned men have written about it, which is a very good plan if anybody happens to be able to answer our question; or else we may adopt the other plan, and by arranging an experiment, try for ourselves."

Here is how it goes:

Observing nature with puzzled curiosity: "I wonder why the stars look reddish tonight."

Perceiving the relations between observations: "Perhaps for the same reason that automobile headlights look reddish on a foggy night."

Constructing hypotheses consistent with both: "Perhaps the red component of the light is less scattered by dust and fog than the blue component."

Contriving experiments to test the hypothesis: "I'll see whether a red light is less dimmed than a blue light when both are viewed separately through a fog."

Three centuries of the "new philosophy" have alerted all of us to the *connectibility* of such natural phenomena, and have provided us with the mental tools required to seek out the connections. Only by exercising those tools can we avoid falling into a modern version of the mediaeval fallacy, in which the guild of professional scientists becomes a priesthood whose utterances are believed but not understood. Already the conspicuous successes of natural science, in its proper province, tempt many to invoke "the scientific method" outside that province. Bertrand Russell, surely no enemy of science, warns us, "Science tells us what we can know, but what we can know is little, and if we forget how much we cannot know we become insensitive to many things of very great importance." Our *value systems*, for example, cannot be examined by the scientific method. "Must the good be eternal in order to deserve to be valued, or is it worth seeking even if the universe is inexorably moving toward death? Is there such a thing as wisdom, or is what seems such merely the ultimate refinement of folly? To such questions no answer can be found in the laboratory." But such questions remain relevant to scientific activity because they embrace and override it.

Thus we ask, in science as elsewhere, good taste and good heart. The advertising copywriter whose man-in-a-white-coat is made to say, "Science tells us that this toothpaste cleans 2.7 percent whiter," is tasteless; he uses a great achievement of man in a trivial context. The Nazi German doctors, who froze patients and studied their behavior as they thawed, were heartless; whatever one may think of their "science," their experiments were humanly impermissible. Surely the fact that an experiment is "scientific" does not automatically justify performing it.

At present we all face a severe test of the interaction of our science and our human values: we possess nuclear weapons capable of destroying the largest part of the race of man. The "scientist" in each of us will ask us not to shrink squeamishly from reading about the enormous destructive power of these weapons. And the good heart in each of us will ask us to work toward denying their use for destruction. Honor those professional scientists, knowing best the destructiveness, who work hardest toward that denial.

The designers of this course have tried to afford you a vista down one of the more fruitful roads traveled by man in recent centuries. If they have succeeded, you can join your fellows in traveling that road as you will—by questioning others, by reading, by observation, by experiment—as the occasions of your life and your tastes suggest. Perhaps you have even gained some confidence in your own ability to answer by yourself some of the questions that natural happenings constantly pose. In any case you have glimpsed a little of the grandeur of a system of "laws" that you did not enact and cannot repeal. And you can experience the inner repose of turning occasionally from the hurly-burly of human affairs to the contemplation of nature's changeless ways.

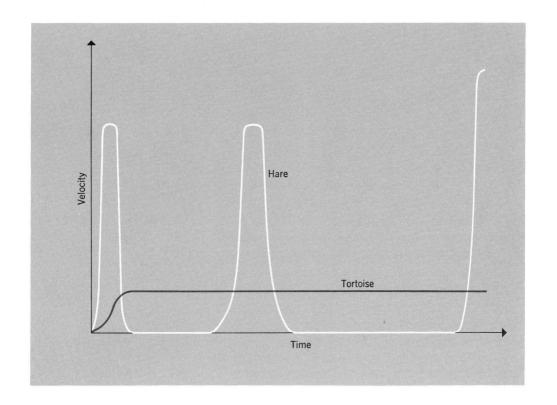

APPENDICES

APPENDIX A

REVIEW OF POWERS OF TEN

In the physical sciences it is frequently necessary to deal with very large and very small numbers. For example,

$$\text{The velocity of light} = 299{,}776{,}000 \text{ meter/sec}$$

The number of molecules in 1 cm³ gas at 1 atmosphere pressure and 0°C is 26,886,000,000,000,000,000.

The mass of an atom of cesium = 0.0000000000000000000000221 g

1. *Expressing a Number in Power-of-Ten Notation*

These large and small numbers can be expressed much more simply in powers of ten. As we will see, this practice will also be of help in keeping track of the decimal point when dealing with these large or small numbers. This appendix is designed to give you some experience in the use of powers of ten. Recall that

$$10 \times 10 = 10^2 = 100$$

The 2 is called the *exponent* of ten and indicates how many 10's are multiplied together. 100 is the "second power" of ten.

Similarly, $10 \times 10 \times 10 = 10^3$, but $10 \times 10 \times 10 = 1000$, so $10^3 = 1000$. Note that the exponent tells you the number of zeros after the digit 1.

In a similar way

$$10^1 = 10$$
$$10^2 = 100$$
$$10^3 = 1000$$
$$10^6 = 1{,}000{,}000$$

A-1 Practice by expressing the following numbers as 10^n, where n is the appropriate exponent.
(a) 10,000   (b) 100,000,000   (c) 1,000,000,000

2. *Multiplication*

$$100 \times 1000 = 100{,}000$$
$$10^2 \times 10^3 = 10^5$$

You can see that to multiply 10^2 by 10^3, we just added $2 + 3$ and got 5 as the exponent in the answer. To generalize:

$$10^n \times 10^m = 10^{n+m}$$

A-2 Practice by solving the following problems, using power-of-ten notation.
(a) $10,000 \times 10^{23}$ (b) 100×100 (c) $10^6 \times 10^4$

3. *Division*

$$\frac{100,000}{100} = 1000$$

$$\frac{10^5}{10^2} = 10^3$$

You can see that to divide 10^5 by 10^2 we just subtracted 2 from 5 and got 3 as the exponent in the answer. To generalize:

$$10^n/10^m = 10^{n-m}$$

A-3 Practice by solving the following problems, using power-of-ten notation.
(a) $1000/100$ (b) $10^6/1000$ (c) $10,000/10$

4. *Numbers less than 10*

$$10/10 = 1 \qquad 1000/10,000 = 0.1$$
$$10^1/10^1 = 10^0 \qquad 10^3/10^4 = 10^{-1}$$

By following the division rule we have already established, you can see that

$$10^0 = 1.0$$
$$10^{-1} = 0.1$$
$$10^{-2} = 0.01$$
$$10^{-3} = 0.001$$
$$10^{-4} = 0.0001$$

Note that, with negative exponents, the number of zeros to the right of the decimal point is *one less than* the value of the exponent. The rules of multiplication and division are the same.

A-4 Practice by doing the following problems, using power-of-ten notation. (a) $0.01 \times 1000 = 10^{-2} \times 10^3 = 10^1 = 10$ (b) 0.0000001×0.01
(c) $\dfrac{100}{10^8}$ (d) $\dfrac{0.01}{100}$

Note that

$$0.001 = 1 \times 10^{-3} = 0.1 \times 10^{-2} = 0.01 \times 10^{-1} = 10 \times 10^{-4}$$

5. *Numbers that are not whole-number powers of ten*

$$5280 = 5.280 \times 1000 = 5.280 \times 10^3$$
$$96{,}734 = 9.6734 \times 10{,}000 = 9.6734 \times 10^4$$
$$0.00095 = 9.5 \times 0.0001 = 9.5 \times 10^{-4}$$

In the examples above we have expressed each number as a number between 1 and 10, multiplied by ten raised to the appropriate power.

A-5 At the beginning of this appendix, values were given for the velocity of light, the number of molecules in 1 cm^3 of gas at 1 atmosphere pressure and 0°C, and the mass of an atom of cesium. Express these values in power-of-ten notation.

You will find that power-of-ten notation greatly simplifies many calculations. Consider the following examples:

(1) How many seconds are there in a day?

60 sec in 1 min $\times$ 60 min in 1 hr $\times$ 24 hr in a day.
$6 \times 10 \times 6 \times 10 \times 2.4 \times 10 = 36 \times 2.4 \times 10^3$
$3.6 \times 10 \times 2.4 \times 10^3 = 3.6 \times 2.4 \times 10^4 = 8.64 \times 10^4$

(2) What fraction of an hour is a second?

$$1 \text{ min} = 1/60 \text{ of } 1 \text{ hr}$$
$$1 \text{ sec} = 1/60 \text{ of } 1 \text{ min}$$
$$1 \text{ sec} = 1/60 \times 1/60 \text{ of } 1 \text{ hr}$$

$1/60 = 0.01666 \ldots \ldots$ for as long as you want to write sixes. In such a case we "round off" the last digit. We choose to write to one number higher if the following digit is greater than 5 but not if the digit is less than 5. (0.2087 is rounded off to 0.209, but 0.2083 is rounded off to 0.208 if we choose to keep only 3 digits after the decimal point.) So we will write $1/60 = 0.0167 = 1.67 \times 10^{-2}$

$$1 \text{ sec} = 1.67 \times 10^{-2} \times 1.67 \times 10^{-2} \text{ hr}$$
$$1 \text{ sec} = 2.79 \times 10^{-4} \text{ hr}$$

(3) If something is moving at 1 mph, how many feet does it move in 1 sec?

$$1\,\frac{\text{mile}}{\text{hr}} = \frac{5280\text{ ft}}{60\text{ min}} = \frac{5280\text{ ft}}{60\times 60\text{ sec}} = \frac{52.80\times 10^2\text{ ft}}{36.0\times 10^2\text{ sec}}$$

$$= \frac{52.80}{36.0}\times 10^0 = 1.461\,\frac{\text{ft}}{\text{sec}}$$

Here we have substituted 5280 ft for its equivalent 1 mile, and 60 min ($= 60 \times 60$ sec) for its equivalent 1 hr. We have also changed 5280 to 52.80×10^2 for convenience in dividing by 36.

A-6   Practice by solving the following problems, using power-of-ten notation.

(a) $\dfrac{86,000,000}{4300}$ (b) $\dfrac{82,000 \times 0.00000012}{0.0021 \times 300,000}$

(c) $\dfrac{4900 \times 12,000}{0.0000004}$

6.   *Addition and Subtraction*

$$490,000 + 25,000 = 515,000$$
$$49.0\times 10^4 + 2.5\times 10^4 = 51.5\times 10^4$$
$$29,000,000 - 10,000 = 28,990,000$$
$$2900\times 10^4 - 1\times 10^4 = 2899\times 10^4$$

In other words, thousands can only be added to thousands ($m\times 10^3 + n\times 10^3$) or millions to millions ($m\times 10^6 + n\times 10^6$), rather like adding only apples to apples and pears to pears.

Note that $60,000,000 = 6.0\times 10^7 = 60\times 10^6 = 600\times 10^5 = 6,000\times 10^4 = 60,000\times 10^3$.

A-7   Practice by solving the following problems in power-of-ten notation.

(a) $250 + 92,000$ (b) $112,000 - 50$
(c) $29,000,000,000 + 16,000,000$

7.   *Exponents of Exponents*

$100 \times 100 = 10,000$

$10^2 \times 10^2 = (10^2)^2 = 10^4$ Note that $2\times 2 = 4$

$1000 \times 1000 = 1,000,000$

$10^3 \times 10^3 = (10^3)^2 = 10^6$ Note that $3\times 2 = 6$

$100 \times 100 \times 100 = 1,000,000$

$10^2 \times 10^2 \times 10^2 = (10^2)^3 = 10^6$ Note that $2\times 3 = 6$

$0.01 \times 0.01 \times 0.01 = 0.000001$

$10^{-2} \times 10^{-2} \times 10^{-2} = (10^{-2})^3 = 10^{-6}$

Note that $-2\times 3 = -6$

To generalize:

$$(10^n)^m = 10^{n \times m}$$

A-8 Practice by solving the following problems in power-of-ten notation.
(a) $(1000)^2 =$
(b) $(1,000,000)^3 =$
(c) $(0.0001)^2 =$

If it is desired to find some power of a number which is not just a whole number power of ten, remember that both parts of the number must be raised to the appropriate power. The following examples should make this clear.

$(3000)^2 = (3 \times 10^3)^2 = 3^2 \times (10^3)^2 = 9 \times 10^6$

$(15,000,000)^3 = (1.5 \times 10^7)^3 = (1.5)^3 \times (10^7)^3 = 3.38 \times 10^{21}$

A-9 Practice by solving the following problems.
(a) $(49,000)^2 =$
(b) $(21,000,000)^2 =$
(c) $(0.000000002)^3 =$

8. *Taking Roots*

$\sqrt{100} = 10$, since $10 \times 10 = 100$, or $\sqrt{100} = \sqrt{10^2} = 10^1 = 10$

These two equations can be read as the square root of 100 is 10 and the square root of 10^2 is 10^1. Similarly $\sqrt{1,000,000} = 1000$ since $1000 \times 1000 = 1,000,000$ or $\sqrt{1,000,000} = \sqrt{10^6} = 10^3 = 1000$. These two equations can be read as the square root of 1,000,000 is 1000, and the square root of 10^6 is 10^3. To generalize:

$$\sqrt{10^n} = (10^n)^{1/2} = 10^{n/2}$$

Thus:

$$\sqrt{100} = (100)^{1/2} = (10^2)^{1/2} = 10^1 = 10$$
$$\sqrt{1,000,000} = (1,000,000)^{1/2} = (10^6)^{1/2} = 10^3 = 1000$$

Cube roots are handled in a similar fashion using the fractional exponent one-third. Consider the following examples.

$$\sqrt[3]{1000} = (1000)^{1/3} = (10^3)^{1/3} = 10^1$$
$$\sqrt[3]{1,000,000} = (1,000,000)^{1/3} = (10^6)^{1/3} = 10^2 = 100$$

You may have noticed that in the examples given, the final exponent always comes out to be a whole number. The examples where chosen so that they would turn out that way. Consider

$$\sqrt[3]{10,000} = (10,000)^{1/3} = (10^4)^{1/3} = 10^{4/3}$$

You can't write this as a one with $4\frac{1}{3}$ zeros. But you can get an answer in the following way:

$$\sqrt[3]{10,000} = (10^4)^{1/3} = (10 \times 10^3)^{1/3} = 10^{1/3} \times 10 = 2.16 \times 10$$

The rule to follow in such cases as this is to always express the number in power-of-ten form in such a way that the fractional exponent times the exponent of ten is a whole number. Thus:

$$\sqrt{250,000} = (25 \times 10^4)^{1/2} = (25)^{1/2} \times (10^4)^{1/2}$$
$$= (5^2)^{1/2} \times (10^4)^{1/2} = 5 \times 10^2$$

But:

$$\sqrt{2,500,000} = (2.5 \times 10^6)^{1/2} = (2.5)^{1/2} \times (10^6)^{1/2}$$
$$= 1.58 \times 10^3$$

A-10   Practice by solving the following problems in power-of-ten notation. If you can't do the square root or cube root by inspection, as in the last example, leave it in the form of the next to the last expression, i.e., $(2.5)^{1/2} \times 10^3$.
(a) $\sqrt{14,400}$ (b) $\sqrt{144,000}$ (c) $\sqrt[3]{125,000}$
(d) $\sqrt{0.0016}$

9.   *Some Large and Small Physical Quantities*

Here are some useful physical quantities expressed in the power-of-ten notation:

$m =$ the mass of an electron $= 9.1066 \times 10^{-28}$ g
$e =$ the charge on an electron $= 1.602 \times 10^{-19}$ coulomb
$M_p =$ the mass of a proton $= 1.672 \times 10^{-24}$ g
$M_n =$ the mass of a neutron $= 1.675 \times 10^{-24}$ g
$c =$ the velocity of light $= 2.9978 \times 10^8$ meter/sec
The mass of the Earth $= 5.983 \times 10^{24}$ kg
Average distance to the Sun $= 1.495 \times 10^{11}$ meters

A-11   Compare quantitatively the mass of an electron and the mass of a proton; that is, which is heavier and what is the ratio of the mass of the heavier one to the mass of the lighter one?

A-12 The time t for something to travel a distance d with a constant velocity v is given by $t = d/v$. How long does it take for light to get to the Earth from the Sun?

A-13 As in Question **A-10**, compare quantitatively the mass of the Earth and the mass of a neutron.

APPENDIX B

GRAPHS

Graphs are diagrams representing information, usually numerical, in a way that makes it more readily grasped by the observer.

Figures B–1 through B–4 show some examples of graphs.

The pie graph, Figure B–1a, is commonly used where the information concerns the proportion of various items that make up the whole. Suppose there are 3600 students in your college distributed among the classes as follows: 1100 freshmen, 900 sophomores, 800 juniors, and 800 seniors. This distribution is shown on the pie graph by measuring a center angle of 110 degrees for the pie-shaped piece representing the freshmen, 90 degrees for the sophomores, 80 degrees for the juniors and 80 degrees for the seniors. (We chose 3600 because it made the arithmetic easy since the whole pie contains 360 degrees.) Note that you cannot tell by looking at Figure B–1a how many students there are in each class or in the whole college. Suppose the pie graph represents a college in which the total population is only 1000 students. Then the number of freshmen would be $110/360 \times 1000 = 11,000/36 = 305$. How many sophomores would there be in the college?

The bar graph, Figure B–1b, is used to show several different quantities that are not necessarily parts of a whole. In Figure B–1b we have used the length of the bar to indicate the number of students. The vertical scale at the left was chosen with a total length representing a number just a little more than the largest number of students and divided into subdivisions that could be conveniently marked off on the graph paper. Since the largest number of students we have to show is 1100, we might choose 1200 as the number to be represented by the total length. To choose the length of a large square to represent 100 students would make the vertical axis too long for our piece of paper. Choosing the length of a small square to represent 100 students gives us a convenient height for the graph on this piece of paper. We mark off the scale, starting with zero at the bottom and increasing, 100 at a time, to 1200 at the top. The horizontal axis, at right angles to the vertical axis, needs no scale

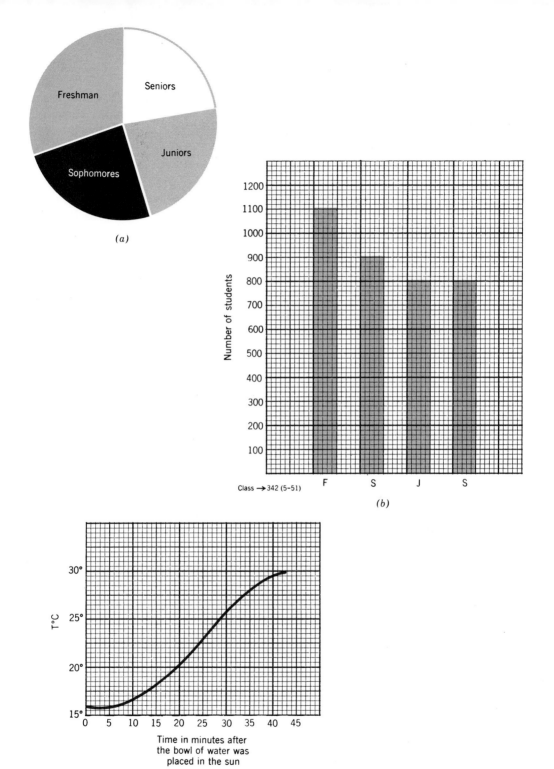

Class → 342 (5-51)

FIGURE B-1   *Examples of various types of graphs.*

such as we made for the vertical axis since nothing is measured along it. It is just the base line or zero line from which the lengths of the bars are measured. The widths of the bars don't represent any quantity, so we can chose any width we like. The bars labeled *F* for freshmen, *S* for sophomores, *J* for juniors, and *S* for seniors show us, at a glance, approximately how many students there are in each class. In this way they give us more information than the pie graph did, but it may not be as easy to see from the bar graph what proportion of the whole each class is.

Figure B–1*c* is a graph showing the temperature of a bowl of water, drawn from a cold water faucet and placed in the Sun on a warm day.

Along the vertical axis (which is called the *ordinate*) we have chosen to show only that part of the temperature scale which we need. There is no need to show temperatures down to zero Celsius because the water was not that cold during the time we are considering. The lengths of divisions were chosen for convenience as they were in the example in Figure B–1*b*. Here the length of each large division represents 5°. Since there are 10 small divisions in each large division, each small division represents 0.5° or half a degree. Along the horizontal axis (the *abscissa*) we have plotted units of time which measure how long the bowl of water stood in the Sun. Here we want to start with zero, the moment that the bowl was first placed in the Sun.

As the bowl of water stands in the Sun, both time and temperature vary. The temperature of the water depends on how long it has stood in the Sun. We say that the temperature is the *dependent variable*. Time, on the other hand, is going to keep on changing inevitably as it always has, independent of the bowl of water and its temperature. Time is the *independent variable*. Customarily the scale for the independent variable is given along the abscissa and that for the dependent variable along the ordinate.

By following the vertical line that is labeled *15 minutes* to the point where the curve crosses it, we see that the temperature of the water after 15 min in the Sun was just a little above 17.5°. By mentally dividing the vertical distance between the 17.5° line and the 18.0° line into parts, we can estimate that the temperature was about 17.7°C. What was the temperature of the water after 20 min in the Sun? after 12 min? after 30 min? 35 min? 40 min? What would you guess the temperature would be after 45 min? 50 min? You can see that the temperature was changing more slowly toward the end of the time. It looks as though the water was becoming about as warm as it was going to get. What would you estimate the temperature of its surroudings to be? Is this in accord with your guess concerning the temperature of the water after 45 min? (The process of guessing values beyond the end of a curve is known as *extrapolation.*)

For any point on the abscissa (i.e., any particular time you wish to name) you can find a point on the curve (directly above the abscissa point) which will correspond to a point on the ordinate directly to the left, that is, a point that will tell you what the temperature of the water was at that particular time. Note that this is not so of the graph in Figure B–1*b*. Which type of graph would be appropriate for each of the following sets of information?

1. A report of the total number of stocks sold on the stock exchange in each day for the month of August.

2. A report of the depth of the water at some point in the Mississippi River from measurements taken each day during the month of August.

3. A report of the temperature of a patient in the hospital, taken three times a day for one week.

4. A report of the maximum temperature reached during each month throughout the year.

5. A report of the price of stocks sold each day on the stock exchange.

6. The apportionment of our tax dollar among various government expenditures.

How is a graph made? Suppose we take report number 2, above, as an example. First we must collect the data. Each day during August we stick a long pole into the water and measure the depth of the water. Since we want to study the changes from one day to the next and not from one part of a day to another, we make the measurement always at the same time of day—10:00 A.M., for example. Each measurement is thus separated from the next by the same time interval.

A question that must always be answered when recording measurements has to do with the precision of the measurement. Should we report the depth of the river as $3\frac{1}{2}$ meter, 3.50 meter, 3.502 meter? The Mississippi River is a swift river. Where the water runs swiftly it will swirl around the measuring stick and will change from one moment to the next by several millimeters, perhaps a couple of centimeters—that is a couple of hundredths of a meter (0.02 meter). But surely we can measure more accurately than to the nearest half a meter. So perhaps the best way to report this measurement would be: 3.50 meter plus or minus 0.02 meter, or 3.50 ± 0.02 meter. This is an important point.

With every physical measurement, no matter how carefully made, goes a certain amount of uncertainty. Sometimes the person making the measurement can make a pretty good estimate of his experimental uncertainty as he makes the measurement. Sometimes the mathematics of statistics is used to determine it, but experimental uncertainty is always present.

Now we will proceed with the river example. The river measurements might turn out to be as given in Table B–1.

Table B–1   Measurements of the Depth of the Mississippi River at Some Point at 10:00 A.M. in August of Some Year

Date	Depth in Meters ±0.02 m	Date	Depth in Meters ±0.02 m
August 1	3.50	August 16	3.62
August 2	3.45	August 17	—
August 3	—	August 18	3.56
August 4	3.40	August 19	3.55
August 5	3.35	August 20	3.50
August 6	3.35	August 21	3.46
August 7	3.40	August 22	3.45
August 8	3.50	August 23	3.71
August 9	3.65	August 24	—
August 10	—	August 25	4.10
August 11	3.80	August 26	4.03
August 12	3.75	August 27	4.01
August 13	3.70	August 28	3.90
August 14	3.70	August 29	3.84
August 15	3.65	August 30	3.75
		August 31	—

How shall we plot these data? Unlike the data of the graph in Figure B–1*b*, the various values in this case represent information about something that was continuously changing—the level of the water. In this way, it is like the information in graph B–1*c* which represented the continuous change of temperature of the water in the Sun. The water-level measurements are samples of information taken at arbitrary intervals of time. (We chose to make the measurements once a day.)

To represent the information on a graph with rectangular coordinates (axes at right angles to each other) we will want to use the abscissa for time, the independent variable, and the ordinate for water level, which changed *as a function of time,* that is, its change was related in some way to change in time. We have 31 points to place at equal spaces along the time scale. Suppose the graph paper has 10 large divisions across the bottom, each subdivided into ten small divisions, making 100 small divisions. If we let each small division represent a day, the graph will only take up a third of the width of the paper. If we let each large division represent a day, we could only get 10 days on the paper. If we let 90 of the small divisions represent 30 days, then 31 days would fit well within the width of the paper, with 3 small divisions representing one day. There is no meaning to "day number zero" so the zero point on the abscissa could be labeled August 1, as in Figure B–2. Note that we do not need

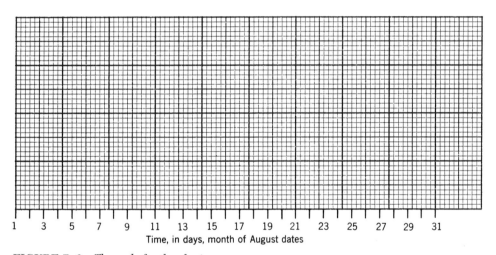

1 3 5 7 9 11 13 15 17 19 21 23 25 27 29 31
Time, in days, month of August dates

FIGURE B-2 *The scale for the abscissa.*

to label every point on the scale. (In many cases, three divisions of the graph paper would be an awkward choice for representing a unit because we would prefer to be able to locate tenths of the unit easily. However, in this case we know we will only need to locate the points on the abscissa that represent whole days, so the three divisions will not be inconvenient.)

The depth of water will be measured along the ordinate. The

smallest depth measured during August was 3.35 meter, and the largest was 4.10 meter. Therefore it is only necessary for the ordinate scale to include this range of depths. The finished graph will go no lower than 3.35 and no higher than 4.10. The whole change in water level is only 0.75 meter. If we spread this over the whole height of the page, then differences so small we could hardly detect them would look very large on the graph. A small division on the graph could represent the smallest division read on our meter stick. With this scale we would use 75 small divisions or 7.5 large divisions for the whole vertical range of our measurements. In order not to let any point get hidden on the abscissa, we could label the ordinate scale so that the lowest point was up half a division, that is, let the ordinate scale begin with 3.30 meter at the bottom, as in Figure B–3.

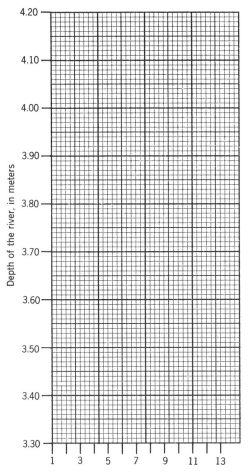

FIGURE B–3
The scale for the ordinate.

Now we are ready to plot the experimental points. The first three are shown in Figure B–4. Since we have no data for August 3 we have to skip that point. Note that it would distort

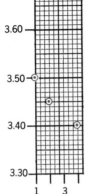

FIGURE B-4
Plot of the first three points.

the picture of the changing level of the river if we moved August 4 over to take the place of August 3. It would look as though the river level had fallen 0.05 meter in one day (since each division along the abscissa represents one day) whereas in fact it took two days to fall by this amount. This is an important point. When you are plotting data, if you are lacking the datum for a particular point, leave that place vacant. You will distort the picture otherwise. Once the scales of your axes have been chosen, they must be strictly adhered to.

Now is there any way that we can show on the graph the fact that we would not want to guarantee our river level measurements more precisely than ±0.02 meter? This is sometimes done by drawing little "flags" which extend outward from the plotted point for a distance which shows the amount of uncertainty of the measurement. On an enlarged picture of our graph, the flags would look like those in Figure B–5. On the scale of our graph, they would look like those in Figure B–6. If we now plot all our data in this manner, we have Figure B–7.

Now what do we know about the level of the river during the month of August? All that we *know* is just how deep it was on certain days at the point where we measured it at 10:00 A.M. But we can infer much more from our knowledge of the nature of the thing we were measuring. From an examination of our data we can be confident that the depth at that point

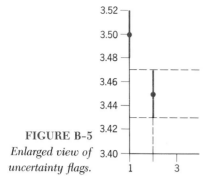

FIGURE B-5
Enlarged view of uncertainty flags.

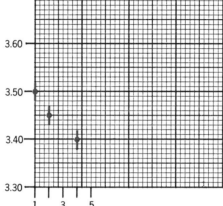

FIGURE B-6
The first three points plotted with uncertainty flags.

in the river did not increase to any value much above 4.10 meters or fall to any value much below 3.40 meter during the month. It would have taken the sudden violent movement of tons of water between two successive times of measurement to achieve this. There were increases and decreases, but you would not expect any sudden change of level. This would be like having a sudden steep slope occur in the surface of the river water. So because of our knowledge of the thing we are measuring, we expect a smooth change of level, and we will proceed to connect our experimental points with a smooth curve shown in Figure B-8.

Note that the curve does not have to go exactly through each point because we know that there is uncertainty about the position of each one. We do not guarantee the value more closely than ±0.02 meter. Note also the kinds of interpretation that go into the drawing of the curve.

Since the level was falling from August 1–5 and rising from

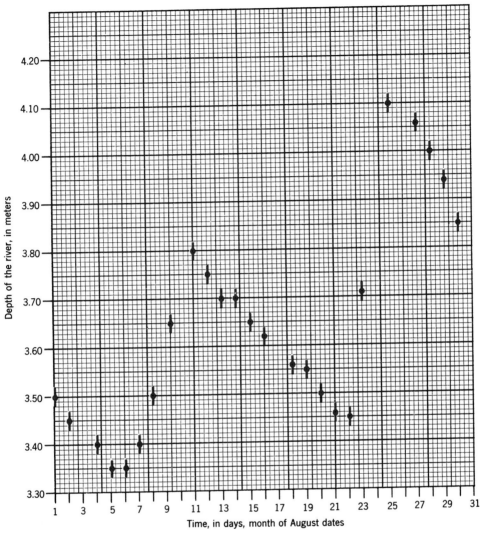

FIGURE B-7   *Plot of the river-depth data for the month of August.*

August 6–11, we can make the curve smoother by letting it dip between August 5 and 6. If it had been rising before the 5th and falling after the 6th, we would have shown an up-curve— a hump—between August 5 and 6. Because we know that the curve must go down right after August 11, we start it into this curve before the 11th. If the reading on August 12 had been 3.95 meter, we would have drawn a straighter line between the August 9 point and the August 11 point.

A mistake has been made in plotting one of the points. Find and correct it, and redraw the curve appropriately. It is a little

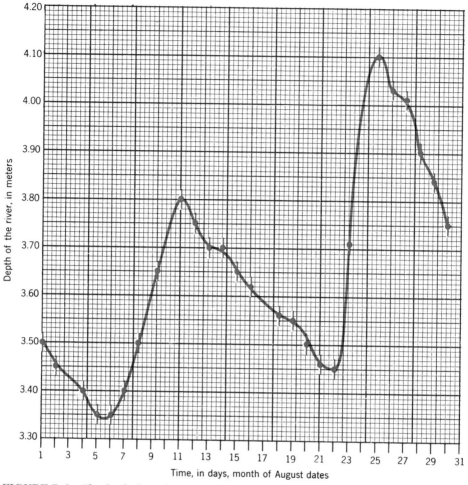

FIGURE B-8   *The finished graph.*

mistake, so it is not very obvious, but suppose we had made a big mistake in recording the data. Suppose, for example, we had mistakenly recorded the August 7 reading as 4.30 instead of 3.40. Note where this point would lie on the coordinate system of the graph paper and see how easy it would be to recognize that a mistake was made. Since you could not be sure that a number switch was the mistake, the only safe thing to do would be to omit the measurement for that day and draw the best curve you could without it, just as we did without all the Sunday points because the bus to the river didn't run on Sunday.

You will recall that the process of guessing beyond the end of a curve drawn from known points is known as extrapolation.

The process of guessing where a curve lies between known points is called *interpolation*.

A further step in interpretation has to do with the causes of the fluctuations of the curve. Beginning about August 11 there must have been a dry spell for a couple of weeks during which time the level of the river fell. Toward the end of this period there must have been a rainstorm to cause quite a sudden increase in the depth of the river. Such steep parts of the curve are hard to draw. The curve must not cross any vertical line (time line) twice. The meaning of this would be that the river had two different depths on our measuring stick at exactly the same moment, which is impossible.

Table B–2 is a table of data and Figure B–9 a piece of graph paper. Plot the data and draw the curve. You will need to go through all the steps that we went through for the water-level problem. The data are the records that a woman kept of her weight over a period of time when she was trying to diet. You might try making a graph from the record of your own weight over a similar period of time.

Table B–2   Weight versus Date

Date		Weight, in Pounds ± 0.2 *lb*
December	9	140.5
	10	142.3
	12	142.5
	17	142.3
	20	141.8
	22	141.0
	26	145.0
	29	146.2
January	3	147.0
	6	147.0
	10	145.8
	15	144.0
	22	142.8
	30	141.7
February	4	141.0
	8	141.5
	10	141.0
	14	140.5
	19	142.0
	25	141.0
	28	140.5

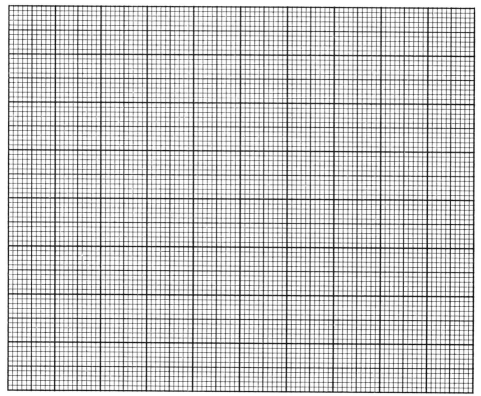

FIGURE B-9

Table B–3 shows data that are somewhat differently related. It shows the density of a liquid which is a mixture of alcohol and water. The density is a function of the composition—that is, when you change the proportion of alcohol to water, the density changes. When you plot the values from this table, should the scale for composition be plotted on the abscissa or the ordinate?

When the service station attendant uses a tube with a floating measuring device in it to determine whether you have enough antifreeze in your car in cold weather, he is determining the composition of the fluid in your radiator by measuring its density.

Graph the data from Table B–3 on an appropriately prepared piece of graph paper. Remember to give the graph a title and to label both scales.

Since no uncertainty limits are given for this table, you cannot draw uncertainty flags on the points. This does not mean that there was no uncertainty in the measurement. However,

Table B–3   Density of Mixtures of Ethyl Alcohol and Water versus
Composition by Volume Percent

Density in Gram/cm³	Percent Alcohol by Volume (cm³ Alcohol × 100/cm³ Mixture)
1.00	0.00
0.99281	5.00
0.98660	10.00
0.98114	15.00
0.97608	20.00
0.97097	25.00
0.96541	30.00
0.95910	35.00
0.95185	40.00
0.94364	45.00
0.93250	50.00
0.92441	55.00
0.91359	60.00
0.90210	65.00
0.89001	70.00
0.87729	75.00
0.86389	80.00
0.84960	85.00
0.83389	90.00
0.81610	95.00
0.79390	100.00

since the results are given to five figures following the decimal
place, we can guess that the uncertainty was in the last figure
as in the case of the river-level measurements. It is thus approximately ±0.00001.

Sometimes we want to graph a relationship for which we have
an equation. A very simple equation would be $x = y$. The
graph for that is shown in Figure B–10. We have shown nega-

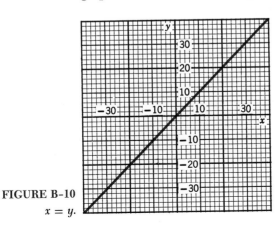

FIGURE B-10
$x = y$.

tive values on both the abscissa and the ordinate here since $x = y$ means $-2 = -2$ just as it means that $+2 = +2$. Negative values would have had no meaning in the water-level graph, the lady's weight graph, or the density graph. By custom the scale for x values is along the abscissa, which is often called the *x-axis* and the scale for y values is along the ordinate, which is often called the *y-axis*. The zero point on both scales is called the *origin*. By custom, the part of the ordinate scale *above* the origin is for *positive* values and the part *below* the origin for *negative* values. On the abscissa scale, the part *to the right* of the origin is for *positive* values, the part *to the left* for *negative* values.

Note that this equation results in a graph which is a straight line. Such an equation is called a *linear equation* and the relationship which it states is called a *linear relationship*.

Make a graph showing the relationship between a length measured in centimeters and the same length measured in inches. One inch equals 2.54 centimeters.

You might want to start by making a table like Table B–4, but you would soon discover that this is a linear relationship as is every relationship expressed by an equation of the form $x = (k)y$, where (k) refers to any constant, in this case, 2.54.

How many points do you need to determine where the straight line should be drawn?

Table B–4   The Relationship Between Inches and Centimeters

Inches	Centimeters
1	2.54
2	5.08
100	254

Will you be able to draw the line more accurately if the points are far apart or if they are close together?

Will negative values have any significance?

Is the completed graph a useful time-saver? Does it make the relationships clearer? How?

Now we will consider an equation that is not quite so simple:

$$p = \frac{1}{s}$$

By substituting some simple values for s we can easily determine p, for example, when $s = 2$, $p = 0.5$. Since these

Table B–5   Values of s and p that Satisfy the Equation $p = 1/s$

s	p
10.000	0.100
8.000	0.125
6.000	0.167
4.000	0.250
2.000	0.500
1.000	1.000
0.500	2.000
0.250	4.000
0.100	10.000
0.010	100.000
0	infinity
−0.010	−100.000
−0.100	−10.000
−0.250	−4.000
−0.500	−2.000
−1.000	−1.000
−4.000	−0.250
−10.000	−0.100

letters do not stand for any physical thing, we have no reason to suppose that the numbers could not be negative as well as positive. Table B-5 shows some values of s and p that satisfy the equation.

You can see that this graph does something quite remarkable. It goes off to infinity on the plus side and comes back from infinity on the minus side. This gives us some food for thought on the subject of the location of infinity with respect to our graph paper.

The values given in Table B–5 are plotted in Figure B–11 connected by a curve. Clearly it is impossible to plot the values when either s or p is close to zero so we have to choose some arbitrary boundaries for the graph.

Note the symmetry of Figure B–11. The upper right looks like the mirror image of the lower left half. Try placing a pocket mirror at right angles to the plane of the paper with its bottom edge on the origin and running from upper left to lower right. Is the mirror image of one curve like the other? The only thing that is different is that, in the mirror, the image of the arrow labeled s would lie along the p axis. It looks as though these two could change places without any change in the shape of the curve. We can go back to the equation and see whether this makes sense:

$$p = \frac{1}{s}$$

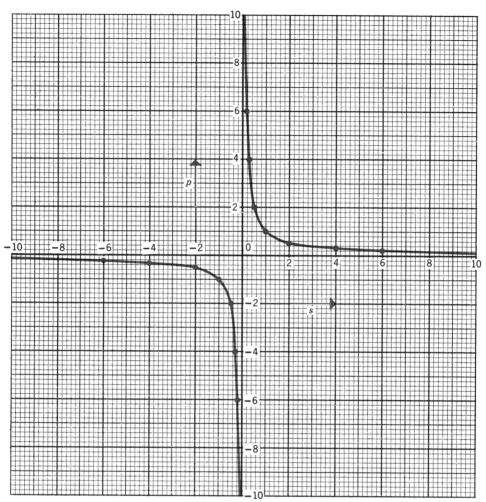

FIGURE B-11 $p = \dfrac{1}{s}$ *graphed as p versus s.*

Multiplying both sides of the equation by s

$$ps = 1$$

Dividing both sides of the equation by p

$$s = \frac{1}{p}$$

It is certainly true that s and p can change places, and the equation will still express the same relationship. This is a *reciprocal relationship*.

In the text, we have the equation

$$f\lambda = v$$

where f = the frequency of a wave
λ = the wavelength of the wave
v = the velocity of the wave

Is the relationship between frequency and wavelength a reciprocal relationship or a linear relationship? Try dividing both sides of the equation by either f or λ.

Make a graph of frequency versus wavelength for visible light. (The velocity of light is 3×10^8 meter/sec.)

Sometimes we can arrange to simplify our graphing job and even make it a more accurate one by choosing a scale that is especially suited to the situation. An example will make this statement more meaningful.

Consider again the equation $p = 1/s$. Suppose we were to lay out the abscissa scale in units of $(1/s)$ instead of s. The equation then says that what is measured in units along the ordinate (namely p) equals what is measured in units along the abscissa (namely $1/s$), and it thus assumes the form of the equation.

In Figure B–12 the data of Table B–5 are plotted in this way, with p versus $1/s$. Note that we only need to plot two points.

Make a graph of $a = s^2$, graphed as a versus s.

Make a graph of $a = s^2$, graphed as a versus s^2.

Make a graph of $F = \dfrac{1}{r^2}$, graphed as F versus r.

Make a graph of $F = \dfrac{1}{r^2}$, graphed as F versus $1/r^2$.

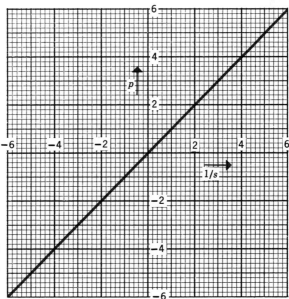

FIGURE B–12

$p = \dfrac{1}{s}$ *graphed as* p

versus $1/s$.

PERIODIC TABLE

1	2	3	4	5	6	7	8	9	10	11	12	13	14	15	16	17	18
1 H 1.008																	2 He 4.00
3 Li 6.94	4 Be 9.01											5 B 10.8	6 C 12.0	7 N 14.0	8 O 16.0	9 F 19.0	10 Ne 20.2
11 Na 23.0	12 Mg 24.3											13 Al 27.0	14 Si 28.1	15 P 31.0	16 S 32.1	17 Cl 35.5	18 Ar 39.9
19 K 39.1	20 Ca 4.01	21 Sc 45.0	22 Ti 47.9	23 V 50.9	24 Cr 52.0	25 Mn 54.9	26 Fe 55.6	27 Co 58.9	28 Ni 58.7	29 Cu 63.5	30 Zn 65.4	31 Ga 69.7	32 Ge 72.6	33 As 74.9	34 Se 79.0	35 Br 79.9	36 Kr 83.8
37 Rb 85.5	38 Sr 87.6	39 Y 88.9	40 Zr 91.2	41 Nb 92.9	42 Mo 95.9	43 Tc (99)	44 Ru 101.1	45 Rh 102.9	46 Pd 106.4	47 Ag 107.9	48 Cd 112.4	49 In 114.8	50 Sn 118.7	51 Sb 121.8	52 Te 127.6	53 I 126.9	54 Xe 131.3
55 Cs 132.9	56 Ba 137.3	see below 57–71	72 Hf 178.5	73 Ta 180.9	74 W 183.9	75 Re 186.2	76 Os 190.2	77 Ir 192.2	78 Pt 195.1	79 Au 197.0	80 Hg 200.6	81 Tl 204.4	82 Pb 207.2	83 Bi 209.0	84 Po 210	85 At (210)	86 Rn (222)
87 Fr (223)	88 Ra (226)	see below 89–															

57 La 138.9	58 Ce 140.1	59 Pr 140.9	60 Nd 144.2	61 Pm (147)	62 Sm 150.4	63 Eu 152.0	64 Gd 157.3	65 Tb 158.9	66 Dy 162.5	67 Ho 164.9	68 Er 167.3	69 Tm 168.9	70 Yb 173.0	71 Lu 175.0
89 Ac (227)	90 Th 232.0	91 Pa (231)	92 U 238.0	93 Np (237)	94 Pu (242)	95 Am (243)	96 Cm (247)	97 Bk (245)	98 Cf (251)	99 Es (254)	100 Fm (253)	101 Md (256)	102	103

Most stable known isotopes are shown in parentheses.

INDEX

6095

T-S
Sc.p-W64
P